Contemporary Issues in Technology Education

Technology education is a developing field, new issues keep arising and timely, relevant research is continually being conducted. The aim of this series is to draw on the latest research to focus on contemporary issues, create debate and push the boundaries in order to expand the field of technology education and explore new paradigms. Maybe more than any other subject, technology education has strong links with other learning areas, including the humanities and the sciences, and exploring these boundaries and the gaps between them will be a focus of this series. Much of the literature from other disciplines has applicability to technology education, and harnessing this diversity of research and ideas with a focus on technology will strengthen the field.

Occasional volumes on a bi-annual basis will be published under the Council for Technology and Engineering Teacher Education (CTETE) inside this series.

For more information, or to submit a proposal, please email Grace Ma: grace.ma@springer.com.

Wendy Fox-Turnbull · P. John Williams
Editors

Locating Technology Education in STEM Teaching and Learning

What Does the 'T' Mean in STEM?

Springer

Editors
Wendy Fox-Turnbull
Division of Education
University of Waikato
Hamilton, New Zealand

P. John Williams
School of Education, STEM Research Group
Curtin University
Perth, WA, Australia

ISSN 2510-0327 ISSN 2510-0335 (electronic)
Contemporary Issues in Technology Education
ISBN 978-981-97-1994-5 ISBN 978-981-97-1995-2 (eBook)
https://doi.org/10.1007/978-981-97-1995-2

This Springer imprint is published by the registered company Springer Nature Singapore Pte Ltd.
The registered company address is: 152 Beach Road, #21-01/04 Gateway East, Singapore 189721, Singapore

Paper in this product is recyclable.

This book is dedicated to all the hard-working technology teachers who have embraced technology education as a student-centred, inquiry-based academic and practical learning area, set within authentic and meaningful contexts. The current STEM movement has the potential to enrich design and technology, placing it at the centre of an interdisciplinary curriculum, assisting the students to learn science and maths in a purposeful context. It also has the potential to relegate it to the use of technical tools and skills—digital or otherwise.

We hope this book will assist and motivate teachers in the former.

Acknoweldgements

Research assistance for data collection in this paper was provided by Sue March (field leader), Fatema Taj Johora, and Junqian Ma and data organisation by Freya Fleer-Stout and Ainslie Holland. Special acknowledgement is made of the two teachers who participated in the study, Rebecca and Oriana. Funding from the Australian Research Council Discovery Grant [DP140101131] and from the Australian Research Council Laureate Fellowship Scheme [FL180100161] the research and writing of this chapter.

Contents

Contributors

Piet Ankiewicz Department of Science and Technology Education, Faculty of Education, University of Johannesburg, Auckland Park, South Africa

Frank Banks The Open University, Milton Keynes, UK

HildaRuth Beaumont University College London, London, United Kingdom

Donal Canty University of Limerick, Limerick, Ireland

Marc J. De Vries University of Technology Delft, Delft, The Netherlands

Andrew Doyle School of Education, University of Waikato, Hamilton, New Zealand

Marilyn Fleer Monash University, Peninsula Campus, Australia

Wendy Fox-Turnbull The University of Waikato, Hamilton, New Zealand

Michael Hacker Hofstra University Center for STEM Research, Hempstead, NY, USA

Jonas Hallström Department of Behavioural Sciences and Learning (IBL), Linköping University, Norrköping, Sweden

Virginia R. Jones Old Dominion University, Norfolk, VA, USA

Kerry Lee University of Auckland, Auckland, New Zealand

Kuen-Yi Lin National Taiwan Normal University, Taipei, Taiwan

Johanna Lätti Tampere University, Tampere, Finland

Sonja Niiranen Tampere University, Tampere, Finland

Per Norström Department of Learning, KTH Royal Institute of Technology, Stockholm, Sweden

Elizabeth Reinsfield Learning Area Lead: Ministry of Education, Wellington, New Zealand

Thomas Roberts Bowling Green State University, Bowling Green, OH, USA

Konrad J. Schönborn Department of Science and Technology (ITN), Linköping University, Norrköping, Sweden

Niall Seery Technological University of the Shannon, University Road, Athlone, Westmeath, Ireland

Yu-Jen Sie National Taiwan Normal University, Taipei, Taiwan

Joachim Svärd Linköping University Sweden, Linköping, Sweden

Sini Teräsahde Tampere University, Tampere, Finland

Hanno van Keulen Delft University of Technology, Delft, The Netherlands

Belinda von Mengersen Australian Catholic University, Strathfield, NSW, Australia

Chapter 1
Locating Technology Education in STEM Teaching and Learning: So, What Does the T Mean in STEM?

Wendy Fox-Turnbull

Abstract The vision for this book is to explore and clearly define technology (T) when embedded in Science, Technology, Engineering and Mathematics (STEM). It grew from a concern that prevailing perceptions of technology when embedded in STEM is merely the use of technological tools to serve the learning needs of science and mathematics. Those who have a clear and deep understanding of technology education know that this is not the case. Interdisciplinary integration is key to this understanding. Burke (Technol Teacher 73:14, 2014) states that an integrative STEM approach recognizes that technology and engineering provide opportunities for students to develop deeper knowledge about science. We would argue that an integrative STEM approach recognizes that technology and engineering provide opportunities for students to develop deeper knowledge about the world and how to intervene in it, drawing on design, science and mathematics knowledge to assist this process.

The vision for this book is to explore and clearly define technology (T) when embedded in Science, Technology, Engineering and Mathematics (STEM). It grew from a concern that prevailing perceptions of technology when embedded in STEM is merely the use of technological tools to serve the learning needs of science and mathematics. Those who have a clear and deep understanding of technology education know that this is not the case. Interdisciplinary integration is key to this understanding. Burke (2014) states that an integrative STEM approach recognizes that technology and engineering provide opportunities for students to develop deeper knowledge about science. We would argue that an integrative STEM approach recognizes that technology and engineering provide opportunities for students to develop deeper knowledge about the world and how to intervene in it, drawing on design, science and mathematics knowledge to assist this process.

W. Fox-Turnbull (✉)
The University of Waikato, Hamilton, New Zealand
e-mail: Wendy.fox@waikato.ac.nz

W. Fox-Turnbull and P. J. Williams (eds.), *Locating Technology Education in STEM Teaching and Learning*, Contemporary Issues in Technology Education,
https://doi.org/10.1007/978-981-97-1995-2_1

The diversity of approaches to STEM complicates its discussion. Not only does the definition of the T in STEM differ according to practitioner perceptions, but the approach to integration within STEM differs between schools and could be multi-disciplinary (concepts and skills separately in each discipline but in reference to a common theme), interdisciplinary integrating knowledge and methods from different disciplines, using a real synthesis of approaches) or transdisciplinary (creating a unity of intellectual frameworks beyond the disciplinary perspectives) and single or multi-teacher-led. The relationship of a STEM activity to the curriculum is also diverse and could be extra-curricular (out of school timetable and outside the school curriculum), intra-curricular (a STEM project delivered within a single subject over a few lessons of weeks, such as Science) or inter-curricular (timetabled as STEM and focused on the pedagogy of fostering the capacity for applied synthesis). Within this diversity, a number of themes emerge in the authors' discussions of STEM within the book, a summary of which follows.

Technological Applications in STEM: This theme highlights the application of technology in STEM education. It covers technology and engineering as disciplinary domains, design pedagogy and the role of food (Reinsfield), textiles (von Mengersen) and makerspaces (de Vries) in the STEM context. The themes emphasize the significance of technology in addressing real-world challenges and fostering innovation (Hallström and Norström). Authors advocate for technology education as a central pillar in STEM, enabling students to apply their knowledge and skills to create practical solutions and technological artifacts (Ankiewicz, Fox-Turnbull, Hacker, Reinsfield, Svärd and von Mengersen).

Philosophy, Theoretical and Historical Perspective of Technology in STEM: This theme explores the philosophical aspects of technology's role in STEM education and includes the development of products or artifacts (van Keulen) and identifies its constructivist theoretical underpinning (Ankiewicz). The historical roots of technology are discussed, and parallels are drawn with the Greek concept of 'techne,' which had a profound influence on European educational practice. By examining the meaning of technology as an artifact and as artful practice, this theme highlights the importance of traditional, tacit practices that cannot solely be taught through theory. The human aspect of technology is emphasized, as it aims to solve problems from a human perspective. This philosophical reflection encourages educators to view technology as an essential component of STEM and recognize its cultural and historical significance, noting that technology can be disruptive whether it be through outcomes or actions (Lee, van Keulen).

Technology in STEM with Integrity: Maintaining the integrity of technology, technological knowledge and technological practice is central to this book. It advocates the avoidance of 'Mucky Brown Paint' syndrome (all learning areas mixed together in busy activity rather than specific learning) (Fox-Turnbull, Lee). Doyle makes a clear distinction between technology and technology tools used for digital learning in education and explores the impact of technology and digital educational tools, especially during the COVID-19 pandemic, on STEM education. He also examines the adoption of digital technology in STEM classrooms and its implications for future STEM activities. The book also advocates for the purposeful addressing of curriculum knowledge in STEM Projects (Banks, Fox-Turnbull, Hacker, Lee, Reinsfield, van Keulen, von Mengersen) to advocate for a focus on the importance of technology teachers having a wide curriculum knowledge to support the ‘T’ in STEM school projects. In addition, it emphasizes cross-curriculum approaches, collaboration and the integration of science and mathematics in technology-focused STEM projects (Hallström and Norström and Lin and Sie) and situates textiles (von Mengersen) and foods (Reinsfield) within the STEM family and acknowledges the importance of play worlds advocating for the bringing together of play and imagination in design to amplify technological learning (Fleer). Learning is not restricted to the classroom but also occurs outside of the curriculum (Lin and Sie) and in community makerspaces (de Vries). Hallström and Norström position the need for experts in the field within technology and engineering practice. Hacker examines the relationship between technology and engineering and the impacts both have on the wider world.

Integration and Curriculum Development: This theme focuses on the integration of STEM education, emphasizing the need to bring together science, technology, engineering and mathematics to create a holistic learning experience drawing on notions of social constructivism (Lee). Various approaches to integration are explored, including conceptual frameworks and models that facilitate the blending of these disciplines (Lee, Fox-Turnbull). The theme highlights the benefits of integrated STEM projects for both teachers and learners while discussing the challenges in implementing such approaches. By addressing curriculum knowledge and cross-curriculum collaboration, this theme advocates for a more comprehensive and interconnected STEM education. The book delves into different approaches and models to integrating STEM education (Banks, Beaumont, Fox-Turnbull, Hacker, Lee,) and explores the role and notions of integration (Lee) and the importance of keeping the academic integrity of distinct knowledge disciplines and the challenges and benefits of implementing integrated STEM projects within authentic learning contexts. (Beaumont, de Vries Lee, Banks, Fox-Turnbull). Jones and Roberts discuss the concept of STEM literacy and its relation to disciplinary literacies. They emphasize technological literacy as a necessary foundation for achieving STEM literacy.

Assessment and Evaluation in STEM: This theme discusses the importance of assessing technology in STEM education. It acknowledges the fluid and non-deterministic nature of technological activities, making assessment a complex task. Seery and Canty explore the role of assessment in informing teaching and learning, aiming to strike a balance between evaluating students’ performance and fostering a student-centered approach. It emphasizes the need for holistic assessment and teachers’ professional judgment to capture the multidimensional aspects of technology in STEM. Fully embedded assessment is for learning rather than of learning (Banks, Seery and Canty). Assessment is a significant component of any technology program, especially in the senior secondary context. Svärd identifies a number of challenges for the senior secondary sector including teacher knowledge, especially across the STEM disciplines, lack of time and teacher preparation and different ways of working technology requires.

Gender and Equity in STEM: Niiranen et al. delved into the issues of gender representation and equity in STEM education and explored gender segregation and unconscious bias in technology-oriented fields. The chapters within this theme discussed the importance of gender-aware approaches to promote equitable participation and representation of genders in STEM (Niiranen et al. Reinsfield). By shedding light on the experiences and perceptions of female students and educators, this theme advocated for creating inclusive and supportive environments for all learners.

Senior Secondary Education in STEM: Svärd focuses on the implementation of STEM education in senior secondary schools. It highlights the significance of interdisciplinary education to prepare students for real-world challenges and encourage them to engage with complex problems. The theme emphasizes the role of technology education in integrating and contextualizing learning, bridging the gap between different STEM subjects.

Overall, the book covers a wide range of themes that collectively contribute to a comprehensive understanding of technology's place in STEM education. It highlights the theoretical, practical and societal aspects of integrating technology within STEM and advocates for a more inclusive, equitable and interdisciplinary approach to teaching and learning (Hacker, Lee, Fox-Turnbull). The exploration of technology's historical and philosophical roots adds depth to the understanding of its significance in shaping human development and problem-solving approaches. Furthermore, the book recognizes the importance of addressing gender disparities and fostering a supportive learning environment for all students. By examining the implementation of STEM at different educational levels, the book offers valuable insights into curriculum development and pedagogical practices. It provides a rich resource for educators, researchers and policymakers interested in advancing STEM education and technology integration in the modern educational landscape.

*ChatGPT was used in the identification of some of the initial themes.

Reference

Burke, B. N. (2014). The ITEEA 6E learning byDeSIGN[TM] model: maximizing informed design and inquiry in the integrative STEM classroom. *Technol Teacher, 73*(6), 14.

Wendy Fox-Turnbull is an Associate Professor at the University of Waikato and Associate Dean-Academic in the Division of Education. Wendy is the editor of Australasian Journal of Technology Education and on Editorial Board of International Journal of Work Integrated Learning and co- editor European Journal of STEM Education. In 2019 she was guest editor for Design and Technology: An International Journal Primary Edition. In 2018 she was the invited section editor for the International Handbook in Teacher Education, Section Technology Teacher Education. Wendy was chair of the Technology Education New Zealand (TENZ Council) from 2006 to 2018 and has convened three TENZ conferencesTENZ 2005, TENZ2015 and TENZ/ICTE2017 and one International Technology Research conference (PATT) in 2013. Research special interests include authentic learning in technology education, the place of women in technology related

STEM careers, the role and nature of effective conversations in learning and teaching and learning approaches for the 21st Century. Wendy is a registered and certified teacher.

Chapter 2
A Philosophy for the Place of Technology in STEM

Hanno van Keulen

Abstract This philosophical reflection focuses on the 'T' of technology in STEM as it appears to stand out as the product or artefact resulting from the practices of science, engineering, and mathematics. This interpretation is enriched by a historical and cultural comparison of technology to 'techne', the Greek word for craft that had a deep influence on European educational practice and draws attention to the value of traditional, tacit practices that are not taught through theory alone. The meaning of technology, as artefact and artful practice, is further analysed from viewpoints of linguistics, cognitive psychology, anthropology, and ethics. Technology is thoroughly human in the sense that it solves problems from a human point of view, and therefore we all should, in principle, be able to understand the technology and recognize its affordances. Technology has shaped us as much as we shape technology: the development of our cognitive system goes hand in hand with making and using technology, both in the development from birth to adult as in the evolution from our prehistoric ancestors to the creative and responsible Homo sapiens we may one day become.

Keywords Philosophy of STEM · Tacit knowledge · Embodied cognition · Craft · STEM education

2.1 Introduction

Do we need to think from a philosophical point of view about STEM? We already have well-established philosophies of science, of technology, and of mathematics. This book focuses on the 'T' of technology in STEM. Does technology change in nature when it becomes a constituent of STEM, requiring additional philosophical reflections? Will the combination reveal new or emerging properties of technology,

---

H. van Keulen (✉)
Delft University of Technology, Delft, The Netherlands
e-mail: hannovankeulen@me.com

W. Fox-Turnbull and P. J. Williams (eds.), *Locating Technology Education in STEM Teaching and Learning*, Contemporary Issues in Technology Education,
https://doi.org/10.1007/978-981-97-1995-2_2

like gelatine can reveal its ability to turn juice and sugar into a tasty dessert, or molecules of H_2O, which are not liquid or wet themselves, can form a drop of water?

A reflection on the origin of STEM shows that STEM was not proposed from a theoretical, let alone a philosophical point of view. The first mention of STEM was in the USA in the last decade of the twentieth century, in the context of workforce shortages related to technical professions. Reportedly, the juggling with nouns towards STEM started with Science, Mathematics, and Technology (SMT), after which Engineering was inserted to make the acronym easier to pronounce (SMET). Apparently, this did not resonate well enough, so the order was changed, and STEM was born. Clearly, STEM is not a natural phenomenon, out there, waiting to be observed, analysed, and defined. It is a social construct, which implies that its meaning is 'in the eye of the beholder' and to be found in the social and cultural practices where STEM is discussed, taught, learned, and applied.

Adding 'engineering' to the mixture of science, technology, and mathematics is an act that potentially changes meaning and opens a venue. It suggests that technology and engineering are different enough to justify the inclusion of both. What could be this difference? Science, engineering, and mathematics can be regarded as practices and not as material things. You cannot pick up science and let it fall on your feet. The word technology, though, is often used for material things, artefacts (Mitcham, 1994), the products of a practice that may be informed by engineering and the application of mathematics and science. So, thinking about the 'T' in STEM is thinking about the special position of human-made artefacts resulting from practices that are informed by the bodies of knowledge and skills of science, mathematics, and engineering.

But perhaps this is true only when and if we speak English. English is the language of academia, but native speakers of other languages will recognize difficulties when translating STEM. For example, in Germany, the acronym most widely used for STEM is 'MINT', which stands for Mathematik (Mathematics), Informatik (Computer Science), Naturwissenschaft (Natural Sciences), and Technik. 'Technik' may be translated into 'technology' but not in the sense that it refers to artefacts. Its meaning is much closer to practice. 'Technik' has an ancient origin in the Greek word 'τέχνη' (techne), which refers to the artful making of useful things. For Germans, 'Technik' and 'Ingenieurwissenchaften' (engineering) are therefore too similar to be used both in a combination like STEM. This is also true for several other related languages that use 'techne', more or less, in its original Greek meaning, like Dutch and Swedish.

2.2 Historical and Cultural Aspects

What this means is that philosophical reflection on STEM and on the 'T' in STEM has historical and cultural connotations. Are we reflecting on technology-as-artefacts that result from scientifically and mathematically informed engineering practices (American style) or on technology-as-artful practice with roots in craftsmanship, rather than in academic science (Greek-German style)? For the ancient Greeks, making things and working with their hands out of necessity was not held in high regard. Craftsmen were respected when their work had quality, but they were artists rather than scientists. They may have had a certain knowledge base, but this was often unarticulated, implicit, tacit, personal, and not the explicit and undeniable, 'scientific', theorems the philosophers sought, to learn about and understand the eternal truths. 'Science', from the Latin 'scientia', means 'knowing' or 'understanding'. Knowing and using your mind was for the Greeks not related to making and using your hands. The school of Aristotle, although he himself a keen of observer of nature and a proto biologist, became known as 'Peripatetic', which means so much as philosophizing by walking around, with hands behind the back. Aristotle's ideas and theories were brilliant, and his influence lasted for centuries, but they were eventually questioned when experimenting became the companion of theorizing and observing. Experimenting implies using your hands; bare hands and the naked eye can be reinforced with tools; and tools and instruments require making. Science and technology, understanding and making, became fundamentally entangled in the age of the scientific revolution with its front runners Galileo, Huygens, and Hooke. But 'Technik', as the practice of non-academic craftsmen, retained a parallel position, still creating artefacts based on tacit approaches that were mastered through imitation, rather than through theory. In 'The Craftsman' (Sennett, 2009), a wonderful book on how craft is taught and learned through the ages, Richard Sennett gives the example of Stradivari, who build violins like never before and after, because he was unable (or unwilling) to articulate his secrets to his pupils. Artefacts made through tradition do not necessarily lack quality; on the contrary, they come into being through pathways that are alien to academia and tend to be ignored or even refuted, precisely for their lack of theoretical transparency and rationale.

Combining academic science, mathematics, and engineering for the sake of producing technology could easily become self-evident and dominant in the US education system since this only came into being well after the seventeenth century. Europeans, however, inherited their school system from classical and mediaeval times. The oldest universities (Bologna, Paris, Oxford) were established around 1200 and were utterly impractical. Just like it was in Greece, the gymnasium (secondary education) and the university were for the elite, who did not need to work for a living but sought intellectual challenge, like reading Plato, writing poetry, excavating Egyptian treasures, or collecting exotic butterflies. Greek, Latin, and History were the important subjects. Medicine was studied through the works of Hippocrates and Galamus, with the result that for the sake of your health, you'd better go to a midwife or herb collector than to a medical doctor.

The Industrial Revolution with its increasing need for workers with advanced knowledge of mechanics, chemistry, transportation, steel production, et cetera, initiated radical changes in the school system. For example, in 1863 the Dutch Government created a new secondary school, the 'Hogere Burger School', with an emphasis on mathematics, physics, and chemistry. Translating the name of the school into American English nicely illustrates the cultural stance towards the 'techne' of Europe. It seems obvious to translate 'Hogere Burger School' into 'high school' or 'higher education for citizens' ('burger' means 'citizen'). But in 1863 'higher' did not refer to the next step in a sequence from primary to tertiary education. Rather, this school was meant for 'higher citizens' as opposed to 'lower citizens': the new elites that made their money through trade and industry and, like the old elites, did not work with their hands. The difference was perhaps that they preferred reading Von Humbold above reading Plato. Theory, and not making, fleshed out the curriculum, and teachers in these schools nowadays still perform experiments mainly to illustrate the theory, not to develop practical skills. The children of the real labourers and craftsmen, on the other hand, did not go to secondary school until well into the twentieth century, and they still are underrepresented in the schools that prepare for academic learning. Many European countries with a Greco-Roman heritage maintain a two-tiered educational system, with a theoretical and a vocational column, with approaches to the material world that are often incompatible.

So, a philosophy for STEM emphasizes a historical and cultural reflection on the place of artefacts, theory, making, and the appreciation of tacit craft in educational settings. If this analysis is adequate, curriculum developers on both sides of the Atlantic may draw different lessons on how to elaborate STEM into their school systems: interest in non-theoretical traditions on the one side and equity for craft on the other. This also has repercussions for the cultures with native craft traditions in Asia, Africa, Oceania, or the Americas. They are exposed to the powerful self-evident views concerning the STEM curriculum from Europe and North America, and the world may lose valuable ways of making artefacts when a one-size-fits-all STEM approach is thoughtlessly implemented. History can inspire reflection on the role of technology in STEM, and there are more points of view that can enrich our understanding of the T in STEM. In the next sections, we will explore viewpoints from linguistics, cognitive psychology, anthropology, and ethics.

2.3 Linguistic Aspects

The reflections above already show how important language is. Greek techne is not identical to American technology. Therefore, what 'STEM' can mean to an American is not what it can mean to a German, a Dutchman, or a Nigerian. The fact that the meaning of a word depends upon historical and cultural contingencies, however, does not imply that we cannot cross the boundaries and understand other viewpoints. The laws and constants of physics require precision up to five decimals and apply everywhere in the world, but cultural and social concepts typically are fuzzier. 'Play', for example, has no definition that includes all instances that everybody agrees on are play; compare tennis to patience. Such constructs are called 'categories': they have instances but no all-encompassing definition.

Words, like STEM, have origins. Words occur in a context. The meaning of words is not a matter of definition from scratch, and teaching is not forcing learners to reproduce these definitions. Rather, students should be taught how to derive meaning from the context. This is a personal act (Polanyi, 1962), not something that is produced by an AI algorithm. This is the domain of hermeneutics, an approach to understanding texts that are not self-evident. The meaning of an unknown word, for example, can be derived from its place and apparent role in a sentence, a paragraph, or a book. Likewise, the meaning of an artefact can be derived from the interaction of the artefact with its material and immaterial context. Interpretation is a key activity: a correct interpretation leads to a fruitful application. The ability to apply correctly implies understanding. This is the hermeneutic circle. For the German hermeneutic philosopher Hans-Georg Gadamer (Gadamer, 1960, 1972), this is how we should also approach technology (in its meaning of artefact): the technology will impose its affordances on me, that is, how it should be used if I encounter it in its proper context, and if I apply all my knowledge of the context to find out what it can do for me.

Technologies that are too new or disruptive, however, cannot exist or come into being in a context that cannot interpret them. That is why the first motorized cars looked like horse-drawn chariots. Imagining a Corvette was beyond imagination. Typically, technology is a solution to a problem, and we cannot understand solutions to problems that have not (yet) been articulated. Currently, e-books have page numbers like printed books, but who knows how we will find our way into texts in the future? Gadamer compares the assessment of technology to the experience we have when we are confronted with a piece of art. Conveniently, the German language uses the same word for a piece of art as for a technological artefact: 'Kunstwerk'. We can recognize what is a true work of art, and the same experience of truth can be obtained when we are confronted with technology. Good technology fits a solution to a problem in a specific context, in our life, experienced as the flow of things and events that makes sense to us. We intuitively appreciate buildings, structures, tools, and utensils when we recognize an applicability that is consistent with the surroundings and our needs, whether they have been created by theoretically informed engineers or by illiterate craftsmen.

To many people, STEM should make room for the 'A' of Arts and evolve into STEAM. What do proponents mean by the A? Does it express the interest of the STEM community in craft and traditional ways of making? Or is meant that painters, sculptors, musicians, dancers, et cetera, are also creative problem solvers whose approaches can inspire STEM professionals? Both interpretations have merit, and linguistic analysis can help to prevent misunderstandings. For a set of articles that are mutually inconsistent in this respect, see the special issue 'STEM and Arts Education' of the European Journal of STEM Education (2021).

Linguistics is also important to understand where the words we use within STEM come from. Living and surviving in the material world means correctly understanding the affordances of this world and effectively applying this understanding to our daily life. To some extent, this can be tacit. But there is a gap from daily life to the highly abstract theoretical language of scientists and engineers, to be bridged by education (Collins, 2010; Sullivan, 2017). A starting point is needed. A new and unknown phenomenon has to be named: deictic pointing and using words like 'this' and 'that' may work on the spot but do not transfer to an audience that does not participate in the practice. Lakoff and Johnson (Lakoff & Johnson, 1980) elaborated the role of metaphor in STEM. Typically, we start with improvising on a concept we know and map its characteristics to the unknown, to develop an initial understanding (Nersessian, 2008). 'Electrical current' is a good example: we recognize the similarities between a river flowing from high to low altitude while transporting useful stuff. We conveniently forget about the differences: electrical current does not evaporate or make you wet. 'Current' has received a new meaning. In this way, well-known phenomena with an explicit vocabulary keep on inspiring scientists to explain the not-yet-well-known (Abrahamson et al., 2012; Browne, 2003). Reflecting on technology in STEM thus is reflecting on language.

2.4 Cognitive Psychology

Technology is material, and so are we. Understanding technology is for a part embodied, in the sense that artefacts like we are subject to similar forces and influences, and artefacts typically are solutions to human problems and challenges to cope with the material world (Niebert et al., 2012). A cup can contain water much better than our hand. Present-day technology is not an incomprehensible black box but a set of babushka dolls: it uses other technology which again uses other technology (Arthur, 2009), but after some peeling-off in the end you reach your body. A cell phone at the bottom is your voice but with the ability to reach others over a greater distance. We may lack the mental and physical facilities to 'really' understand string theory, quantum coupling, or four dimensions that are perpendicular to each other. But we are fundamentally related to artefacts and every child, student and citizen should in principle be able to understand the devices we are confronted with, however complicated they may seem at first sight. A challenge for education!

We experience the world through our senses, and we interact with the world with our hands. We feel gravity pulling us down from the moment we are born, and the struggle to stand on our feet is a yearlong intense effort, so when we feel 'down', we can notice that we also keep a low body posture. Therapists know this and advise you to actively keep your head up and lift the corners of your mouth to automatically feel better. More important people sit in a higher place, suggesting that they were stronger than others in the struggle to move up against gravity. Our understanding of this metaphorical use of such words is embodied (Lakoff & Johnson, 1999). In contrast to feeling high or low, we never feel 'left', because there is no force acting on the horizontal as there is a force like gravity acting on the vertical, and consequently moving left or right is emotionally irrelevant.

For many psychologists, cognition and language, therefore, are embodied (Abrahamson et al., 2012; Gibson & Pick, 2000; Glenberg & Kaschak, 2002; Zwaan & Kaschak, 2009; Marechal et al., 2013). We develop our concepts through perceiving affordances and acting on these. What fits our bodily experiences is easier to understand. Technology is meant to solve our material problems, and therefore we should always be able to connect with it. It should fit like a glove. Surely, a dolphin won't understand this metaphor: dolphins concentrate on other affordances. For children, is far easier to understand the meaning of 'up' and 'down' and apply this knowledge when opening a drawer than to understand the meaning of 'left' and 'right'.

When we learn to walk, we become aware of the meaning of shades, recognizing what we have to do with our feet to prevent stumbling. When the sun is behind you and the shade is behind the obstacle, you must lift your feet because it is an elevation. When the shade is before the obstacle, there is a hole to step over. Through learning, our brain and our cognition become adapted to this specific material world (Dehaene, 2020). We are accustomed to a world in which the sun is above us and not below us, otherwise, our automatic interpretation of shades would have led to the opposite behaviour. Also, we do not experience a change in our clothes when we move from the dark into the light: we assume that these are the same clothes and thus must retain their colour. Computer scientists have only recently mastered the animation of shades in such a way that the animation appears somewhat realistic to the human observer because our brain has learned to create an awareness of what makes sense to us, not what is objectively visible (Ware, 2013; Wolfe et al., 2009). So, in a sense, reflecting on technology in STEM is reflecting on early childhood education and cognitive psychology.

2.5 Anthropological Aspects

Our body allows us to work with technology, especially with artefacts that elaborate on our body. A hammer is like a fist, but heavier and less vulnerable. A file is like nails but harder and, again, less vulnerable. The brain we are born with allows us to understand and operate these tools. The first learning takes place at a very early age before we can speak or think with words. We shape technology, or so we think. Developmentally, however, there is a case for technology shaping us and thinking as preceded by making. Anthropologists have studied how we learn to make stone tools (Malafouris, 2013). It was two million years ago that the first stone tools were made, by ancestors that were not yet evolved into Homo sapiens ('wise', 'understanding', 'thinking'). Was the tool designed and developed from the imagination of a prehistoric mastermind? Anthropologists think not. With *f*MRI techniques, they studied how expert tool makers use their brains when tool-making and compared this to novices learning to make stone tools. Experts use the brain regions associated with conscious thinking, especially the prefrontal cortex. They have learned what to do and what to observe, and they act on purpose, doing what they want to do. Novices, however, when trying to make stone tools, primarily use the motor regions of the brain. They act randomly at first, and they are unable to repeat a lucky strike at will. Eventually, when the body tacitly understands what to do, the prefrontal cortex enters the stage, and novices slowly start to become conscious about what they are doing and what they see, feel, and hear. It is through this sustained doing that stable connections between neurons in the brain will be made, and these will connect the motor region with perception, emotion, and language, allowing us to remember, describe, explain, rejoice in success, and repeat and improve the action till perfection (Hurford, 2007). Learning is a permanent change in the brain, and the development of the prefrontal cortex is correlated with the advent of the 'maker movement' (Feinberg & Mallat, 2016). Between the first primitive stone tools and conscious technological advancements lies a vast amount of time, enough to evolve a brain suitable for tool-making, speech, and thinking. Within our lifetime we will not notice how technology shapes us, but our successors will. Reflecting on technology is reflecting on human development (Bickerton, 2014; McGinn, 2015).

2.6 Ethical Aspects

Tool making and tool using shaped us. Tools and technology shaped the world. This process has accelerated enormously, especially since the nineteenth century. And not just for the good. Kevin Kelly advocates the view that if technology is 51% beneficial, in the long run, humanity will benefit (Kelly, 2011), but who pays for the collateral damage? Maybe these trade-offs remain inevitable when thinking about the impact of technology on society is restricted to scientists, engineers, and mathematicians and predominantly plays a role in the introductory and closing paragraphs of grant proposals. Scientists and engineers nowadays know they can use help from ethics, various religions, and other sources of wisdom (Goorden, 2017). Jacques Ellul, both a philosopher of technology and a Christian theologist, pointed out that it all started in a Garden but, at least for those willing to believe the Book of Revelations, it will end with a City, requiring architects, constructors, and craftsmen (Ellul, 1970). Perhaps, optimists like Kelly are right and technology, used wisely, will help create the sustainable and safe society we all hope for. What attracted and seduced Eve and Adam in the Garden may not have been an apple but a stone tool, perhaps with a warning tag 'to be used creatively but responsibly'. Thinking about the consequences is, yet, not hard-wired in our DNA, unfortunately. There is a resemblance with the problem of obesity: our body can handle scarcity much better than abundance. Likewise, we are well-equipped to recognize the affordances of new technologies, but we fail to recognize the trade-offs. Will this capacity ever evolve? Reflecting on the 'T' of technology in STEM thus is reflecting on how we should live.

2.7 Conclusion

Technology is the central concept of STEM. Technology-as-artefact complements science, engineering, and mathematics, as the product of practices. Technology-as-techne reminds us that not all practices that create valuable products have a rational foundation in evidence-based academic theories. Technology-as-art has the double connotation of valuing traditional craftsmanship and present-day Art School creativity. Technology-as-embodiedness reminds us that technology is not a black box but reflects our material selves. Situating technology in both a developmental and an ethical context appeals to us to apply our creativity and problem-solving capacities freely to design technology as a solution to a problem, but always in a responsible way.

References

Abrahamson, D., Gutiérrez, J. F., & Baddorf, A. K. (2012). Try to see it my way: The discursive function of idiosyncratic mathematical metaphor. *Mathematical Thinking and Learning, 14*, 55–80. https://doi.org/10.1080/10986065.2012.625076

Arthur, W. B. (2009). *The nature of technology—What it is and how it evolves*. Free Press.

Bickerton, D. (2014). *More than nature needs. Language, mind, and evolution*. Harvard University Press.

Browne, T. L. (2003). *Making truth. Metaphor in science*. The University of Illinois Press.

Collins, H. (2010). *Tacit and explicit knowledge*. The University of Chicago Press.

Dehaene, S. (2020). *How we learn*. Viking.

Ellul, J. (1970). *The meaning of the city*. Eugene, Wipf and Stock.

European Journal of STEM Education. (2021). *Special issue on STEM and arts education*. https://www.lectitopublishing.nl/european-journal-of-stem-education/stem-and-arts-education

Feinberg, T. E., & Mallat, J. M. (2016). *The ancient origins of consciousness*. MIT Press.

Gadamer, H.-G. (1960). *Wahrheit und Methode [Truth and Method]*. Frankfurt, J.C.B. Mohr.

Gadamer, H.-G. (1972). Theorie, Technik, Praxis. In *Neuere Philosophie II* (pp. 243–266). J. C. B. Mohr (Paul Siebeck).

Gibson, E. J., & Pick, A. D. (2000). *An ecological approach to perceptual learning and development*. Oxford University Press.

Glenberg, A. M., & Kaschak, M. P. (2002). Grounding language in action. *Psychonomic Bulletin & Review, 9*(3), 558–565.

Goorden, L. (2017). *De sprong in de techniek. Nadenken over wat we doen [Jump into technology. Thinking about what we do]*. ISVW Uitgevers.

Hurford, J. R. (2007). *The origins of meaning: Language in the light of evolution*. Oxford University Press.

Kelly, K. (2011). *What technology wants*. Penguin Books.

Lakoff, G., & Johnson, M. (1980). *Metaphors we live by*. University of Chicago Press.

Lakoff, G., & Johnson, M. (1999). *Philosophy in the flesh*. Basic Books.

Malafouris, L. (2013). *How things shape the mind*. The MIT Press.

Mareschal, D., Butterworth, B., & Tolmie, A. (Eds.). (2013). *Educational neuroscience*. Chichester.

McGinn, C. (2015). *Prehension*. The MIT press.

Mitcham, C. (1994). *Thinking through technology*. The University of Chicago Press.

Nersessian, N. (2008). *Creating scientific concepts*. MIT Press.

Niebert, K., Marsch, S., & Treagust, D. F. (2012). Understanding needs embodiment: A theory-guided reanalysis of the role of metaphors and analogies in understanding science. *Science Education, 96*(5), 849–877. https://doi.org/10.1002/sce.21026

Polanyi, M. (1962). *Personal knowledge*. University of Chicago Press.

Sennett, R. (2009). *The craftsman*. Yale University Press.

Sullivan, K. (2017). Conceptual metaphor. In B. Dancygier (Ed.), *The Cambridge handbook of cognitive linguistics* (pp. 385–406). Cambridge University Press.

Ware, C. (2013). *Information visualization. Perception for design*. Morgan Kaufmann.

Wolfe, J. M., Kluender, K. R., & Levi, D. M. (2009). *Sensation and perception*. Sunderland, Sinauer.

Zwaan, R. A., & Kaschak, M. P. (2009). Language in the brain, body, and world. In P. Robbins, & M. Ayede (Eds.), *The Cambridge handbook of situated cognition* (pp. 368–381). Cambridge University Press

Hanno van Keulen received his PhD from the faculty of Chemistry of Utrecht University on 'Making Sense—How chemistry students can learn to do research in organic synthesis education'. He has tried to make sense of the material world ever since, in various positions from higher education staff and educational development to STEM in primary and early childhood care and education. He currently is Director of the MSc Science Education and Communication, that is the Teacher Training College of Delft University of Technology in the Netherlands. He is Editor-in-Chief of the European Journal of STEM Education.

Chapter 3
STEM as Integration-Maximising Learning Opportunities

Kerry Lee

Abstract This chapter will outline the history, theories and approaches pertaining to the notion of integrated learning, before providing an overview of different types of integrated STEM education. Comparisons and connections will be made between various notions of integration including Kelley and Knowles' four-pulley conceptual framework, Bybee's nine perspectives of STEM, Nadelson and Seifert's STEM Continuum and Vasquez's four forms of STEM integration. This will provide the reader with a deep understanding of the levels and approaches to STEM integration. The role and confusion relating to the inclusion of the T (technology) in the STEM acronym will be provided. The benefits of STEM integration for teachers and learners will be outlined, followed by the key challenges of implementation. Prior to the conclusion is recommendations for future research.

Keywords Integration · Interdisciplinary · Multidisciplinary · Transdisciplinary · Models

3.1 Introduction

This chapter builds on the notions of the philosophy of STEM introduced in the previous chapter. It outlines the many facets of STEM integration. The chapter will begin with a brief overview of integration of learning before narrowing down and delving into STEM integration; the foundations; and the approaches before finally reflecting on the implications for teachers and learners.

K. Lee (✉)
University of Auckland, Auckland, New Zealand
e-mail: k.lee@auckland.ac.nz

W. Fox-Turnbull and P. J. Williams (eds.), *Locating Technology Education in STEM Teaching and Learning*, Contemporary Issues in Technology Education,
https://doi.org/10.1007/978-981-97-1995-2_3

3.2 History Behind the Notion of Integrated Learning

The origins of the term integration in the field of education have been recorded by a number of authors including Barber (2012, 2014), Huber (2005) and Perez and Barber (2018). For a detailed account of the progression of the notion of integration see Klein (1990). Klein stated the earliest notable uses of the term integration appeared in a book by Herbert Spencer in 1855 but it was not until 50 years later, in 1898, that Alexis Bertrand wrote about the history of integrated instruction. It was at this time that integration was first recognised as a pedagogy (Goldberg, 2012). Initially the term referred to natural relations among the disciplines (Ciccorico, 1970). By the 1920s the term was linked with a project approach to learning, with an emphasis on the students' personal and social concerns. It was associated with the core curriculum movement in the 1930s and problem-solving in the 1940s. Since this time the focus has been on content integration and process integration (Klein, 2005).

Integration of learning has increased in popularity and has now become a buzzword (Leadbeatter, 2021). With many believing that the "integration of learning is a central goal for all professions" (Clapton et al., 2008, p. 334), including that of education (Jamison et al., 2014), and is a superior and ideal type of learning (Leadbeatter, 2021). The increase in popularity of integrated learning is also evidenced by the increase of research in the field, with approximately 1,470,000 research articles written between 2005 and 2010 and an impressive 2,030,000 between 2010 and 2015 (using a Google Scholar Search). Initially the general focus during these times was on approaches and knowledge drawn together from more than one discipline in an integrated unit, thematic study or a social-problem approach to science. However latterly the focus has been more on interdisciplinary study using communities of learning and collaborative approaches.

Although integration of learning is now routinely part of a teachers' lexicon it is rarely explained, with many incorrectly presuming a common and shared understanding of the term. However integration of learning can imply an outcome (Booth, 2011; Huber & Hutchings, 2004) or an action (Gill, 2007; Goldman & Schroth, 2012). Or it can focus on learner dispositions (Welsh & Dehler, 2013; Wolff & Luckett, 2013). It is also referred to as a process of learning, where the learner pulls together experience, knowledge and skills, gained from a range of contexts. This enables the learner to see similarities and *connect* the knowledge, *apply* the knowledge to a new context or *synthesise* existing knowledge to create new knowledge (Barber, 2014). Others refer to the integration of learning as a curriculum process (Goldman & Schroth, 2012); a goal of education (Clapton et al., 2008; Jamison et al., 2014); or any combination of these. Unfortunately these many interpretations of the term have proven to be problematic (Leadbeatter, 2021; Saito et al., 2016). The notion of STEM integration is also problematic, with commonly used terms including: STEM integration, integrated STEM, integrative STEM and interdisciplinary STEM (Moore et al., 2020).

3.3 Theories and Approaches Pertaining to Integration of Learning and STEM Learning

In order to understand the foundation of STEM integration it is important to first understand the foundation of all integrated learning; as one is a subset of the other. These will only be discussed briefly in order to provide a reference point, as they are expanded upon in Section B of this book. The conceptualisation of integration of learning has been founded on many theories. These include the famous works by Lao Tzu, Buddha, Heraclitus, Kant and Vico (Prince & Felder, 2006), followed by constructivist theorists such as Dewey, Piaget and Vigostsky, as well as Rogers, Montessori and Maslow (Ültanir, 2012).

The works by these theorists underpin well-known approaches used in STEM teaching and learning, which include: discovery methods (Bruner, 1960); inquiry-based learning (Schwab, 1960); project methods (Wingert et al., 2011); problem-based learning (Barrows, 1996); Bloom's taxonomy (Boom, 1956); Fischer's skill theory (Fischer, 1980); SOLO taxonomy (Biggs & Collis, 1982); planned behaviour, (Ajzen, 1985); just-in-time teaching (Novak et al., 1999); multicultural pedagogies (Jamison et al., 2014); transfer of learning (Barber, 2012); constructive developmental perspectives (Barber, 2014; Perez & Barber, 2018); as well as the use of tools as inquiry, discussion methods and peer-teaching. Because STEM learning has a practical nature it is often associated with cross-curricular learning (Barnes, 2011), theme-based learning (Mumford, 2000), generative learning (Wittrock, 1989) and real-life learning (Steele & Ashworth, 2013).

The majority of the above theories and approaches are underpinned by the notions of social constructivism. Learning is expected to be driven by the learner, and the teacher is the facilitator or supporter alongside. The traditional transmission teacher with the teacher standing in front of the classroom is incompatible with these approaches. Learning is strongly influenced by prior knowledge. Motivation is derived from authentic real-world professionally relevant situations and problems (Moore et al., 2020) which provide context for learning content and skills (Prince & Felder, 2006). Teaching is expected to move from a teacher-centred pedagogy to a learner-centred teaching and learning approach: a content-based paradigm, to an outcomes-based paradigm.

3.4 STEM Integration

Initially teachers used STEM to holistically refer to the four separate and distinct fields of science, mathematics, engineering and technology (Kelley & Knowles, 2016; Sanders, 2009; Wang et al., 2011); subjects which had historically been taught in isolation (Banks & Barlex, 2014b). However, the teaching of siloed subjects seemed to show a disconnection to real-world situations and problems where multi-disciplinary teams are needed (Wang et al., 2011). In this way disciplines are seen as interdependent (Bybee, 2013a; Kelley & Knowles, 2016; Wang et al., 2011). Until the late 2000s, researchers and practitioners emphasised the connections between subjects and advocated new *integrative* approaches to design and STEM (Honey et al., 2014; Kennedy & Odell, 2014).

Although disciplinary crossing is one of the most significant features of integrative STEM, the approaches of integration and the extent of crossing vary considerably (English, 2016). For example, some researchers (e.g. Honey et al., 2014; Wang et al., 2011) emphasise that the context or the problem should be complex enough to require students to apply more than one subject's knowledge. Honey et al. (2014) defined integration as "working in the context of complex phenomena or situations on tasks that require students to use knowledge and skills from multiple disciplines". Other researchers propose merging the content of one or more given subject(s) into other subjects' courses (e.g. Kennedy & Odell, 2014; Sanders, 2009), such as when students gain science learning outcomes in a technology class (Kelley & Knowles, 2016). Due to the diverse interpretations of integration, more comprehensive perspectives and systems have been proposed, in which different types of integrated STEM are organised by certain rules. Most agree that STEM education integration requires at least two disciplines, with many requiring more (Moore et al., 2020).

This chapter provides an overview of a number of categorizations that define different types of integrated STEM education, including Kelley and Knowles (2016)'s four-pulley conceptual framework, Bybee (2013b)'s nine perspectives of STEM, Nadelson and Seifert (2013)'s STEM Continuum and Vasquez (2014)'s four forms of integration. The four notions of STEM integration presented by Vasquez (2014) are now commonly used, and for this reason they will be overlayed on the models developed by Kelley and Knowles, Bybee and Nadelson and Seifert to show the cohesion and commonality between the various interpretations.

3.4.1 Levels of Integration

Vasquez (2014) provided four labelled and clearly differentiated forms of integration: disciplinary, multidisciplinary, interdisciplinary and transdisciplinary. In this model the extent of integration and the interconnection amongst the subjects increases along the continuum (as shown in Fig. 3.1). The initial STEM acronym highlighted the importance of each of the respective subjects, however, a preference quickly developed for interdisciplinary and transdisciplinary approaches (English, 2016; Vasquez, 2014). Fox-Turnbull in her chapter identifies a number of issues and innovations with each approach as it relates to the primary and lower secondary classroom.

As the teachers increase the level of integration the distinction between disciplines becomes more blurred (Vasquez, 2014). In a multidisciplinary approach to STEM the topic is the only thing that ties the learning together. For example with the topic "Change", the children could: learn about states of matter and heating and cooling in science; look at how various cultures preserve food using dehydration, before making their own dehydrated snack in technology; investigate the how the wind and

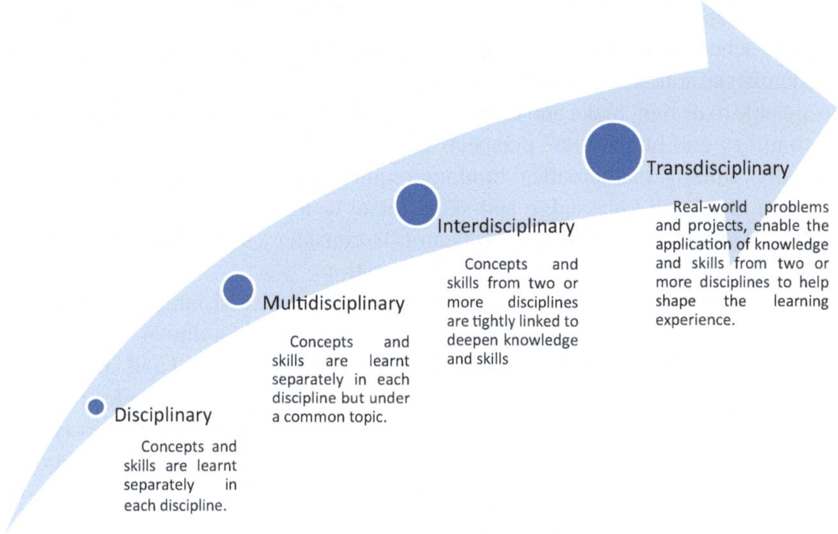

Fig. 3.1 Increasing levels of STEM integration. Adapted from English (2016) and Vasquez et al., (2013)

sun are used to make electricity before applying this to their own engineering design; and regularly measure and record the wind speed and direction as well as rainfall in mathematics. In an interdisciplinary STEM approach, teachers support students to undertake a cross-disciplinary topic, with the disciplinary knowledge and skills connected and interdependent. In transdisciplinary STEM, students apply knowledge and skills across disciplines to solve a real problem or a project. For example, to help bring birds back to the cities, students can record the number of birds shown in a park near the school on a daily or hourly basis and make plots to describe the main trends (mathematics); to do this they will need to have undertaken research on bird species for identification, which they would later build on and include research on bird species' habits and habitats (science), in order to make objects which might attract specific birds, e.g. feeding tables, sugar water holders, insect attractants, nesting boxes and bird baths (technology and engineering).

3.4.1.1 Kelley and Knowles's (2016) Four-Pulley Conceptual Framework

In Kelley and Knowles's (2016) conceptual framework, the four pulleys, representing STEM subjects, are bound by the rope of community practices and work cooperatively to lift a load of situated STEM learning. The framework requires the four subjects to work in harmony to ensure the whole system powers and activates authentic and relevant learning in a situated context. The engineering design works as the entry point and incorporates engineering practice into the curriculum and provides students with a systematic method to solve problems. The scientific inquiry requires teachers to encourage students' curiosity and motivate them to think and act like scientists conducting scientific practices. Technological literacy refers to students who are able to design, make and use technology, including engineering perspectives of technology and humanities' perspectives of technology (see Herschbach, 2009, for details). Finally, mathematical thinking requires students to learn maths content as well as how such knowledge and skills relate to life and to solving real-world problems. This framework prepares educators to consider a simplistic and integrated STEM and pedagogical approach to implementation.

When referring to Vasquez et al. (2013)'s levels of STEM integration Fig. 3.2 has the boxes representing the disciplinary level of integration drawn separately to show that at this level the disciplines are important but are isolated and siloed from each other. Multidisciplinary integration involves the subjects working independently but on the same related topic. Whilst interdisciplinary integration involves two or more of the STEM disciplines working closely together, utilising content and procedural knowledge when needed. Transdisciplinary is drawn to cover all of the STEM disciplines, which work together in a cohesive way.

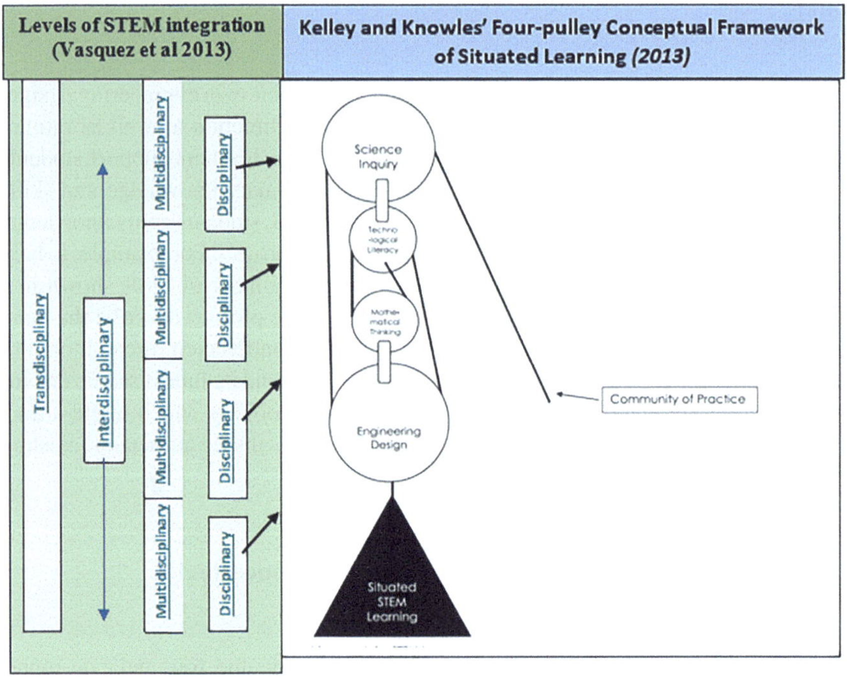

Fig. 3.2 Kelley and Knowles' four-pulley conceptual framework of situated STEM learning. Adapted from Kelley and Knowles (2016) and Vasquez et al., (2013)

3.4.1.2 Bybee's (2013b) Nine Perspectives of STEM

Based on numerous resources (e.g. discussions, projects and articles), Bybee (2013b) identified nine perspectives of STEM education, each of which was provided with an analogy and figure to visually demonstrate the relationships between each of the STEM disciplines. In this chapter the reader has been provided with a condensed summary of these perspectives via a table and accompanying text (see Table 3.1).

Bybee identified that some people although referring to STEM were actually only referring to science or occasionally mathematics and only one branch of the discipline such as physics or biology. This single-discipline reference was Bybee's first perspective. It is noted that such use of the STEM acronym violates the multiple disciplinary orientations of the term itself. The second perspective of STEM referred to the two disciplines of science *and* mathematics. Although the two disciplines were identified they were seen as individual entities and treated as silos. Technology

Table 3.1 Perspectives and features of STEM education

Levels of STEM integration (Vasquez et al., 2013)	Perspectives and features of STEM education (Bybee, 2013b)	
	Perspectives	Features
Disciplinary	1. Single-discipline reference	STEM equates to science (or mathematics)
Multidisciplinary	2. STEM as a reference for science and mathematics	Separate disciplines of science and mathematics
	3. Separate science disciplines that incorporate other disciplines	Separate science disciplines that incorporate technology, engineering or mathematics as appropriate
	4. Separate disciplines	Separate disciplines including science, technology, engineering and mathematics
Interdisciplinary	5. Science and math connected by technology or engineering programme	Connection between science and mathematics made by technology and engineering programme
	6. Coordination across disciplines	Concepts, processes and resources are coordinated across boundaries to separate disciplines
	7. Combining two and three disciplines	Combine two (or three) disciplines to form a course
	8. Integrated disciplines	Integrated disciplines through overlapping and sequencing
Transdisciplinary	9. STEM as a transdisciplinary course or programme	Transdisciplinary

Adapted from Bybee (2013b) and Vasquez et al., (2013)

and engineering were acknowledged but never visible or included in the discussion. Bybee refers to these as two disciplines as postholes (you know they are there, but you can't see them). In the third perspective, one or more other disciplines is incorporated into science lessons. For example, a science lesson might include technology or engineering when deemed appropriate. Although this perspective keeps science as the dominant discipline, it represents the first step towards integration. The fourth perspective is where "STEM equals a quartet of separate disciplines" (Bybee, 2013a, 2013b, p. 76). This perspective is similar to the second perspective but includes all four subjects, each treated as individual silos. This perspective may incorporate separate units within a course or cover four courses. Although the four disciplines are evident, they are not generally all seen as equal.

The fifth perspective refers to science and mathematics courses which connect to programmes that emphasise technology or engineering. For example, a science course may undertake STEM when it links to a career and technical education programme (CTE) which involves biomedical and agriculture projects, where students use technology or engineering methods to connect science and math information to solve problems. The sixth perspective refers to teachers from different disciplines working together: one teaching concepts to support the learning in another discipline's course. For example, graphing techniques are introduced in a mathematics class before being applied in an engineering course. Ideally all four would be coordinated, but in reality, concepts and processes of two disciplines are usually transferred. The seventh perspective refers to the integration of two or three disciplines where each disciplines has equal emphasis (e.g. creating a new course on science and technology with each having equal weight). In the eighth perspective, STEM is integrated by sequencing disciplines in a series of units or courses, for example, students investigate a scientific question by progressing through a sequence of lessons focussing on the four disciplines. Finally, the ninth perspective is a transdisciplinary approach where all four disciplines and possibly others are utilised to understand an issue such as a "wicked problem". Often this approach is only adopted when students have undertaken traditional siloed courses for the disciplines. Bybee notes he does not intend to identify one "true STEM perspective" but show there are numerous perspectives which need to be considered when considering or engaging in education reform.

Table 3.1 highlights how Vasquez et al. (2013)'s levels of STEM integration is easily linked with Bybee's (2013b) perspectives and features of STEM education. Vasquez et al. (2013)'s disciplinary level of integration, links with Bybee's first "single-discipline" perspective, and the multidisciplinary level is associated with Bybee's perspectives 2–4. Whilst perspectives 5–8 are seen as interdisciplinary, and perspective 9 is clearly transdisciplinary.

3.4.1.3 Nadelson and Seifert's STEM Continuum

This model identified what was occurring in local business and industry as well as the classroom. It did not place judgement as to which facet was the most desired or approved but rather stated that "all regions of the STEM continuum have value and are essential for teaching and learning" (Nadelson & Seifert, 2013 para. 7). As shown in Fig. 3.3, the first column has the STEM disciplines separated. Content is delivered by the teacher to the student generally in an instructional manner. This engages the learner in vocabulary building and memorisation and is what is deemed necessary for STEM literacy. Vasquez et al. (2013) would refer to this as the disciplinary level. The middle column has some disciplines overlapping in applied and acquired learning. Students are provided foundational content via a top-down approach but are also able to build skills by a bottom-up approach with the teacher or mentor working alongside. Vasquez et al. (2013) would refer to this as the multidisciplinary level. The far right column has STEM fully integrated, with learners solving complex, ill-structured problems found in industry and business. Knowledge is developed from the bottom-up, with the teacher acting as a facilitator, and due to the complexity of the problem there is no "right answer", but rather multiple solutions which need to be evaluated for feasibility. Vasquez et al. (2013) would refer to this as interdisciplinary or transdisciplinary depending on how the integration occurred and how much student ownership there was.

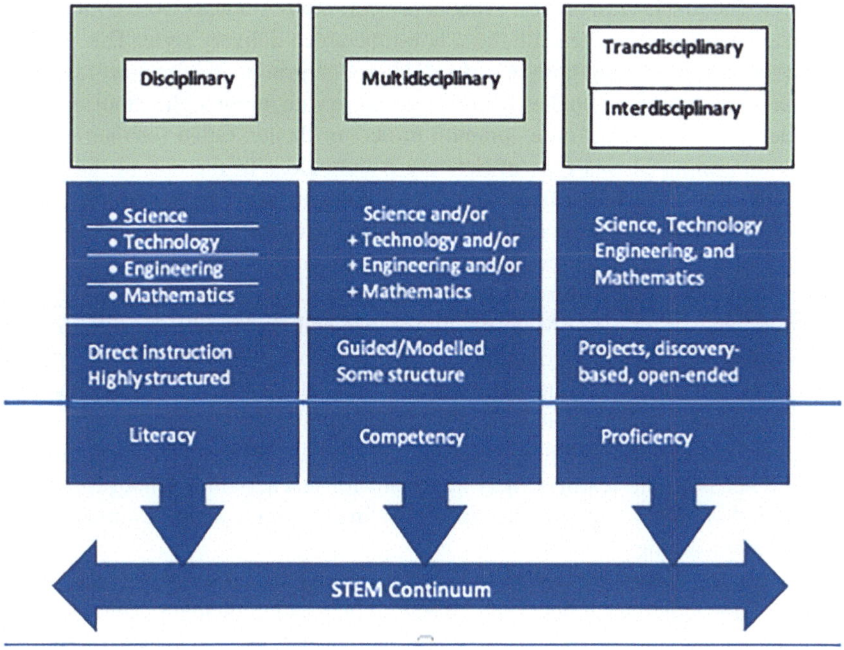

Fig. 3.3 STEM Continuum. Adapted from Nadelson and Seifert (2017) and Vasquez et al., (2013)

3.4.2 Approaches to Integration

The various types and levels of integration can be achieved in many ways. Banks and Barlex (2014a) are well recognised for their three approaches to working across the STEM disciplines: these being the coordinated approach, the collaborative approach and the integrated approach.

The coordinated approach requires the teacher to be aware of not only what is being taught in the classrooms of other disciplines, but also when this teaching is occurring. This enables a coordinated approach to ensure that language, principles and analogies are shared and reinforced across the disciplines in a timely fashion. Learning is still siloed but reference is made to learning occurring in other disciplines. The collaborative approach requires teachers to plan together so that activities support learning across the disciplines. Each activity or series of activities is able to be applied to the next discipline's classes. This approach often involves investigating real-world problems such as the need for renewable energy. The lessons are still teacher-centred and well structured, but the knowledge and skills developed by the learner are easily transferrable.

Banks and Barlex (2014a) suggest the integrated approach can be undertaken in two ways. The first is via a one-off event or project, where students work with a range of teachers who are team teaching in a non-timetabled delivery style. This is often more easily achieved with those in early childhood or primary (elementary) settings. At the secondary/high school level, this occurs more with junior high school students: all students work together on a common project or theme. Often for management reasons the timing of this type of delivery occurs close to the end of the school year when sessions are less formal, as senior students are busy with examination preparation. The second approach is a full integration of the STEM disciplines. This approach is more easily achieved at the primary/elementary level as it requires one teacher to follow a themed project across a number of lessons.

In their chapter on a STEM curriculum Banks and Barlex (2014a) refer to "the three ways of viewing the curriculum: Specified, Enacted and Experienced" (p. 45). The curriculum directly influences what a teacher does and what is experienced by pupils. In most countries the specified curriculum is the legal document which underpins teaching and learning. Teachers are able to interpret the curriculum to meet the needs of the learners. It is important for teachers in a school to unpack what STEM is, to develop a shared understanding of what constitutes good STEM teaching and learning.

The enacted curriculum refers to what teachers do. This is dependent on the teacher's knowledge (what they bring to bear to plan and implement their teaching). Often teachers will focus only on their personal discipline knowledge to achieve a specified learning outcome. Yet by utilising the strengths of a teacher from another discipline their own teaching and student learning can be enhanced. The experienced curriculum incorporates both the specified and the enacted curriculum and in effect it is what the students learn. Students gain and transfer knowledge from one subject to another often incidentally. Rarely are links overtly made, and the transfer of skills and knowledge planned and supported.

The myriad of interpretations and approaches reflect that the problem is not only a terminology issue, but also a conceptual one. Prince and Felder (2006) reviewed the effectiveness of many of the inductive teaching methods, commonly used in STEM teaching and learning, and identified that there was conflicting evidence on the effects of integration of STEM, with research finding the approach to be both positive and negative for teachers and learners.

3.5 Technology and STEM Integration

Technology education has undergone a metamorphosis over the last century. It has moved from a focus on technical, manual and life skills, to vocational, craft and industrial arts, through to its current form with its focus on design and innovation (Ferguson, 2009; Jones et al., 2013). The word technology has been used in many science curricula, generally to refer to the use of machines, tools, engineering concepts and/or digital devices. This may have been one of the reasons there has been confusion between science and technology (Constantinou et al., 2010; Sanders, 2009), engineering and technology (Wicklein, 2006), applied science and technology (Layton, 1993) and technical and technology education (Lawal, 2014). Often technology is synonymous with construction and practical activities even if no design is involved. Some think "using technology" is the same as "doing technology". In fact when a technology educator is asked what they teach they spend more time explaining what they don't teach (Kelley, 2010).

It is therefore not surprising that the confusion continued when acronyms arose such as: S&T (science and technology); STS (science, technology and society); SMET (science, mathematics, engineering and technology); TAS (technology as applied science); SET (science, engineering and technology); ETE (engineering and technology education); MST (mathematics, science and technology); STEM (science, technology, engineering and mathematics); and more recently STEAM (science, technology, engineering, art and mathematics) (Hacker & Barak, 2017; Lee & Granshaw, 2018; Wells, 2019). Whilst a new term (i-STEM) that has recently been coined already has multiple meanings: integrative STEM (Chang & Chen, 2020); integrated STEM (Gunbatar et al., 2022); integrative instructional blending of the individual STEM elements (McComas & Burgin, 2020); and interdisciplinary STEM education (Saray, 2022).

As a stand-alone discipline, technology education is often confused with educational technology (Brown & Brown, 2010), e-learning, digital technology, information and communication technology (ICT) (Compton & Compton, 2013; Lee, 2019). The confusion is compounded when reading about STEM as many authors refer to technology as digital devices, robotics and high-tech gadgetry. However in *technology education* these digital devices are seen as tools to help the students undertake technological practice.

Technological practice is generally practical, focusses on design, innovation and creativity, builds tacit knowledge by working with expert others and ideally is authentic (Lee et al., 2019). Authenticity is very important and can refer to the initial problem posed to the learner, the process and selected materials, the interest and ability of the learner and the modelling of real-life practice (Lee, 2011; Snape & Fox-Turnbull, 2013). Hacker & Barak (2017) wrote "the National Institute of Science and Technology, differentiates between engineering and technology by explaining that engineering is oriented toward the solution of specific problems, whereas technology is oriented toward development of new capability" (pp. 31–32). Their research showed that technology closely aligns with engineering as both disciplines involve design, modelling, systems, resources and human values (Hacker & Barak, 2017).

3.5.1 Advantages of STEM Integration

Integration of learning can simultaneously connect theoretical foundations, (Huber & Hutchings, 2004), applied skills (Hassan et al., 2019), tacit knowledge, as well as social knowledge and knowledge of self and others (Booth, 2011; Weurlander et al., 2016).

STEM integration has numerous advantages to teachers and learners (Honey et al., 2014). It has been shown to produce increased numbers in the STEM professions (Reeve, 2021), as well as providing career awareness (John et al., 2018; Ryu et al., 2018; Sullivan et al., 2013) and an ever-changing workforce (Woods, 2020). It facilitates experience in real-world problems (Burrows et al., 2018; Lowrie & Larkin, 2020; Weng & Li, 2020), whilst making connections within the STEM field (Reeve, 2021; Ryu et al., 2018).

Rather than siloed teaching and learning there is an infusion of STEM concepts which often assists with transference and linkages between concepts (Myers, 2015) and shared specialised vocabulary, representation symbols and measurement instruments (Nathan et al., 2017). An integrated STEM curriculum can promote the development of STEM literacy (Sullivan et al., 2013) which is crucial for student success (Fayer et al., 2017, January). STEM integration has consolidative qualities that also incorporate general literacy skills (Welsh & Dehler, 2013).

An integrated STEM curriculum has been shown to improve educational outcomes for children. It has been found to support students with twenty-first century skills (Bybee, 2013a; Honey et al., 2014) such as problem-solving (Tank et al., 2018; Weng & Li, 2020), higher order thinking (Honey et al., 2014), critical thinking, resilience (Kewalramani et al., 2020), persistence (Kewalramani et al., 2020), self-regulation (Aldemir & Kermani, 2017), as well as questioning, analysing and working with others (Aldemir & Kermani, 2017). Evidence shows that a comprehensive integrated STEM curriculum can promote cognitive development (Fogarty, 2009; Klein & Knitzer, 2006; Zan & Geiken, 2010) and curiosity (Lin et al., 2021). Both formal and informal integrated STEM education have been shown to boost children's metacognitive and social knowledge construction (Kewalramani et al., 2020). It can increase student performance (Aldemir & Kermani, 2017); specifically their knowledge (John et al., 2018; Lin et al., 2021), skills (Lin et al., 2021; Weng & Li, 2020), understanding (Tank et al., 2018) and higher order thinking (Wells, 2016).

In the younger years it provides a solid foundation for the STEM disciplines (Vasquez et al., 2020a). "STEM integration in ECE [early childhood education] in developmentally appropriate ways helps to close the achievement gap that places ethnically and economically diverse children at a disadvantage from the beginning of their lives" (Aldemir & Kermani, 2017, p. 1704).

STEM integration strongly enables and encourages learner-centred approaches (Aldemir & Kermani, 2017; Kewalramani et al., 2020; Lin et al., 2021; Tank et al., 2018; Tippett & Milford, 2017; Weng & Li, 2020). It provides learning based on students' interests (Tippett & Milford, 2017) and has the ability to change attitudes (John et al., 2018), increase engagement (Lin et al., 2021; Tank et al., 2018; Vasquez et al., 2013), motivation (Aldemir & Kermani, 2017; Dilek et al., 2020; Lin et al., 2021), enthusiasm (Smith & Cline, 2016) and a sense of confidence (Vasquez et al., 2020a). Students are encouraged to develop independence of thought and critical analytical mindset (Booth et al., 2009), whilst the effectiveness of STEM has also been evidenced by an increase in persistence in completing activities (John et al., 2018).

Many studies have shown that students of low socio-economic and ethnic minority groups have already fallen behind when they enter school and that STEM programmes have the ability to close the gap (Haskins & Rouse, 2005; Howes et al., 2008; Mehalik, 2008). By providing carefully considered contexts and content, a wider and more diverse group of students can be engaged (Wasserman & Rossi, 2015) to undertake activities which seem more relevant and applicable to their lives (Lou et al., 2011). Moreover, Tank et al. (2018) reported that STEM integration promotes equity for girls through early education.

Well planned STEM activities can provide powerful and meaningful learning experiences (Aldemir & Kermani, 2017; Huber & Hutchings, 2004; Steele & Ashworth, 2013) and connections (Smith & Cline, 2016) which develop a sense of community (Dilek et al., 2020; Vasquez et al., 2020a) and agency (Dilek et al., 2020). Rich-task learning (Lowrie & Larkin, 2020) develops collaboration skills (Huber & Hutchings, 2004; Kewalramani et al., 2020) and helps students apply knowledge and concepts in meaningful ways (Hassan et al., 2019). Students are able learn content from multiple perspectives (Cheng & So, 2020; Huber & Hutchings, 2004; Welsh & Dehler, 2013; Wingert et al., 2011), make linkages (English, 2016; Weng & Li, 2020), develop interconnections (Cheng & So, 2020) and interrelationships between STEM disciplines (English, 2016), as well as undertaking peer-teaching (Kewalramani et al., 2020). A recent literature review of 109 articles focussing on STEM integration justified the need for integration because STEM disciplines were connected by big ideas and common skills and practices (Moore et al., 2020).

The integration of theory and practical work can optimise delivery time (Huber & Hutchings, 2004), as well as helping teachers to challenge themselves and recognise their crucial role in supporting prerequisite skills necessary for children to learn more advanced STEM concepts for the future education (Aldemir & Kermani, 2017). As a result teachers often feel more empowered (John et al., 2018) when being involved in STEM projects.

3.6 Cautions for STEM Integration

Although STEM integration has many noted advantages, it is not easy to implement and there is mixed evidence on its efficacy (Honey et al., 2014). It generally utilises non-traditional teaching techniques and therefore requires the buy-in of teachers, to challenge their previous teaching practices for new techniques. This requires considerable support, professional development and time (Reeve, 2021). Many students and teachers are underprepared to engage and implement STEM education (National Research Council, 2012).

Many believe that integration has resulted in modularised programmes (Booth et al., 2009), a sense of loss of teacher's personal relationships with students (Booth et al., 2009), larger class sizes (Booth et al., 2009) and the difficulty of covering subjects in depth (McDowall & Hipkins, 2019).

Those considering STEM implementation need to consider if the benefits outweigh the issues (Banks & Barlex, 2014a). Many believing that it takes time away from each of the STEM disciplines resulting in key concepts and procedures not being taught and prescribed targeted learning not being achieved (English, 2016).

There are a number of challenges associated with the integration of learning. "While many programs (sic) are teaching undergraduates, they may not be preparing professionals. Integrative disciplinary knowledge is essential for professional preparation" (Gill, 2007, p. 260). Many perceive that subjects will lose their status if their knowledge is merged with that of a "soft" subject (Banks & Barlex, 2014a; Steele & Ashworth, 2013). The teachers from different STEM disciplines and their academic backgrounds have a major influence on STEM integration (Banks & Barlex, 2014a) and what can and should be conceived as possible (Stubbs & Myers, 2016).

Integration is not always easy. Initially integration of content material can slow down the learning process, often requiring a higher cognitive load (Samson, 2014). Teachers have commented on the lack of curriculum materials (Banks & Barlex, 2014a), the importance of logistics such as the space, access to supplies, classroom management, timing of integration of subjects (Steele & Ashworth, 2013), addressing potential timetabling issues (Cheng & So, 2020) and the lack of resources (Prendergast et al., 2019). It requires teachers and learners to be able to successfully transfer knowledge and skills, from one discipline to another (Hudson et al., 2015).

Teacher self-efficacy plays an essential role in STEM pedagogy integration (John et al., 2018). Generally teachers new to STEM need to learn "new concepts, content, or practices that may be "outside their area of expertise" (Reeve, 2021, p. 96). A survey in 2012 identified low levels of STEM content knowledge for STEM teachers, with only 14% of high school science teachers having taken one or more engineering courses (Banilower et al., 2013). Not surprisingly then, research has found that most teachers were apprehensive of teaching and integrating engineering design

(Cunningham (2009), and they believe they have limited content knowledge and confidence to implement STEM integration (Berlin & White, 2012; Honey et al., 2014).

A national investment in supporting STEM education needs to be a priority to build knowledgeable skilled and innovative members of our workforce (Carnegie Corporation of New York, 2009). However, one challenge many teachers face is the constant expectation for a strong and heavy emphasis on literacy and numeracy rather than STEM education (Aldemir & Kermani, 2017).

There is a lack of personal experience for teachers and student teachers of STEM (Ryu et al., 2018). This means that student teachers have limited experiential knowledge of what STEM lessons should and could look like, do not have role models, a community of practice or the opportunity to witness quality STEM teaching on practicum (Ryu et al., 2018). The long-term effect of STEM pre-service education programmes is compromised as existing school cultures and structures enculturate pre-service teachers into traditional teaching practices and discourage new ways of teaching (Putnam & Borko, 2000; Ryu et al., 2018, p. 505). Even after undertaking STEM pre-service education, many student teachers feel ill-prepared to implement fully integrated STEM units (Berlin & White, 2012).

Integration of STEM has been shown to capture, increase and build on students' learning interests; however rarely are STEM programmes student initiated or developed as a result of student interest (Ng et al., 2022).

3.7 Future Recommendations

More examples of fully integrated STEM topics which have been successful in engaging children in authentic STEM tasks (Ng et al., 2022) are needed for teachers. More articles such as Banks and Barlex (2014a), Vasquez et al. (2020b) and websites such as https://www.pta.org/docs/defaultsource/files/programs/stem/2017/math-nighttoolkit/stem-families-resources-stem-bystate.Pdf are needed.

Although STEM teaching and learning has been relatively well researched, more research is needed around: the role of the teachers (Hooper et al., 2014); challenges teachers are facing; learners who expect contemporary methods of education; and learners being moved beyond simply regurgitating and recreating the familiar to unexpected pathways, which can be very unsettling (Dev & Kumar, 2015) and even overwhelming for both teachers and learners (McDowall & Hipkins, 2019).

More research is needed to develop standardised instruments to assess students learning in STEM (Aldemir & Kermani, 2017).

3.8 Conclusion

Although there are many advantages of STEM integration, it is not necessarily an easy way of teaching. Considerable time and effort is required of the teaching team to develop successful programmes which contain flexible, authentic, integrative STEM activities that target the interests and learning needs of students. Every approach is a compromise of time, resources and discipline content. More research is needed as to the effectiveness and value of STEM integration topics, approaches and integration strategies to determine if the effort is worthwhile for all teachers and their students.

References

Ajzen, I. (1985). From intentions to actions: A theory of planned behavior. In J. Kuhl & J. Beckmann (Eds.), *Action control: From cognition to behavior* (pp. 11–39). Springer-Verlag.

Aldemir, J., & Kermani, H. (2017). Integrated STEM curriculum: Improving educational outcomes for Head Start children. *Early Child Development and Care, 187*(11), 1694–1706. https://doi.org/10.1080/03004430.2016.1185102

Banilower, E. R., Smith, P. S., Weiss, I. R., Malzahn, K. A., Campbell, K. M., & Weis, A. M. (2013). *Report of the 2012 National Survey of Science and Mathematics Education*. Horizon Research.

Banks, F., & Barlex, D. (2014b). What is STEM? In Teaching stem in the secondary school: Helping teachers meet the challenge. In *Teaching stem in the secondary school: Helping teachers meet the challenge* (Vol. 1). Taylor & Francis.

Banks, F., & Barlex, D. (2014a). *Teaching stem in the secondary school: Helping teachers meet the challenge*. Routledge/Taylor & Francis.

Barber, J. P. (2014). Integration of learning model: How college students integrate learning. *New Directions for Higher Education, 2014*(165), 7–17. https://doi.org/10.1002/he.20079

Barber, J. P. (2012). Integration of learning: A grounded theory analysis of college students' learning. *American Educational Research Journal, 49*(3), 590–617. https://doi.org/10.3102/0002831212437854

Barnes, J. (2011). *Cross-curricular learning* (2nd ed.). Sage.

Barrows, H. S. (1996). Problem-based learning in medicine and beyond: A brief overview. *New Directions for Teaching and Learning, 68*, 3–12. https://doi.org/10.1002/tl.37219966804

Berlin, D. F., & White, A. L. (2012). A longitudinal look at attitudes and perceptions related to the integration of mathematics, science, and technology education. *School Science and Mathematics, 112*(1), 20–30.

Biggs, J., & Collis, K. (1982). *Evaluating the quality of learning: The SOLO taxonomy (Structure of the Observed Learning Outcome)*. Academic Press.

Boom, B. S. (1956). *Taxonomy of educational objectives: The classification of educational goals*. David McKay.

Booth, A., McLean, M., & Walker, M. (2009). Self, others and society: A case study of university integrative learning. *Studies in Higher Education, 34*(8), 929–939. https://doi.org/10.1080/03075070902773818

Booth, A. (2011). Wide-awake learning: Integrative learning and humanities education. *Arts and Humanities in Higher Education, 10*(1), 47–65. https://doi.org/10.1177/1474022210388399

Brown, R. A., & Brown, J. W. (2010). What is technology education? A review of the "Official Curriculum." *The Clearing House: A Journal of Educational Studies, 83*(2), 49–53. https://doi.org/10.1080/00098650903505449

Bruner, J. S. (1960). *The process of education*. Harvard University Press.

Burrows, A., Lockwood, M., Borowczak, M., Janak, E., & Barber, B. (2018). Integrated STEM: Focus on informal education and community collaboration through engineering. *Education Sciences, 8*(4). https://doi.org/10.3390/educsci8010004

Bybee, R. W. (2013a). *What are the challenges for STEM education?* (Vol. 1). National Science Teachers Association.

Bybee, R. W. (2013b). *What is your perspective of STEM education?* (Vol. 8). National Science Teachers Association.

Carnegie Corporation of New York. (2009). *The opportunity equation—Transforming mathematics and science education for citizenship and the global economy*. Retrieved September 23, 2009, from https://www.carnegie.org/media/filer_public/80/c8/80c8a7bc-c7ab-4f49-847d-1e2966f4dd97/ccny_report_2009_opportunityequation.pdf

Chang, C.-C., & Chen, Y. (2020). Cognition, attitude, and interest in cross-disciplinary i-stem robotics curriculum developed by thematic integration approaches of webbed and threaded models: A concurrent embedded mixed methods study. *Journal of Science Education and Technology, 29*, 622–634. https://doi.org/10.1007/s10956-020-09841-9

Cheng, Y. C., & So, W. W. M. (2020). Managing STEM learning: A typology and four models of integration. *International Journal of Educational Management, 34*(6), 1063–1078.

Ciccorico, E. W. (1970). Integration in the curriculum. *Main Currents in Modern Thought, 27*(November/December), 60–62.

Clapton, G., Cree, V. E., Allan, M., Edwards, R., Forbes, R., Irwin, M., MacGregor, C., Paterson, W., Brodie, I., & Perry, R. (2008). Thinking 'outside the box': A new approach to integration of learning for practice. *Social Work Education, 27*(3), 334–340. https://doi.org/10.1080/02615470701379826

Compton, V., & Compton, A. (2013). Teaching the nature of technology: Determining and supporting student learning of the philosophy of technology. *International Journal of Technology and Design Education, 23*(2), 229–256. https://doi.org/10.1007/s10798-011-9176-2

Constantinou, C., Hadjilouca, R., & Papadouris, N. (2010). Students' epistemological awareness concerning the distinction between science and technology. *International Journal of Science Education, 32*(2), 143–172. https://doi.org/10.1080/09500690903229296

Cunningham, C. M. (2009). Engineering is elementary. *The Bridge, 30*(3), 11–17.

Dev, S., & Kumar, J. (2015). Teacher's perception towards integration of learning disabled students into regular class room – A study in Dubai & Abu Dhabi Schools. *Procedia-Social and Behavioral Sciences, 211*, 605–611. https://doi.org/10.1016/j.sbspro.2015.11.079

Dilek, H., Taşdemir, A., Konca, A. S., & Baltaci, S. (2020). Preschool children's science motivation and process skills during inquiry-based STEM activities. *Journal of Education in Science Environment and Health, 6*(2), 92–104. https://doi.org/10.21891/jeseh.673901

English, L. D. (2016). STEM education K-12: Perspectives on integration. *International Journal of STEM Education, 3*(1), 1–8. https://doi.org/10.1186/s40594-016-0036-1

Fayer, S., Lacey, A., & Watson, A. (2017, January). *STEM occupations: Past, present, and future U.S.* Bureau of Labor Statistics Retrieved September 23, 2017, from https://www.bls.gov/spotlight/2017/science-technology-engineering-and-mathematics-stem-occupations-past-present-and-future/pdf/science-technology-engineering-and-mathematics-stem-occupations-past-present-and-future.pdf

Ferguson, D. (2009). *Development of technology education in New Zealand schools 1985–2008*. Ministry of Education.

Fischer, K. W. (1980). A theory of cognitive development: The control and construction of hierarchies of skills. *Psychological Review, 87*(6), 477–531. https://doi.org/10.1037/0033-295X.87.6.477

Fogarty, R. (2009). *How to integrate the curricula*. Corwin.

Gill, D. L. (2007). Integration: The key to sustaining kinesiology in higher education. *Quest, 59*(3), 269–286. https://doi.org/10.1080/00336297.2007.10483552

Goldberg, M. (2012). *Arts integration: Teaching subject matter through the arts in multicultural settings* (4th ed.). Pearson.

Goldman, E., & Schroth, W. S. (2012). Deconstructing integration: A framework for the rational application of integration as a guiding curricular strategy. *Academic Medicine, 87*(6), 729–734. https://doi.org/10.1097/ACM.0b013e318253cad4
Gunbatar, S., Kiran, B., Boz, Y., & Roehrig, G. H. (2022). A closer examination of the STEM characteristics of the STEM activities published in NSTA journals. *Research in Science & Technological Education.* https://doi.org/10.1080/02635143.2022.2121692
Hacker, M., & Barak, M. (2017). Important engineering and technology concepts and skills for all high school students in the United States: Comparing perceptions of engineering educators and high school teachers. *Journal of Technology Education, 28*(2), 31–52.
Haskins, R., & Rouse, C. (2005). *Closing achievement gaps. The future of children policy brief.* Princeton Brookings.
Hassan, M. N., Abdullah, A. H., Ismail, N.,Suhud, S. N. A., & Hamzah, M. H. (2019). Mathematics curriculum framework for early childhood education based on science, technology, engineering and mathematics (STEM). *International Electronic Journal of Mathematics Education, 14*(1), 15–31. https://doi.org/10.12973/iejme/3960
Herschbach, D. (2009). *Technology education: Foundations and perspectives.* American Technical Publishers.
Hooper, B. R., Greene, D., & Sample, P. L. (2014). Exploring features of integrative teaching through a microanalysis of connection-making processes in a health sciences curriculum. *Advances in Health Sciences Education, 19*(4), 469–495. https://doi.org/10.1007/s10459-013-9481-5
Honey, M., Pearson, G., & Schweingruber, H. (2014). *STEM integration in K-12 education: Status, prospects, and an agenda for research.* The National Academies Press.
Howes, C., Burchinal, M., Pianta, R., Bryant, D., Early, D., Clifford, R., & Barbarin, O. (2008). Ready to learn? Children's pre-academic achievement in pre-kindergarten programs. *Early Childhood Research Quarterly, 23*(1), 27–50. https://doi.org/10.1016/j.ecresq.2007.05.002
Huber, M. T., & Hutchings, P. (2004). *Integrative learning: Mapping the terrain.* The academy in transition. *Association of American Colleges and Universities.*
Huber, M. T. (2005). Integrative learning as an intellectual art: Panel, closing plenary. In *Association of American colleges and universities network conference on integrative learning.*
Hudson, P., English, L., Dawes, L., King, D., & Baker, S. (2015). Exploring links between pedagogical knowledge practices and student outcomes in STEM education for primary schools. *Australian Journal of Teacher Education, 40*(6). https://doi.org/10.14221/ajte.2015v40n6.8
Jamison, A., Kolmos, A., & Egelund Holgaard, J. (2014). Hybrid learning: An integrative approach to engineering education. *Journal of Engineering Education, 103*(2), 253–273. https://doi.org/10.1002/jee.20041
John, M.-S., Sibuma, B., Wunnava, S., Anggoro, F., & Dubosarsky, M. (2018). An iterative participatory approach to developing an early childhood problem-based STEM curriculum. *European Journal of STEM Education, 3*(3), 7. https://doi.org/10.20897/ejsteme/3867
Jones, A., Buntting, C., & de Vries, M. (2013). The developing field of technology education: A review to look forward. *International Journal of Technology and Design Education, 23*(2), 191–212. https://doi.org/10.1007/s10798-011-9174-4
Kelley, T. (2010). Staking the claim for the 'T'in STEM. *Journal of Technology Studies, 36*(1), 2–11. http://scholar.lib.vt.edu/ejournals/JOTS/v36/v36n1/kelley.html
Kelley, T. R., & Knowles, J. G. (2016). A conceptual framework for integrated STEM education. *International Journal of STEM Education, 3*(1), 11. https://doi.org/10.1186/s40594-016-0046-z
Kennedy, T. J., & Odell, M. R. L. (2014). Engaging students in STEM education. *Science Education International, 25*(3), 246–258.
Kewalramani, S., Palaiologou, I., & Dardanou ,M. (2020).Children's engineering design thinking processes: The magic of the ROBOTS and the power of BLOCKS (electronics). *EURASIA Journal of Mathematics, Science and Technology Education, 16*(3), em1830. https://doi.org/10.29333/ejmste/113247
Klein, J. T. (1990). *Interdisciplinarity: History, theory, and practice.* Wayne State University Press.

Klein, L. G., & Knitzer, J. (2006). Effective preschool curricula and teaching strategies. In *Pathways to early school success*, issue brief no. 2. Columbia University, National Center for Children in Poverty.

Klein, J. T. (2005). Integrative learning and interdisciplinary studies. *Peer Review, 7*(11), 8–10.

Lawal, A. W. (2014). Technical and vocational education, a tool for national development in Nigeria. *International Letters of Social and Humanistic Sciences (ILSHS), 3*, 53–59.

Layton, D. (1993). *Technology's challenge to science education*. Open University Press.

Leadbeatter, D. (2021). What is integration of learning? *Teaching in Higher Education, 26*(1), 1–19. https://doi.org/10.1080/13562517.2019.1632824

Lee, K. (2011). Looking back, to look forward: Using traditional cultural examples to explain contemporary ideas in Technology Education. *Journal of Technology Education, 22*(2), 42–52. http://scholar.lib.vt.edu/ejournals/JTE/v22n2/lee.html

Lee, K., & Granshaw, B. (2018). STEM education. In B. Frey (Ed.), *The SAGE encyclopaedia of educational research, measurement, and evaluation* (pp. 1619–1620). SAGE. https://doi.org/10.4135/9781506326139.n669

Lee, K. (2019). Technology teacher education: Issues. In *Encyclopedia of teacher education*.

Lee, K., Courtney, M., McGlashan, A., Neveldsen, P., & Toso, M. (2019). Initial teacher education students' perceptions of technology and technology education in New Zealand. *International Journal of Technology and Design Education*, 1–22. https://doi.org/10.1007/s10798-019-09516-6

Lin, X., Yang, W., Wu, L., Zhu, L., Wu, D., & Li, H. (2021). Using an inquiry-base science and engineering program to promote science knowledge, problem-solving skills and approaches to learning in preschool children. *EaRly Education and Development, 32*(5), 695–713. https://doi.org/10.1080/10409289.2020.1795333

Lou, S.-J., Shih, R.-C., Diez, C. R., & Tseng, K.-H. (2011). The impact of problem-based learning strategies on STEM knowledge integration and attitudes: An exploratory study among female Taiwanese senior high school students. *International Journal ofTechnology & Design Education, 21*, 195–215. https://doi.org/10.1007/s10798-010-9114-8

Lowrie, T., & Larkin, K. (2020). Experience, represent, apply (ERA): A heuristic for digital engagement in the early years. *British Journal of Educational Technology, 51*(1), 131–147. https://doi.org/10.1111/bjet.12789

McDowall, S., & Hipkins, R. (2019). *Curriculum Integration: What Is Happening in New Zealand Schools?*. New Zealand Council for Educational Research. https://www.nzcer.org.nz/system/files/Curriculum%20Integration%202018-2019.pdf

McComas, W., & Burgin, S. (2020). A critique of "STEM" education: Revolution-in-the-making, passing fad, or instructional imperative? *Science & Education, 29*, 805–829. https://doi.org/10.1007/s11191-020-00138-2

Mehalik, M. M., Doppelt, Y., & Schunn, C. D. (2008). Middle school science through design-based learning versus scripted inquiry: Better overall science concept learning and equity gap reduction. *Journal of Engineering Education, 97*(1), 71–85.

Moore, T. J., Johnston, A. C., & Glancy, A. W. (2020). STEM integration: A synthesis of conceptual frameworks and definitions. In *Handbook of research on STEM education* (pp. 3–16). Routledge.

Mumford, L. (2000). *Art and technics*. Columbia University Press.

Myers, A. (2015). *The STEM shift: A guide for school leaders*. Corwin.

Nadelson, L. S., & Seifert, A. (2013). Perceptions, engagement, and practices of teachers seeking professional development in place-based integrated STEM. *Teacher Education and Practice, 26*(2), 242–266.

Nadelson, L. S., & Seifert, A. L. (2017). Integrated STEM defined: Contexts, challenges, and the future. *The Journal of Educational Research, 110*(3), 221–223. https://doi.org/10.1080/00220671.2017.1289775

Nathan, M. J., Wolfgram, M., Srisurichan, R., Walkington, C., & Alibali, M. W. (2017). Threading mathematics through symbols, sketches, software, silicon, and wood: Teachers produce and maintain cohesion to support STEM integration. *The Journal of Educational Research, 110*(3), 272–293. https://doi.org/10.1080/00220671.2017.1287046

National Research Council. (2012). *A framework for k-12 science education: practices, crosscutting concepts, and core ideas*. The National Academies Press.

Ng, A., Kewalramani, S., & Kidman, G. (2022). Integrating and navigating STEAM (inSTEAM) in early childhood education: An integrative review and inSTEAM conceptual framework. *EURASIA Journal of Mathematics, Science and Technology Education, 18*(7), em2133. https://doi.org/10.29333/ejmste/12174

Novak, G. M., Patterson, E. T., Gavrin, A. D., & Christian, W. (1999). *Just-in-time teaching: blending active learning with web technology*. Prentice-Hall.

Perez, R. J., & Barber, J. P. (2018). Intersecting outcomes: Promoting intercultural effectiveness and integration of learning for college students. *Journal of Diversity in Higher Education, 11*(4), 418–435. https://doi.org/10.1037/dhe0000067

Prendergast, M., Harbison, L., Miller, S., & Trakulphadetkrai, N. V. (2019). Pre-service and in-service teachers' perceptions on the integration of children's literature in mathematics teaching and learning in Ireland *Irish Educational Studies, 38*(2), 157–175. https://doi.org/10.1080/03323315.2018.1484302

Prince, M. J., & Felder, R. M. (2006). Inductive teaching and learning methods: Definitions, comparisons, and research bases. *Journal of Engineering Education, 95*(2), 123–138. https://doi.org/10.1002/j.2168-9830.2006.tb00884.x

Putnam, R. T., & Borko, H. (2000). What do new views of knowledge and thinking have to say about research on teacher learning? *Educational Researcher, 29*(1), 4–15.

Reeve, E. M. (2021). The need for STEM education: Now more than ever! *Southeast Asian Journal of STEM Education, 2*(1), 94–110.

Ryu, M., Mentzer, N., & Knobloch, N. (2018). Preservice teachers' experiences of STEM integration: Challenges and implications for integrated STEM teacher preparation. *International Journal of Technology and Design Education, 29*, 493–512. https://doi.org/10.1007/s10798-018-9440-9

Saito, T., Anwari, I., Mutakinati, L., & Kumano, Y. (2016). A look at relationships (Part I): Supporting theories of STEM integrated learning environment in a classroom-A historical approach. *K-12 STEM Education, 2*(2), 51–61. https://www-learntechlib-org.ezproxy.auckland.ac.nz/p/209575/

Sanders, M. (2009). Stem, stem education, stemmania. *The Technology Teacher*, December/January, 20–26.

Samson, G. (2014). From writing to doing: The challenges of implementing integration (and interdisciplinarity) in the teaching of mathematics, sciences and technology. *Canadian Journal of Science Mathematics and Technology Education ,14*(4), 346–358. https://doi.org/10.1080/14926156.2014.964883

Saray, H. R. (2022). *Peer mentorship in interdisciplinary STEM education: The underrepresented groups experience* San Francisco State University. https://www.proquest.com/docview/2666475071?pq-origsite=gscholar&fromopenview=true

Schwab, J. (1960). Inquiry, the science teacher, and the educator. *The School Review, 68*(2), 176–195.

Smith, B. A., & Cline, J. E. (2016). Zebras and jaguars, ohmy! Integrating science and engineering standards with art during prekindergarten block time. *Journal of STEM Arts, Crafts, and Constructions, 1*(1), 62–74.

Snape, P., & Fox-Turnbull, W. (2013). Perspectives of authenticity: Implementation in technology education. *International Journal of Technology and Design Education, 23*(1), 51–68.

Steele, A., & Ashworth, E. L. (2013). Walking the integration talk: An Artsci project. *Canadian Journal for the Scholarship of Teaching and Learning, 4*(2), 6.

Stubbs, E. A., & Myers, B. E. (2016). Part of what we do: Teacher perceptions of STEM integration. *Journal of Agricultural Education, 57*(3), 87–100.

Sullivan,A., Kazakoff, E. R., & Bers, M. U. (2013). The wheels on the bot go round and round: Robotics curriculum in pre-kindergarten. *Journal of Information Technology Education, 12*, 203–219. https://doi.org/10.28945/1887

Tank, K. M., Rynearson, A. M., & Moore, T. J. (2018). Examining student and teacher talk with in engineering design in kindergarten. *European Journal of STEM Education, 3*(3), 1–15. https://doi.org/10.20897/ejsteme/3870

Tippett, C. D., & Milford, T. M. (2017). Findings from a pre-kindergarten classroom: Making the case for STEM in early childhood education. *International Journal of Science and Mathematics Education, 15*(1), 67–86. https://doi.org/10.1007/s10763-017-9812-8

Ültanir, E. (2012). An epistemological glance at the constructivist approach: Constructivist learning in Dewey, Piaget, and Montessori. *International Journal of Instruction, 5*(2), 1308–1470.

Vasquez, J. (2014). STEM beyond the acronym. *Educational Leadership, 72*(4), 10–15.

Vasquez, J., Comer, M., & Gutierrez, J. (2020a). *Integrating STEM teaching and learning into the K-2 Classroom*. ERIC.

Vasquez, J., Comer, M., & Gutierrez, J. (2020b). Pioneering into STEM intergration. In R. Ledbetter (Ed.), *Integrating STEM teaching and learning into the K-2 classroom* (pp. 11–22). NSTA Press.

Vasquez, J., Sneider, C., & Comer, M. (2013). *STEM lesson essentials, Grades 3–8: Integrating science, technology, engineering, and mathematics*. Heinemann.

Wang, H.-H., Moore, T. J., Roehrig, G. H., & Park, M. S. (2011). STEM integration: Teacher perceptions and practice. *Journal of Pre-College Engineering Education Research (J-PEER) Journal of Pre-College Engineering Education Research (J-PEER), 1*(2), 1–13. https://doi.org/10.5703/1288284314636

Wasserman, N. H., & Rossi, D. (2015). Mathematics and science teachers' use of and confidence in empirical reasoning: Implications for STEM teacher preparation. *School Science and Mathematics, 115*(1), 22–34. https://doi.org/10.1111/ssm.12099

Wells, J. G. (2019). STEM education: The potential of technology education. In M. K. Daugherty, & V. Carter (Eds.), *The Mississippi valley conference in the 21st century: fifteen years of influence on thought and practice* (pp. 195–229). Council on Technology and Engineering Teacher Education. http://hdl.handle.net/10919/93963

Wells, J. G. (2016). Efficacy of the technological/engineering design approach: Imposed cognitive demands within design-based biotechnology instruction. *Journal of Technology Education, 27*(2), 4–20.

Welsh, M. A., & Dehler, G. E. (2013). Combining critical reflection and design thinking to develop integrative learners. *Journal of Management Education, 37*(6), 771–802. https://doi.org/10.1177/1052562912470107

Weng, J., & Li, H. (2020). Early technology education in China: A case study of Shanghai. *Early Child Development and Care, 190*(10), 1574–1585. https://doi.org/10.1080/03004430.2018.1542383

Weurlander, M., Scheja, M., Hult, H., & Wernerson, A. (2016). The struggle to understand: Exploring medical students' experiences of learning and understanding during a basic science course. *Studies in Higher Education, 41*(3), 462–477. https://doi.org/10.1080/03075079.2014.930122

Wicklein, R. C. (2006). Five good reasons for engineering as the focus for technology education. *The Technology Teacher, 65*(7), 25–29.

Wingert, J. R., Wasileski, S. A., Peterson, K., Mathews, L. G., Lanou, A. J., & Clarke, D. (2011). Enhancing integrative experiences: Evidence of student perceptions of learning gains from cross-course Interactions. *Journal of the Scholarship of Teaching and Learning, 11*(3), 34–57.

Wittrock, M. C. (1989). Generative processes of comprehension. *Educational Psychologist, 24*(4), 345–376. https://doi.org/10.1207/s15326985ep2404_2

Wolff, K., & Luckett, K. (2013). Integrating multidisciplinary engineering knowledge. *Teaching in Higher Education, 18*(1), 78–92. https://doi.org/10.1080/13562517.2012.694105

Wood, C. (2020). *Daily STEM: How to create a STEM culture in your classrooms & communities*. https://dailystem.com/books

Zan, B., & Geiken, R. (2010). Ramps and pathways: Developmentally appropriate, intellectually rigorous, and fun physical science. *Young Children, 65*, 12–17.

Kerry Lee is a member of the National Council of Technology Education New Zealand (TENZ), a senior lecturer and discipline leader for STEM, at the University of Auckland, and the Managing Editor for the International Journal of Adult Education and Technology (IJAET). She has lead roles on many national and international bodies and has published widely in the fields of technology education, supporting indigenous technology education, digital technologies, linking with industry and employers, sustainability, enterprise, and entrepreneurship education. Kerry has won numerous awards including a national Teacher Educator in Technology Education Award; ISSFAL Congress New Investigator Award; and an International Award for Excellence in the world universities field. Kerry is also a registered primary school teacher.

Chapter 4
Technology and Digital Learning Tools: Technology Education and Educational Technology

Andrew Doyle

Abstract With the COVID-19 pandemic, many educators and students were conscripted to online learning. This represented a monumental shift in how to prepare for, teach and assess learning across all levels of education. Unsurprisingly, with the limited preparation time and availability of resources, teachers and learners found this migration challenging. Educators had to find the balance between learners' needs, at the same time navigating a novel context for learning. This chapter sets out to explore this novel context and put forward some recommendations for adopting digital educational technology in STEM classrooms. After setting the scene regarding the terminology associated with technology and education, key findings from adopting digital educational technology are presented. The challenges associated with adopting digital education technology during the pandemic and more generally are reviewed, and following this, opportunities for learning in STEM are considered. STEM education as a context for learning is held in focus, as implications for the adoption (by choice) of digital educational technology for future STEM education lessons are put forward.

Keywords Technology education · Digital educational technology · Educational technology · STEM education · Information and communication technology (ICT)

4.1 Introduction

The distinction between technology education as a school subject and educational technology became very apparent in 2020. With the COVID-19 pandemic, many educational institutions around the world were forced to "go online". This rather abrupt migration to digital education was facilitated through various forms of educational technology. In this space, Microsoft Teams®, Google Classroom®, Zoom®

A. Doyle (✉)
School of Education, University of Waikato, Hillcrest, Hamilton 3216, New Zealand
e-mail: andrew.doyle@waikato.ac.nz

W. Fox-Turnbull and P. J. Williams (eds.), *Locating Technology Education in STEM Teaching and Learning*, Contemporary Issues in Technology Education,
https://doi.org/10.1007/978-981-97-1995-2_4

and Moodle® usage at all levels of education significantly increased (Callimaci & Fortin, 2022). Educational technology (sometimes referred to as Ed-Tech) solutions providers flooded the market with a variety of innovative platforms for e-learning, ICT solutions and digital learning. For many educators, new terms associated with online learning emerged, asynchronous learning or synchronous learning was an issue for planning remote learning—which approach should educators adopt? And why? E-readiness was touted at staff meetings as the key goal in the weeks, or some instances, days preceding national and local lockdowns.

Although various forms of *distance education*, *remote teaching* and *online instruction* have been around since the 1980s (Casey, 2008), for the majority of educators and learners, this was a new context for learning. In the opening paragraph, I attempted to represent how difficult the digital learning space is to navigate, as research (Magomedov et al., 2020; Selvaraj et al., 2021) has shown that the pandemic and the associated shift to online learning was a challenging time for teachers and learners. While this may not be surprising given the limited time for preparation, it provides an opportunity to investigate the opportunities and limitations of adopting educational technology in STEM education. This chapter will explore the challenges of using educational technology in the traditional classroom scenario as well as a tool to facilitate learning in remote or distant scenarios. In the broader context of this book, first, there is a need for clarity surrounding the "T" in STEM education. What do we mean by the term "technology"?

4.2 Technology Education

The word *technology* has different meanings in different contexts. Technology is used to describe new products, techniques and systems. Mostly associated with artefacts or "things" the term technology is now inextricably linked with computers. Information and communication technology (ICT) is just one way in which technology can manifest. Understandings from the philosophy of the technology field show us that technology is much more than new products or systems and that technology does necessarily have a physical materiality. Carroll (2017) in their etiological analysis of technology developed a three-fold definition, encompassing

> (a) something that is always inherently intelligent enough either to function, to be used to function, to be imbued with, or to be interpreted as having, a function that only intelligent beings (human or otherwise) have the ability to comprehend; (b) something devised, designed (i.e., primary intention), or discovered (i.e., secondary intention) that serves a particular purpose from a purely secular standpoint, without requiring that mankind be responsible for it, though he may be (i.e., the aspect of reflexivity through purpose in that salt doesn't inherently "elevate" or do anything deliberately, but it does "elevate" the boiling point of water, which it has been found to do and can be considered to serve a purpose); (c) a significant beneficiary of rationally-derived knowledge that is "used for" a purpose, without itself necessarily being translated into something physical or material that "does" (e.g., instructional methodologies in education, processes, ideas).
>
> (Carroll, 2017, p. 18).

As can be seen from this definition, technology is elevated beyond materiality to include an extensive variety of artefacts, systems, ideas and solutions that we find around us in the made world. As a result of this, De Vries (2018) noted that the impact of technology on our lives has been so significant that we speak of "technological literacy" as a requirement for every citizen that should be learned at school. This understanding of technology is represented in many school curricula internationally (Banks & Williams, 2022).

As outlined, general perceptions of the word technology have become intertwined with electronics and computers. In the academic sense (i.e. as a component of STEM education), technology has been studied to a great extent. For example, Mitcham (1994) presented a fourfold classification of technology, noted by de Vries (2018) as having been widely adopted by philosophers of technology and in the context of technology education. Mitcham delineated four manifestations of technology: (1) technology as artefact, (2) technology as knowledge, (3) technology as process and (4) technology as volition. The most common conception of technology is that of an artefact. Artefact here refers to the vast array of material things in the made world. When the word "technology" is used to describe transportation, communication, etc. we are typically talking about technological artefacts (de Vries, 2018). The second manifestation of technology proposed by Mitcham is technology as knowledge, something that may be studied and learned. The delineation of technological knowledge from scientific knowledge has been a long endeavour in the philosophy of technology field (Houkes, 2009). However, it is now accepted that technology does have its own content knowledge, and importantly, this knowledge goes beyond the application of scientific knowledge as was previously conceived (McCormick, 2004). Technology can also manifest as a process, technological activity as something that is done, where there are significant parallels with "design" as an activity. The overlap is so significant that some national contexts have adopted design and technology as a school subject. The final manifestation of technology identified by Mitcham is technology as volition. This perspective is more abstract, pointing towards the actor behind technological decisions. Who are you as a technologist? Mitcham (1994) identified artefacts as material manifestations of technology, but immaterially the knowledge associated with developing objects and activities, and the volition, or decision-making behind technological development can also be considered as technology. This is critical in understanding the basis of technology education. Another perspective on how to explore technology is presented by Pacey (1983). In considering conceptions of technology, in particular, endorsing a more holistic understanding of the term, Pacey delineated three different aspects of technology, namely technology as having a technical, cultural and organisational aspect.

Technology education (capitalised) as a curricular area deals with this multitude of manifestations of technology. Although the subject has in many instances stemmed from a vocational predecessor, with a significant technical aspect, today's emphasis on technology capability (Doyle et al., 2019) and technological literacy (Dakers, 2018) seeks to incorporate the cultural and organisational aspects of technology. The technical context for technology education can differ internationally (Banks & Williams, 2022); for example, if a country has a specific craft heritage, technology education can be enacted in this context. Likewise, technology education may be enacted in new or evolving contexts, such as autonomous cars or genetically modified crops. The key distinction to note is that technology education is not restricted to its technical aspect. With this and the breadth of manifestations of technology outlined earlier, there is significant room for divergence in how technology education is conceived and enacted. The variability of contexts for technology facilitates a wide variety of different manifestations of the subject. For example, the focus may be placed on learners acting in a technological way, as evidenced in the English national curriculum with a subject entitled design and technology (DfE, 2013). In other contexts, in Sweden and New Zealand, for example, the Nature of Technology itself is emphasised (Ministry of Education, 2017; Skolverket, 2011). Such an approach places an understanding of the Nature of Technology at the forefront of teaching and learning. Irrespective of the focus on a particular aspect of technology to use Pacey's terminology, or a manifestation to use Mitcham's, the object of study is technology.

This may seem very straightforward for someone familiar with the domain of technology education. However, evidence of a reductive perspective on technology identified within the STEM education literature (Ellis et al., 2020; McGarr & Lynch, 2017) suggests the need for clarification. The tensions here are between technology education, as described above, and educational technology (note the small "t").

4.3 Educational Technology

Much like how the term "technology" is now commonly associated with anything that runs on electricity, educational technology is now almost exclusively associated with anything that runs on electricity, for the purpose of facilitating teaching and learning. Again, this conceptualisation of (educational) technology presented is reductive. There are two significant challenges with this.

Firstly, there are connotations associated between technology and digital. Educational technology can not only be taken as digital artefacts and systems but also as non-digital artefacts and systems. Technology, given the definition presented previously (Carroll, 2017) can be devised, designed or discovered to serve a particular purpose from a secular standpoint. So long as the "technology" is intelligent enough to function, or to be used to function, it may be considered technology. This results in a vast array of artefacts and systems being considered technological in nature. Resultantly, if any of these artefacts or systems function can be categorised as serving an educational purpose, we can then consider that technology as a form of "educational technology". This means that any human intervention in the made world that affects education may be considered a form of educational technology. Consider anything that has been invented to facilitate communication between teachers and learners. The blackboard, the chalk used to write on a blackboard and even the language(s) used to communicate in written text fall under the broadest definition of technology.

Thus, although technology has become intertwined with ICT in everyday understandings of the term, it is important to remember that this is just one way in which educational technology may manifest. This perspective is supported by the Association for Educational Communications and Technology (AECT) definition, whereby "Educational technology is the study and ethical practice of facilitating learning and improving performance by creating, using, and managing appropriate technological processes and resources" (Januszewski & Molenda, 2008, p. 1).

Secondly, and perhaps more pertinently, the emphasis on technology as an artefact. Here we again draw on Pacey's delineation between the technical aspect of technology, which forms a restricted view, and the cultural and organisational aspects of technology. The materiality of physical (touch and feel/use) technologies is straightforward to explain. Beyond this, there are values which influence how such technologies are adopted and used. Pacey (1983) described how these values, together with the various beliefs and habits of thinking, form the cultural aspect of technology. The emergence of "pandemic pedagogies" (Williamson et al., 2020, p. 108) serves as a useful example. Taking the organisational component of Pacey's model of technology, pedagogy can be considered technology. A pedagogical theory or model that has been developed for a specific purpose, for example, flipped learning, may be considered a manifestation of technology. This may not be true for all pedagogical approaches, as it could be argued that some pedagogical approaches appear in the world without human intervention. There are several heuristics which are exhibited by multiple different species that fall into this category. However, in the instances where specific approaches have been developed by humans (technological), serving a pedagogical goal (educational), they may be considered educational technology.

4.4 Digital (Educational?) Technology

4.4.1 What Do We Mean by "Digital"?

If we take the example of clocks, we can begin to unpack the difference between a *digital* object and a *non-digital* object. I use "non-digital" here as there are a variety of terms used to describe artefacts that are not digital in nature, this largely depends on the contexts in which "digital" is being discussed. For example, in the context of clocks the discussion is about digital vs. analogue, and in other contexts terms such as technical, mechanical, print or natural may be used. What is the commonality between a digital clock and a digital thermometer, and how does this help us understand digital learning? The commonality lies in the use of electronic memory. If a device records, stores or processes electronic data it may be considered digital in nature (Cambridge University Press, 2023). A traditional pendulum wall clock uses the momentum of the swinging weight as its timekeeping element. As the pendulum swings back and forth, gears control a falling weight, which ultimately powers the clock. In mechanical wristwatches, the power comes from a coiled mainspring. Both of these examples use potential energy, which when converted to kinetic energy powers the clock/watch and ultimately turns the hands. It is important to note that the lack of electrical energy here is not sufficient to define something as non-digital. If we are to simply replace the potential energy with chemical energy, for example using a battery to power a wristwatch, there is no memory being generated, stored or processed. A clock/watch may only be considered digital where electronic memory is engaged. For example, a digital clock may process memory to show time on a digital interface such as a liquid crystal display, leap years or the number of days in a month are held in memory or the clock that has a snooze function that can be set to different intervals. With each of these examples, the clock has been or can be programmed and when in operation, the clock accesses this programmed memory in executing a function.

4.4.2 Digital Educational Technology

This definition of digital assists in specifying educational technology. As noted, educational technology can be taken as much broader than how it is represented in everyday discussions and the media. If we consider the perspective on "digital" presented above, framing education as a purposeful manifestation of digital technology emerges as a useful refinement. This shifts the generalised discussions surrounding educational technology to a specific focus on digital technology. As such, the term digital educational technology will be used throughout this chapter, whereby digital educational technology is taken to be any digital technology that has been developed for educational purposes or digital technology that has been adopted for educational purposes.

4.5 Digital Educational Technology: Challenges for STEM Education

The migration to online learning, mandated by the COVID-19 pandemic and facilitated by digital educational technology, showed how prepared the education community was for a wholly digital education. Although the expedited timeframe is not representative of an online learning experience that had significant time for planning, findings have emerged from research that may inform future adoptions of digital educational technology.

4.5.1 Access—Digital Inequalities

Fundamental to a learner's ability to engage with online or remote learning is access to digital education technologies. In the case of synchronous learning, a high-speed internet connection, as well as the appropriate hardware (e.g. a tablet, computer or smartphone), is essential to facilitating online learning. The implications here are immediately apparent. However, when the home environment is investigated further, the facilities available to learners highlight less obvious challenges. An investigation of the English national context identified that 58% of primary school students from the least affluent families did not have access to a private study space (Andrew et al., 2020). Similarly, it was identified that students from better-off families not only had a better environment at home but also had access to more individualised resources and time with teachers (2020). Parental support for academic studies when learners study from home also emerged as a source of tension for some students during the COVID-19 pandemic. Family obligations in the form of household chores and expectations to work emerged as challenges for many learners (Ahmadon et al., 2020). A similar trend to Andrew et al. was noted here, as families with a lower socio-economic status reported higher instances of expectations to work and engage with chores, highlighting the importance of teachers understanding the fundamental needs of learners when studying from home.

4.5.2 Assumptions—Digital Natives?

The notion of *digital natives* was first put forward by Prensky (2001). It is an alluring idea, Prensky suggests that digital natives constitute the ever-growing group of people born post-1984. As a result of being born into a technologically advanced world, the relative ubiquity of digital technologies was theorised to affect the ways in which students think and process information. Prensky suggested that "our students' brains have physically changed" (p. 1), and resultantly, the challenge was laid down to adapt how we teach to better facilitate learning for *digital natives*. Inherent to the concept of digital natives is the assumption of familiarisation and, in turn, expertise in engaging with digital technology.

Despite the widespread adoption of the idea of digital natives, research has continually shown that there are significant variations in the ways in which today's learners access, navigate and use digital education technology (Kirschner & De Bruyckere, 2017). Studies of learner engagement suggest that not all learners are as digitally savvy as this rhetoric would suggest (Beckman et al., 2018). De Bruyckere et al. (2016) highlight recent findings from the field of neuroscience that suggest the implications for teaching *digital natives* may be very different from Prensky's initial hypothesis:

> [G]rowing up with Internet technologies, "Digital Natives" gravitate toward "shallow" information processing behaviors characterized by rapid attention shifting and reduced deliberations. They engage in increased multitasking behaviors that are linked to increased distractibility and poor executive control abilities. Digital natives also exhibit higher prevalence of Internet-related addictive behaviors that reflect altered reward-processing and self-control mechanisms. Recent neuroimaging investigations have suggested associations between these Internet-related cognitive impacts and structural changes in the brain.
>
> (Loh & Kanai, 2015, as cited in De Bruyckere et al., 2016)

There are several important implications for STEM education as a result of this. First, there is no immediate need for significant changes to how teachers teach. Learners today are not so different from those born pre-1984, and the emergence of digital educational technologies does not warrant a reconsideration of how learning happens, although such theories are present in the literature, c.f. Bell (2011). The second, related point here is more nuanced. There are new and emerging digital educational technologies that teachers and learners alike will engage with, highlighting the need for specific education on how to engage with and best utilise such technologies. The non-existence of *digital natives* may be viewed as an impetus to tailor inductions to new digital education technologies and not make assumptions about learners' capacity to engage with or navigate emerging technologies. To assist with this, the concept of "Digital Competence" is being used and will be discussed in the next section.

4.6 Adopting Digital Educational Technology in STEM Education

The non-existence of *digital natives* does not warrant a reconsideration of how learning happens; however, it does warrant serious consideration of the introduction of digital educational technologies in STEM education classrooms. There are several approaches that can be directly adopted by teachers, teacher educators and policymakers to better understand and facilitate the introduction of digital educational technology.

4.6.1 RAT Model

The RAT model is an assessment framework for understanding technology's role in teaching, learning and curricular practices. Comprised of three components, technology as Replacement, Amplification and Transformation (hence RAT), the model is used to determine how digital technology informs practice. Application of the model is proposed across teaching, spanning instruction methods, student learning processes and curriculum goals. Technology as replacement theorises that the incorporated technology is used to replace, and "in no way change established instructional practices, student learning processes, or content goals" (Hughes et al., 2006, p. 1616). An example of this may be the adoption of e-readers or tablets to replace conventional textbooks in a lesson. Although the medium through which learners engage with content changes here, the purpose of the technology is to merely replace existing technology. With technology as amplification, the applied technology increases the efficiency, effectiveness and productivity of instructional practices, student learning practices or content goals. While the tasks remain largely the same, technology in some way extends the capabilities of the teachers. An example of this may be the adoption of a Learning Management System (LMS). Although a teacher prepares for teaching the same topic every term, the LMS may serve as an archive through which information, tasks and resources may be shared with the next cohort of learners. Lastly, technology as transformation holds the ideal of shifting established educational patterns or goals. While the newly incorporated technology may share similarities with amplification, the key difference here is that the instruction, learning process and/or content are fundamentally different. An example of technology as transformation would be changing how students solve problems. For instance, consider the introduction of a shared online platform for modelling concepts. It is important to note that as the RAT model analyses practices and changes in practices, the same practices may be considered differently depending on the teacher. In other words, what may be considered transformative for one teacher may be only amplification for another (Blanchard et al., 2016). This focus is reflective of the initial focus of the framework, which was developed to introduce self-assessment for pre-service and in-service teachers to "increase critical decision-making concerning the integration of technology" (Hughes et al., 2006, p. 1616). Educators are encouraged to use the framework to analyse how they incorporate digital educational technology into their practices, importantly, comparing that to prior practices. For readers who want to consider their existing practices or indeed offer a critical perspective to any future integrations of technology, Hughes (2023) presents a set of resources to analyse how technology is currently being integrated into classrooms.

4.6.2 Technological Pedagogical Content Knowledge (TPACK)

Technological Pedagogical Content Knowledge was introduced by Mishra and Koehler (2006) as an extension of Pedagogical Content Knowledge (Shulman, 1986, 1987). Pedagogical Content Knowledge was introduced by Shulman in an attempt to articulate the unique knowledge required by teachers. Proposing that "Pedagogical Knowledge" (PK) is about how to teach, "Content Knowledge" (CK) is knowledge about the subject matter to be taught, and thus, "Pedagogical Content Knowledge" (PCK) is knowledge of how to teach specific subject matter. Shulman's thesis was that knowledge of the subject matter and pedagogical approaches, in general, was insufficient to facilitate good teaching. PCK was presented as the relevant knowledge to facilitate effective teaching, inclusive of "the ways of representing and formulating the subject that make it comprehensible to others" (Shulman, 1986, p. 9).

In adding "Technological Knowledge" (TK) as teachers' knowledge about digital educational technologies to the PCK framework, Technological Pedagogical Content Knowledge [TPCK, more commonly referred to as TPACK (Thompson & Mishra, 2007)] is presented as the synthesis of TK, PK and CK, illustrating the importance of how teachers integrate technology into their subject matter specific pedagogy (Koehler & Mishra, 2009). While acknowledging the idea that knowledge about technology cannot be treated as context-free and should be understood and applied is not new, the TPACK model serves as a framework to introduce this idea.

Recent studies have begun to unpack the role that technological knowledge and TPACK specifically have in understanding teachers' practices. One such study identified the different ways in which pre-service and in-service teachers conceptualise the use of technology. In their study of technology-enhanced lesson plans, Wekerle and Kollar found that "pre-service teachers' lesson plans envision teachers as managers and in-service teachers' lesson plans envision teachers as learning guides" (2022, p. 15), highlighting the importance of holding the students as the agent of learning. This perspective mirrors that of the RAT model of learning, in that technology should be used to enhance learning, and not shift the focus from learning.

4.6.3 Digital Competence

Lastly, and from the macro-perspective of goals for teaching in the twenty-first century, digital literacy or digital competence has begun to permeate curricula and steering documents. The terminology here varies between contexts, but "digital competencies are generally defined as a multi-layered and complex construct, in which higher-level thinking skills are supported by technical and practical digital knowledge and skills" (Godaert et al., 2022, p. 2). Recognising the multifaceted nature of the construct, the European Commission (2018) developed the DigiComp 2.1 framework to articulate digital competency for citizens. The framework consists of five competence areas: information and data literacy, communication and collaboration, digital content creation, safety and problem-solving. While the framework itself is abstract, it is accompanied by descriptions of eight proficiency levels for the competencies which serve as useful prompts for discussion in the STEM classroom.

4.7 Using Digital Educational Technology to Facilitate STEM Education

4.7.1 Online Interactive Learning

Despite the challenges of adopting digital educational technology, there are some opportunities that should also be considered. Primary among these is the opportunity to engage external stakeholders. Although this has always been possible in education, digital educational technology affords a new form of engagement between educators and the wider community. With the COVID-19 pandemic and associated lockdowns, habits of working have changed (Babapour Chafi et al., 2022) and industries have had to adapt how they offer services, communicate and engage with the community. For example, the school trip to the local processing plant, farm, zoo or industry will always have a place in education, however in the instances where a trip is too costly to be feasible, or perhaps an industry can only cater to small groups, virtual tours may serve as a viable alternative. When the interdisciplinary nature of STEM education is considered, digital educational technology provides an opportunity to engage stakeholders from the foundational STEM education disciplines of Science, Technology, Engineering and Mathematics. This may be in the form of team-teaching larger groups with multiple domain-specific teachers or the form of external consultations or online visits with relevant experts from industry is at the discretion of teachers. In addition, the potential for individual or small groups of learners to contact an expert from the field may be facilitated through digital educational technology.

4.7.2 Simulation

The second significant opportunity for digital educational technology lies in the area of resourcing, in particular, the power of computers to simulate real-world responses in a virtual space. In terms of vocational education, augmented reality has been used for simulating environments from welding (Papakostas et al., 2022) to operating rooms focusing on spinal surgery (Umebayashi et al., 2018). The advantages of using a simulation over conventional training methods are predominantly related to cost, but also safety. Papakostas et al. outline the costs of welding infrastructures such as gas extraction and enhanced electrical installations. Further, there are direct consumables such as materials to be welded and possibly gases depending on the specific welding process. The suggestion here is not to replace traditional hands-on welding education, but to augment this, particularly in the initial stages of learner progression. It should be noted that simulation not only relates to manipulative skills but there are also opportunities for cognitive development and understanding of abstract concepts. Take for example how equations are graphed using the Cartesian coordinate system. GeoGebra® and similar applications provide interactive simulations where a variety of mathematical and scientific concepts can be investigated.

The significant advantage here is the simultaneous interconnection between algebraic and geometric representations, providing learners with the opportunity to infer the formal representations of physical phenomena (Solvang & Haglund, 2021).

4.7.3 Gaming

Gaming for educational purposes has been around for some time, however since the early 2000s specifically, the focus has been paid to the idea of digital games informing teaching and learning in STEM classrooms. Bekebrede et al. (2011) cite Prensky's (2001, 2003) work as a foundation for the idea of using digital games for educational purposes. The approach adopts specifically designed educational video games as learning tools for schools and other educational institutions. Advocates for the approach cite the ability of digital games to teach and reinforce important skills, such as collaboration, communication and problem-solving. Although this research area is still developing, the latest research has shown significant potential for gaming within the STEM education community (Wang et al., 2022). Some findings have begun to emerge as to how and when gaming may be most effective in STEM education. For example, findings highlighting the learning effectiveness at different levels of schooling, where early years, in particular, have been identified as having a significant impact on learning gains as compared to later years of school and indeed university (Riopel et al., 2019). In addition to this, Wang et al. identified the potential of digital games being integrated with additional technologies (e.g. mobile, AR/VR and 3D). It was reported that the substitution of digital games improved students' learning experience and encouraged learning enthusiasm and motivation, with emerging technologies viewed as a medium to further enhance such opportunities.

4.8 Conclusion

The relationship between technology education and STEM education is not always made clear as there are often different perspectives on the place and role of the "T" in STEM education. Given the trend of reductive perspectives on the "T" within STEM being conceived as educational technology (Pleasants et al., 2019), it is important to reemphasise the distinction between technology education and the manifestation of technology as outlined previously. Technology, and in turn, technology education can manifest is a variety of different ways. Mitcham's (1994) classification identified artefact, process, knowledge and volition as four ways that technology can manifest. To view a singular manifestation of technology, namely a digital educational technology, as technology education's contribution to STEM not only grossly misrepresents the nature of technology, it also limits STEM education. In moving beyond this representation, technology education (as a discipline, note the big "T"), educational technology (technology used for educational purposes, technology as a tool), digital technology (as a specific manifestation of technology) and, ultimately, digital educational technology (as a manifestation of technology for a specific purpose to enhance learner's experiences) have been outlined in this chapter. In the same way that Science and Mathematics have established learning areas, with the goal of scientific literacy and mathematics literacy being held as a goal for the learning area, technology education holds technological literacy as the goal (Tang & Williams, 2019). It is envisioned that clarity surrounding the relationship between education and technology will serve to enhance STEM education.

References

Ahmadon, F., Ghazalli, H. I. M., & Rusli, H. M. (2020). Studying during Pandemic: A review of issues from online learning in the middle of COVID-19. In *2020 6th International Conference on Interactive Digital Media (ICIDM)* (pp. 1–4). https://doi.org/10.1109/ICIDM51048.2020.9339644

Andrew, A., Cattan, S., Dias, M. C., Farquharson, C., Kraftman, L., Krutikova, S., Phimister, A., & Sevilla, A. (2020). Inequalities in Children's experiences of home learning during the COVID-19 Lockdown in England*. *Fiscal Studies, 41*(3), 653–683. https://doi.org/10.1111/1475-5890.12240

Babapour Chafi, M., Hultberg, A., & Bozic Yams, N. (2022). Post-Pandemic office work: Perceived challenges and opportunities for a sustainable work environment. *Sustainability, 14*(1), Article 1. https://doi.org/10.3390/su14010294

Banks, F., & Williams, P. J. (2022). International perspectives on technology education. In A. Hardy (Ed.), *Debates in design and technology education* (2nd ed., pp. 26–44). Routledge. https://doi.org/10.4324/9781003166689-4

Beckman, K., Apps, T., Bennett, S., & Lockyer, L. (2018). Conceptualising technology practice in education using Bourdieu's sociology. *Learning, Media and Technology, 43*(2), 197–210. https://doi.org/10.1080/17439884.2018.1462205

Bekebrede, G., Warmelink, H. J. G., & Mayer, I. S. (2011). Reviewing the need for gaming in education to accommodate the net generation. *Computers & Education, 57*(2), 1521–1529. https://doi.org/10.1016/j.compedu.2011.02.010

Bell, F. (2011). Connectivism: Its place in theory-informed research and innovation in technology-enabled learning. *International Review of Research in Open and Distributed Learning, 12*(3), 98–118. Érudit. https://doi.org/10.19173/irrodl.v12i3.902

Blanchard, M. R., LePrevost, C. E., Tolin, A. D., & Gutierrez, K. S. (2016). Investigating technology-enhanced teacher professional development in rural high-poverty middle schools. *Educational Researcher, 45*(3), 207–220. https://doi.org/10.3102/0013189X16644602

Callimaci, A., & Fortin, A. (2022). Intended use of educational technology after the COVID-19 pandemic. *Journal of Further and Higher Education*, 1–15.https://doi.org/10.1080/0309877X.2022.2079970

Cambridge University Press. (2023). Digital. In *Cambridge Dictionary*. Cambridge University Press. https://dictionary.cambridge.org/dictionary/english/digital

Carroll, L. S. L. (2017). A comprehensive definition of technology from an ethological perspective. *Social Sciences, 6*(4). https://doi.org/10.3390/socsci6040126

Casey, D. M. (2008). A journey to legitimacy: The historical development of distance education through technology. *TechTrends, 52*(2), 45. https://doi.org/10.1007/s11528-008-0135-z

Dakers, J. R. (2018). Philosophy of technology and engineering. In M. J. de Vries (Ed.), *Handbook of technology education* (pp. 3–6). Springer. https://doi.org/10.1007/978-3-319-44687-5_67

De Bruyckere, P., Kirschner, P. A., & Hulshof, C. D. (2016). Technology in education: What teachers should know. *American Educator, 40*(1), 12–18. https://eric.ed.gov/?id=EJ1094203

de Vries, M. J. (2018). Philosophy of Technology: Themes and topics. In M. J. de Vries (Ed.), *Handbook of technology education* (pp. 7–16). Springer International Publishing. https://doi.org/10.1007/978-3-319-44687-5_1

DfE. (2013). *National curriculum in England: Design and technology programmes of study [online].* GOV.UK. https://www.gov.uk/government/publications/national-curriculum-in-england-design-and-technology-programmes-of-study/national-curriculum-in-england-design-and-technology-programmes-of-study

Doyle, A., Seery, N., Canty, D., & Buckley, J. (2019). Agendas, influences, and capability: Perspectives on practice in design and technology education. *International Journal of Technology and Design Education, 29*(1), 143–159. https://doi.org/10.1007/s10798-017-9433-0

Ellis, J., Wieselmann, J., Sivaraj, R., Roehrig, G., Dare, E., & Ring-Whalen, E. (2020). Toward a Productive Definition of Technology in Science and STEM Education. *Contemporary Issues in Technology and Teacher Education, 20*(3), 472–496. https://eric.ed.gov/?id=EJ1267643

European Commission, Joint Research Centre, Carretero, S., Vuorikari, R., & Punie, Y. (2018). *DigComp 2.1: The digital competence framework for citizens with eight proficiency levels and examples of use.* Publications Office. https://doi.org/10.2760/38842

Godaert, E., Aesaert, K., Voogt, J., & van Braak, J. (2022). Assessment of students' digital competences in primary school: A systematic review. *Education and Information Technologies.* https://doi.org/10.1007/s10639-022-11020-9

Houkes, W. (2009). The nature of technological knowledge. In A. Meijers (Ed.), *Philosophy of technology and engineering sciences* (pp. 309–350). North-Holland.

Hughes, J. (2023). *Replacement, Amplification, and Transformation: The R.A.T. Model.* https://techedges.org/r-a-t-model/

Hughes, J., Thomas, R., & Scharber, C. (2006). Assessing technology integration: The RAT—replacement, amplification, and transformation—framework. In C. Crawford, R. Carlsen, J. McFerrin, J. Price, R. Weber & D. Willis (Eds.), *Proceedings of SITE 2006—Society for Information Technology & Teacher Education International Conference* (pp. 1616–1620). Association for the Advancement of Computing in Education (AACE). https://www.learntechlib.org/primary/p/22293/

Januszewski, A., & Molenda, M. (2008). *Educational Technology: A Definition with Commentary* (2nd ed.). Routledge.

Kirschner, P. A., & De Bruyckere, P. (2017). The myths of the digital native and the multitasker. *Teaching and Teacher Education, 67*, 135–142. https://doi.org/10.1016/j.tate.2017.06.001

Koehler, M., & Mishra, P. (2009). What is technological pedagogical content knowledge (TPACK)? *Contemporary Issues in Technology and Teacher Education, 9*(1), 60–70.

Magomedov, I. A., Khaliev, M.S.-U., & Khubolov, S. M. (2020). The negative and positive impact of the pandemic on education. *Journal of Physics: Conference Series, 1691*(1), 012134. https://doi.org/10.1088/1742-6596/1691/1/012134

McCormick, R. (2004). *Issues of learning and knowledge in technology education. 1996.* http://oro.open.ac.uk/818/

McGarr, O., & Lynch, R. (2017). Monopolising the STEM agenda in second-level schools: Exploring power relations and subject subcultures. *International Journal of Technology and Design Education, 27*(1), 51–62.

Ministry of Education. (2017). *Technology in the New Zealand Curriculum.* https://nzcurriculum.tki.org.nz/content/download/168478/1244184/file/NZC-Technology%20in%20the%20New%20Zealand%20Curriculum-Insert%20Web.pdf

Mishra, P., & Koehler, M. J. (2006). Technological pedagogical content knowledge: A framework for teacher knowledge. *Teachers College Record, 108*(6), 1017–1054. https://doi.org/10.1111/j.1467-9620.2006.00684.x

Mitcham, C. (1994). *Thinking through technology.* The University of Chicago Press.

Pacey, A. (1983). *The culture of technology.* MIT press.

Papakostas, C., Troussas, C., Krouska, A., & Sgouropoulou, C. (2022). User acceptance of augmented reality welding simulator in engineering training. *Education and Information Technologies, 27*(1), 791–817. https://doi.org/10.1007/s10639-020-10418-7

Pleasants, J., Clough, M. P., Olson, J. K., & Miller, G. (2019). Fundamental issues regarding the nature of technology. *Science & Education, 28*(3), 561–597. https://doi.org/10.1007/s11191-019-00056-y

Prensky, M. (2001). Digital natives, digital immigrants. *On the Horizon NCB University Press, 9*(5), 6.

Prensky, M. (2003). Digital game-based learning. *Computers in Entertainment, 1*(1), 21. https://doi.org/10.1145/950566.950596

Riopel, M., Nenciovici, L., Potvin, P., Chastenay, P., Charland, P., Sarrasin, J. B., & Masson, S. (2019). Impact of serious games on science learning achievement compared with more conventional instruction: An overview and a meta-analysis. *Studies in Science Education, 55*(2), 169–214. https://doi.org/10.1080/03057267.2019.1722420

Selvaraj, A., Radhin, V., KA, N., Benson, N., & Mathew, A. J. (2021). Effect of pandemic based online education on teaching and learning system. *International Journal of Educational Development, 85*, 102444 https://doi.org/10.1016/j.ijedudev.2021.102444

Shulman, L. S. (1986). Those who understand: Knowledge growth in teaching. *Educational Researcher, 15*(2), 4–14.

Shulman, L. S. (1987). Knowledge and teaching: Foundations of the new reform. *Harvard Educational Review, 57*(1), 1–23.

Skolverket. (2011). *Curriculum for the compulsory school, preschool class and the leisure-time centre 2011*. Skolverket.

Solvang, L., & Haglund, J. (2021). How can GeoGebra support physics education in upper-secondary school—A review. *Physics Education, 56*(5), 055011. https://doi.org/10.1088/1361-6552/ac03fb

Tang, K.-S., & Williams, P. J. (2019). STEM literacy or literacies? Examining the empirical basis of these constructs. *Review of Education, 7*(3), 675–697. https://doi.org/10.1002/rev3.3162

Thompson, A. D., & Mishra, P. (2007). Editors' Remarks. *Journal of Computing in Teacher Education, 24*(2), 38–64. https://doi.org/10.1080/10402454.2007.10784583

Umebayashi, D., Yamamoto, Y., Nakajima, Y., Fukaya, N., & Hara, M. (2018). Augmented reality visualization–guided microscopic spine surgery: Transvertebral anterior cervical foraminotomy and posterior foraminotomy. *Journal of the American Academy of Orthopaedic Surgeons. Global Research & Reviews, 2*(4), e008. https://doi.org/10.5435/JAAOSGlobal-D-17-00008

Wang, L.-H., Chen, B., Hwang, G.-J., Guan, J.-Q., & Wang, Y.-Q. (2022). Effects of digital game-based STEM education on students' learning achievement: A meta-analysis. *International Journal of STEM Education, 9*(1), 26. https://doi.org/10.1186/s40594-022-00344-0

Wekerle, C., & Kollar, I. (2022). Using technology to promote student learning? An analysis of pre- and in-service teachers' lesson plans. *Technology, Pedagogy and Education, 31*(5), 597–614. https://doi.org/10.1080/1475939X.2022.2083669

Williamson, B., Eynon, R., & Potter, J. (2020). Pandemic politics, pedagogies and practices: Digital technologies and distance education during the coronavirus emergency. *Learning, Media and Technology, 45*(2), 107–114. https://doi.org/10.1080/17439884.2020.1761641

Dr Andrew Doyle is Lecturer in Technology Education at the University of Waikato, where he teaches undergraduate and master's level teacher education programmes. Before New Zealand, Andrew worked in or studied technology education in the Irish, Swedish and English national contexts. His research is focused on better understanding the nature of technology for education, including teachers' representations of the purpose(s) of teaching technology. More generally, he is interested in representations of STEM education and the role that technology and engineering education are perceived to play within STEM education.

Chapter 5
Addressing Curriculum Knowledge in STEM Projects

Frank Banks

Abstract This chapter sets out the curriculum knowledge that technology teachers need in order to support the 'T' in STEM school projects. If schools move from the STEM subjects being taught in separate isolated subject 'silos', to teaching STEM in a loose coordination of the subjects, then to teachers collaborating on school projects that require a progressive integration of the STEM area of the curriculum there is clearly a need for the teachers to develop increasingly wide curriculum knowledge. Not only do STEM teachers need to gain a deeper understanding of their own specific curriculum knowledge but they also must take notice of what, when and how STEM is being taught by their colleagues. In short, they also should 'Look-Sideways' and collaborate for the mutual benefit of their learners. Using four school STEM curriculum examples based on UNESCO sustainability goals, this chapter highlights the benefits of adopting a cross-curriculum approach to STEM education, both through the contribution that technology brings to school STEM projects whatever the main 'subject focus' of that project and through the contributions that science and mathematics can make to technology-focussed STEM projects. The different aspects of curriculum knowledge required of technology teachers are considered through the 'personal subject construct' that teachers have of their role in teaching and in supporting STEM projects. This dispositional knowledge influences what discipline knowledge they consider is important to possess. In conclusion, assessment and how STEM student learning can be mapped are explored.

Keywords Subject knowledge · Pedagogical knowledge · STEM Curriculum collaboration · Assessing STEM

F. Banks (✉)
The Open University, Milton Keynes, UK
e-mail: frank.banks@open.ac.uk

W. Fox-Turnbull and P. J. Williams (eds.), *Locating Technology Education in STEM Teaching and Learning*, Contemporary Issues in Technology Education,
https://doi.org/10.1007/978-981-97-1995-2_5

5.1 Introduction

What a teacher knows in terms of their teaching strategies, their understanding of the place of their subject in the school system—which is particularly important in their school's structure of STEM education—and the teacher's own specific subject curriculum knowledge all come together to contribute to their effectiveness as a teacher. Some years ago, when observing student-teachers in the classroom, David McNamara noted:

> Teachers' subject matter knowledge influences the way in which they teach, and teachers who know more about a subject will be more interesting and adventurous in their methods and, consequently, more effective. Teachers with only a limited knowledge of a subject may avoid teaching difficult or complex aspects of it and teach in a manner which avoids pupil participation and questioning, and which fails to draw upon children's experience (McNamara, 1991:115).

Deborah Winn when considering links across the curriculum pointed out that the 'silo' nature of the school curriculum in many countries makes the relevance of school lessons obscure to many students, yet in contrast:

> If appropriate links are clear, pupils are able to see context to what they are learning and as such, knowledge becomes embedded, and they are more confident to be creative in all subject areas. Providing a context prevents the belief some students have that they cannot see the point of learning a topic as they will not use the knowledge in their future careers and life (Winn, 2021:273)

I intend to adopt a straightforward approach to what I mean by curriculum knowledge: the combination of both subject knowledge and of pedagogy that a teacher needs in order to be competent; and this chapter looks at the necessary curriculum knowledge that can be brought to school STEM projects to support the 'T' in STEM education. However, school STEM projects can be wide-ranging, and so what is reasonable for a teacher to know depends on how a school organises its STEM curriculum. Clearly, the fewer teachers involved in teaching across school STEM projects, the greater the professional knowledge—subject knowledge and pedagogical content knowledge—is required of those teachers. The approach to the teaching of STEM school projects can be Coordinated, Collaborative or Integrated.

5.1.1 A Coordinated Approach

In a properly coordinated approach, teachers in each contributory STEM subject become familiar with the work carried out by their colleagues in the others and plan their curricula so that the *timing* of topics within each subject is sensitive to others' needs and taught in a way that supports the students' developing understanding, rather than one that causes confusion. For example, proficiency with the use of measuring in millimetres, and ways of collating data from respondents, would benefit technology if covered in mathematics; and if electricity is explained using similar analogies and terminology in both technology and science, the students' developing ideas can be reinforced. (See example 3 below.)

5.1.2 A Collaborative Approach

In a collaborative approach, some activities within each subject are *designed and planned by teachers working together* to establish an effective relationship. For example, in lessons on alternative energy in Scotland, students undertake four 'learning journeys': 'From fossil fuels to wind' meets some of the science requirements, 'Wind, wave and tidal' meets some of the technology requirements, and 'Calculating the wind' meets some of the mathematics requirements of the curriculum with links across to science and to technology. At the end, in 'This island is going renewable', students are challenged to make the case for the use of renewable energy by the small island community (ES, 2022). (See also Food Technology example 4.)

5.1.3 The Integration of STEM Subjects

There are two ways of considering strategies for the full integration of STEM subjects. One is by organising synchronous inputs from a range of staff for an off-timetable event or project. Here, all the educators support the activities through team-teaching and students turn to a particular member of staff for advice and support when they are available. Around the world, students of all ages take part in STEM competitions and challenges or attend workshops in science museums and higher-education institutions. In Taiwan, for example, robot-building from using Lego through to full combat 'robot wars' models is a common out-of-school activity. In Japan, 'STEM-inars' occur early in the school year. School students are encouraged to attend a university for an intensive one-week 'deep dive' into a STEM area of interest.

The second way is a full integration of the STEM subjects in school so that one teacher follows a themed project across a number of lessons, as is often the case in primary schools (see Chapters by Fox-Turnbull and Lee for further consideration of levels of integration). Adopting this at the secondary level assumes that considerable expertise is available in the one teacher, or that staffing resources are available for a STEM curriculum team-teaching approach.

This chapter will look at four aspects when teaching the T in STEM:

1. The contribution that technology brings to school STEM projects whatever the main 'subject focus' of that project.
2. The contribution that science and mathematics can make to technology-focussed STEM projects.

3. The 'personal subject construct' that teachers have of their role in teaching and in supporting STEM projects which influences what knowledge they consider important.
4. The way that STEM student learning can be mapped.

5.2 Technology's Contribution to STEM Education Whatever the Main Subject Focus

I think most would agree that young people want answers to the type of question that seeks knowledge of the 'knowing how' variety—how a thing works, how it is used, how it is possible to improve the function of something or the way something is done, or how to create something which has a new purpose. This can be thought of as technological knowledge. It is the practical knowledge of *application*, i.e., 'know-how'. The other type of question seeks knowledge of the 'knowing why' variety—why the world is the way it is, firstly to help us understand the rules that confirm generally accepted agreement about what we know, and secondly to help us rationalise the experience of our senses. This type of knowledge can be thought of as scientific knowledge. In a nutshell, as Per Norström would summarise it:

> Technology is about creating artefacts and solving problems, while science is primarily about describing, [understanding] and explaining phenomena in the world. (Norström, 2013:378)

In technology, if the knowledge is 'useful' then it continues to be exploited until it is no longer of use. In science, a concept that is not 'correct' in that it does not match experimental results or other related theories is discarded. However, the rejection of certain established scientific ideas such as phlogiston, and the caloric theory of heat, and acceptance of energy as quanta, took many years!

What are Technology's 'Big Ideas' of subject content knowledge that teachers need to consider when being responsible for covering the T in STEM education? Following the lead of Harlen (2010) who set out the 'Big Ideas' needed when teaching Science, Barlex and Steeg (2017) considered what they regarded as the 'Big Ideas' in technology education. They suggested:

knowledge of design—engaging in problem-solving to develop design ideas through identifying peoples' needs and market opportunities;

knowledge of systems thinking—choosing appropriate mechanisms and designing electronic circuits to solve specific problems. Controlling using microprocessors, in particular, is now commonplace (see Banks & Barlex, 2021:198);

knowledge of materials—including sources, properties, its environmental footprint and longevity especially of those in short supply;

knowledge of manufacturing—which can be considered, generally, as four main methods: subtraction, addition, forming and assembly or construction;

knowledge of functionality—most STEM education projects have to 'work' and the function is therefore crucial. This could be set out as powering, controlling and structuring;

understanding of evaluating design ideas—how do they affect the lives of those who use them and those that make them? How do they affect the planet? 'Critiquing for stewardship' involves considering life cycle analysis, and an appreciation of how to critique an intended outcome in terms of ethics and justice.

It is clear from this list that the term 'Technology', unlike in common usage (particularly in the USA) means much *more* than just 'computing'. 'Big Tech' companies such as Google or Meta who have traditionally considered 'Technology' as being equipment, laptops and phones now also centre on Artificial Intelligence (AI) or the 'Internet of Things'. But important though these may be technology is *not* merely 'the appliance of science'. The Big Ideas suggested above show that the T in STEM education can be clearly distinguished from science, and indeed, the assumption that science knowledge always precedes technology knowledge can be challenged through some wide-ranging examples. For instance, refining copper, canning food using 'tins' and the first commercial steam engines by Thomas Savery and Thomas Newcomen are all examples of technologies that were developed without any understanding of the underlying science. Conversely, some 'Blue Skies' science research such as the search for life on Mars, atom-smashing machines at CERN and radio-telescope synthesised images of a black hole depend on technology for their success. The principal point, therefore, is that technology has its own Big Ideas and is more than the application of fully understood scientific knowledge; a point acknowledged by the economist Nathan Rosenburg:

> It is knowledge of techniques, methods, and designs that work, and that work in certain ways and with certain consequences, even when we cannot explain exactly why. It is [...] a form of knowledge which has generated a certain rate of economic progress for thousands of years. Indeed, if the human race had been confined to technologies that were understood in a scientific sense, it would have passed from the scene long ago. (Rosenberg, 1982:143)

Let us consider some examples of the contribution the 'T' for technology education can bring to aspects of Science, Maths and Engineering in school and in doing so focus especially on the specific Big Idea of 'critiquing for stewardship' by addressing the concerns of many young people for Climate Action, Clean Water and Sanitation and Responsible Consumption and Production which are some of the UNESCO sustainability goals (UNESCO, 2022).

Example 1—Obtaining Clean Water The importance of the context of this school STEM project is stark. According to the World Health Organisation, approximately 2 billion people around the world do not have access to clean and safe water. Contaminated water is responsible for the transmission of diseases such as diarrhoea, cholera, dysentery, typhoid and polio, with an estimated 485,000 deaths from diarrhoea alone each year.

In a science context, students would need to consider the following questions:

- What is in water?
- Why does water need to be treated before we can drink it? What engineering techniques can be used to clarify and purify water at scale?
- How can we reduce the quantity used and the environmental impact of obtaining water?

The Welsh examination board for their Applied Science examination for those aged 16 points out that obtaining clean water allows learners to explain everyday and *technological* applications of science; evaluate associated personal, social, economic and environmental implications; and make decisions based on the evaluation of evidence and arguments (WJEC, 2022).

Sedimentation, filtration and the use of chlorination and ultraviolet light can be explored in the school laboratory, but the technology department can support the application of those techniques by modelling how they can be implemented at scale. Alternatively, the science can be applied to smaller-scale filtration techniques to purify scarce domestic water wells or pools in developing world contexts. Just One Africa (2022), use filters designed after the technology employed in kidney dialysis. Hollow fibre membrane removes 99.99999% of all bacteria such as Salmonella, E. coli, Vibrio cholerae and Salmonella typhi (which cause Cholera and Typhoid) and all protozoa such as Giardia and Cryptosporidium.

By constructing units in technology lessons, the filters which are just 100 mm (4 inches) long and less than 225 g (8 oz) can deliver 20 L (5 gallons) of clean water in about 30 min. The high flow rate eliminates the need to store water, which reduces the chances of water being contaminated after it is filtered. The only external force required for the system to function is gravity. The project requires a knowledge of Design, Systems Thinking and Functionality.

Banks and Barlex (2021:60) point out:

> There is a growing movement to encourage designers to tackle the problems encountered in the developing world. Books such as Design for the other 90 per cent (Smith, 2008), Design like you give a damn (Architecture for Humanity, 2006) and Design Revolution (Pilloton & Chochinovd, 2009) are written to inform the general public about the way design can be a powerful means to improving the situation and to provoke a response from the design community. There is no reason why this approach cannot be extended to students in schools through the design & technology curriculum and build on the work carried out in science.

Example 2—Group Problem-Solving of a Real-Life Problem in Mathematics Mathematics, traditionally, has been somewhat resistant to group work and so problem-solving projects that can be done in a group, and which are constructed, like in technology, around ill-structured, open-ended and ambiguous real-life problems deserve careful consideration as, in this example, it leads to an understanding and evaluation of different design ideas.

An approach to mathematics common in schools uses Model-Eliciting Activities (MEAs) which are designed to help students apply the mathematical procedures they have learned, to create mathematical models (Table 5.1). To initiate MEAs, typically the teacher sets up a context for the students about the real-life topic to be considered and students respond to some questions based on the article. The problem is then posed to the students who work in groups to model possible solutions. Often such problems have multiple solutions, and like in technology lessons, the students are working towards and evaluating a 'best fit'. An example (adapted here) from an idea suggested by Chan Chun Ming (2008) in Banks and Barlex (2021:139) is as follows:

> Your group is in charge of hiring some workers to help create a new wildflower area, including a pond, for your school in an environmentally sensitive way. These workers must complete the job within four days.

Conditions

1. You can hire only from one company once, and you have to accept the number of workers for that company.
2. You have a worksite supervisor who can only supervise at most 12 workers per day, so try to hire as many as 12 per day. Assume that each worker to be hired works the same amount of time and produces the same amount of work per hour.
3. You need at least: 14 workers for clearing the land, 14 workers for planting and 14 workers for digging and preparing the pond

The productivity index is calculated as (Total no. of workers/Total Cost) × 100 calculated to 3 decimal places. Are you getting value for money? The larger the index the better.

When the group has made the best fit, they present their suggestion(s) to the rest of the class.

Table 5.1 Data for environmental model-eliciting activity

Data					
Land clearing					
Company	A	B	C	D	E
No. of workers	4	2	6	3	5
Cost $	160	78	270	120	175
Planting					
Company	F	G	H	R	J
No. of workers	3	6	7	4	5
Cost $	114	240	315	160	210
Pond work					
Company	K	L	Q	N	T
No. of workers	7	4	3	6	4
Cost $	245	160	135	225	1140

5.2.1 Science and Maths' Contribution to Technology-Focussed STEM Projects

As suggested above, a technology teacher's required specific curriculum knowledge depends on the way that a school organises its STEM curriculum: Coordinated, Collaborative or Integrated. In all cases, however, it is important that teachers work together in their teaching for the benefit of their students. Teachers need to 'Look-Sideways' at how their colleagues are contributing to STEM education in their school.

Example 3—Science and Technology Coordination—Emergency Lighting in Tents Installing temporary lighting for refugee tents following an emergency such as conflict, flooding, earthquake or some other natural disaster requires knowledge of functionality and a clear understanding of current and series and parallel circuits. As Torben Steeg, (Banks & Barlex, 2021:76) points out:

> If the light-up elements are in series then the number of elements that can be used is quite small, but if they are arranged in parallel then a greater number can be used. If the students want to work out how long the light can be used for on a single battery, then the scientific understanding required here is interesting. If the explanation of the circuit is couched in terms of current flow, then the explanation will revolve around understanding that 'more components' leads to 'higher current flow' and therefore 'the less time the battery will last'. Hence Torben was inclined to use current as the pertinent factor (which can either be measured or looked up from a component's datasheet) and use the fact that most batteries provide a mAh value; this reduces the science understanding demand significantly as there is no need to consider voltage or power. The use of the unit of 'mAh' allows different power sources to be easily compared and matched to the current demand of a circuit.

However, it is not only the 'subject knowledge' that is needed by the technology teacher but also a similar Pedagogical Content Knowledge (PCK) to that of the science teacher, so that students in science and technology are not confused by different analogies but rather have their understanding reinforced.

> Within the category of pedagogical content knowledge, I include, for the most regularly taught topics in one's subject area, the most useful forms of representation of those ideas, the most powerful analogies, illustrations, examples, explanations and demonstrations – in a word, the ways of representing and formulating the subject matter that make it comprehensible to others. (Shulman, 1986:9)

If both technology and science describe electricity as conventional current flow from positive to negative, for example, understanding is reinforced. However, if one describes current but the other electron flows in the opposite direction then it can easily lead to confusion. However, as we will discuss later, PCK as set out by Shulman implies a *fixed* set of understandings 'for the most regularly taught topics in one's subject area, the most useful forms of representation of those ideas' which does not sit well in the constantly changing STEM curriculum.

A further step from coordinating the STEM curriculum as shown in Example 3 could be the more unified collaborative approach where teachers in each subject plan their curricula so that activities within each subject are designed using the expertise of each contributing teacher either at separate times or together in a team-teaching approach as is demonstrated in the next Food Technology example. Alternatively, if Example 4 is configured as a project for the earlier years at High School, one teacher could support and integrate all the suggested activities.

Example 4—Food Technology, Science and Maths—Bangladesh 'Pumpkins Against Poverty' In Bangladesh, 160 million people still live on less than $1 a day. Every year, monsoon rains cause the rivers of Bangladesh to swell, resulting in devastating floods. The rains in 2022 were particularly severe washing away homes, submerging land and destroying possessions. Consequently, families face months of hunger and malnutrition when they are forced to find a new place to live and a new means of earning a living without land to cultivate. When the floods recede, new sandbars are revealed but they are too infertile to cultivate. Practical Action (2022a), however, introduced a new way to grow crops through a project called 'Pumpkins Against Poverty' where pumpkins are grown on the sandy, barren soil left behind when flood waters recede. By digging holes in the sand and filling them with manure, compost and pumpkin seeds together in a jute bag, these crops can thrive. Pumpkins give a high yield, are packed full of nutrients and can be stored for up to a year, meaning people have a crop both for their families and also to sell, easing them out of poverty.

Practical Action has made available free classroom resources including a teacher's guide (Practical Action, 2022b) which has been adapted for this example. After some preliminary work finding out about the geography and climate of Bangladesh, a series of school worksheets draw on Food Technology along with Science and Maths to help students investigate the nutrition of Pumpkins, the ways they can be best prepared for growing and eating and the economics of cultivation and sale of surplus.

Activity 1: Students can cut up pictures of different food types and their definitions, to match the titles: Protein, Vitamin C, Fats, Carbohydrate, Minerals and Vitamin A.

Activity 2: Working in pairs and using a knowledge of functionality, the students decide which, for a Bangladeshi family involved in the project, are the most important factors to consider in the plants that they grow, putting the most important at the top, equal in the middle and the least important at the bottom from the following list:

- Is full of nutrients to keep people healthy
- Grows in sandy soil
- Grows on top of the ground
- Looks good
- Grows and is harvested between November and April
- Has seeds that are easily preserved for another year
- Is a food that people like to eat
- Can be sold for a profit
- Can be stored then used a long time after harvesting
- Does not need too much water and care when growing
- Can be eaten raw

- Smells good.

Activity 3: Getting pumpkin seeds to germinate in different conditions.

(Science investigation), choosing different parameters: dark warm place, light warm place, in the fridge, cold but in the light, different amounts of water and different soil types.

Activity 4: Getting cash from pumpkins to support the family's ambitions.

This contribution from mathematics considers what might be a typical annual surplus that can be sold, and using an understanding of evaluation ideas, make some suggestions as to different benefits for the family. The yield of 100 seeds is four pumpkins from each seed. Each group works out how many pumpkins they might grow. If they keep 10% and sell the rest at 10 taka each, what could they buy? 1 Taka = 1 US cent. See Table 5.2 for approximate values of potential commodities.

A supplementary part of this activity is to investigate what animal-based alternative fuels to charcoal are used by some Bangladeshi families.

Activity 5: Making a Bangladeshi Snack.

Using your knowledge of the following ingredients, design your own fried snack from a quarter of a pumpkin:

Vegetable oil
Turmeric
Some salt
Chilli
What other spices might you use? What other ways of cooking?

Each of the above examples shows the complexity of tackling real-world problems. Real life is no respecter of school subject areas, and it is only by bringing together subject contributions across STEM that a satisfactory solution can be attempted. The T in STEM can lead the problem such as in Examples 3 and 4 where the context is a technological problem, or it can be a significant contributor as in Examples 1 and 2. In all cases, and particularly in Example 4, the way that a school organises its curriculum so that the appropriate knowledge is available to ensure students are properly supported is the key to success. It is clear, therefore, that a context that brings together the many aspects of STEM needed to tackle the problem needs an integrated approach to curriculum design.

Table 5.2 Adapted data for the possible sale of pumpkins in Bangladesh—2022

Items	Cost in taka
Cow	81,000
School uniform	550
Rice for a week	199
Charcoal for cooking for a week	210
Repair the roof	5100
Children/s clothes	550

5.3 A Teacher's Personal Subject Construct for Teaching the T in STEM

Applebee (1989:217) notes that when 'when we start to teach a new subject, one of the most powerful influences on what we do is our memory of how we were taught'. But STEM as an integrated whole and technology as a specific subject do not have a history long enough to have shared memory. But central to the dynamic process of constructing and drawing on curriculum knowledge for teaching STEM are the 'personal constructs' of both teacher and students: a complex amalgam of past experiences of learning, a personal view of what constitutes 'good' teaching and, perhaps most crucially, a belief in the purposes of the subject. This is the root of dispositional knowledge which underpins a teacher's professional knowledge as it shapes what they *believe* they should 'know' and how they should 'act' in response to feedback from students and colleagues.

The diagram below (Fig. 5.1) is an example of a diagram constructed as part of the international DEPTH study—'Developing Professional Thinking for Technology Teachers' (Banks et al., 2004 and 2008). It is not a Venn Diagram; it is rather a 'Graphical Tool' which was presented to teachers for them to complete from their personal perspective in their country context. The blank framework was completed by teachers in Australia, Canada, Finland, New Zealand and the UK, as a tool to support reflection on their own professional knowledge. The DEPTH studies discovered that the graphical tool was equally useful in helping them to become more self-aware as a technology teacher when reflecting on their practice whatever their home country across the world.

This DEPTH diagram I have constructed in line with the ideas explored in this chapter includes the 'Big Ideas' discussed above and techniques for assessment for learning as discussed below. Note that Pedagogic Content Knowledge (PCK) had been re-configured as the more general 'Pedagogical Knowledge'. Shulman's view quoted above that PCK is 'the most useful forms of representation of *those ideas*, the most powerful analogies, illustrations, examples, explanations and demonstrations' suggests a constrained and fixed set of understandings that teachers needed to know, but the DEPTH team's position was that technology and STEM outside the classroom is changing rapidly and that STEM teachers across a school need to grow in their understanding of pedagogical knowledge together. Finally, the idea of 'School Knowledge' takes account of the way that successful teachers appreciate that the prevailing school context both enables and restricts the teaching of technology and so also the teaching of STEM—for example, the different resources, tools and equipment available and the specific role of the students' design portfolio in assessment.

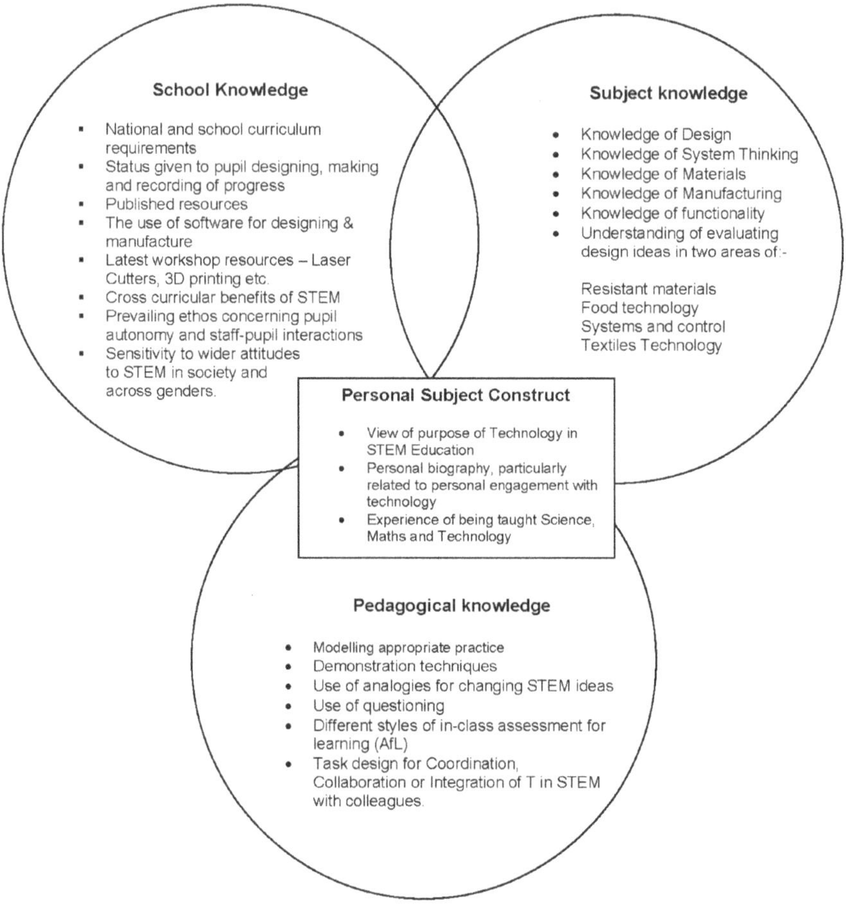

Fig. 5.1 DEPTH diagram for teaching the T in STEM

5.4 The Way that STEM Student Learning Can Be Mapped

If STEM is a collaborative process, then it presents a practical difficulty in high-stakes assessment regimes such as those which are prevalent in England, the USA and many countries in South Asia to assess the particular contribution of an individual. In England, for example, the examination system currently only looks at summative assessment of the contributory separate STEM subjects examinations taken at age 16 (GCSE) and at age 18 (A-levels), Coe et al. (2008) noted:

> Differences between subjects appear to be extremely stable over time. […] Given this level of consistency across methods, time and subgroups, it seems appropriate to examine the relative difficulties of science and non-science subjects. At A-level, the STEM subjects are not just more difficult on average than the non-sciences, they are without exception among the hardest of all A-levels. The other sciences are also generally more difficult than the

> non-sciences, though the difference is not as extreme. At GCSE the STEM subjects are a little more difficult on average than the non-sciences, though the difference is less than at A- level.

Currently, all technical qualifications must have at least 40% of subject knowledge assessed through an end-of-course written test. Are there better ways to map learning across STEM subjects? In Sweden, teachers grade students at both ages 16 and 18 without any, or just minor, written exams and in Wales, Science and Technology is a combined area of learning and experience for all students aged 3 to 16 years, and there has also been a fundamental change in the attitude to assessment:

> Working within the Curriculum for Wales framework, this change recognises that schools/settings are best placed to design the most appropriate curriculum and assessment arrangements for their learners and their contexts. Schools design a curriculum, they select the knowledge, skills and experiences that best support their learners to progress in the ways described in the Curriculum for Wales – they will need to develop assessment arrangements to support each individual learner to progress in relation to their school/setting's curriculum. The focus will be on identifying where the learner is in their learning, their next steps and the support or challenge needed to move forward in their learning. (Hwb, 2022)

Student learning across STEM can best be mapped by focussing on assessment that is *for* learning (AfL), as opposed to merely *of* learning, looking forward to what they need to focus on in their future learning as well as considering what they have already achieved looking back. This kind of assessment has a 'formative' purpose: it helps to structure the future rather than simply to gauge and record past achievements, and it was developed from the seminal work of Black and Wiliam (1998).

I summarise some key points here:

Questioning: Teachers found that giving more time for students to reply to their questions elicited more thoughtful answers. Moving from questions focussed on knowledge recall to exploring understanding enriched student thinking. For example, a science teacher who had begun lessons with questions such as 'What is this instrument and where would you find it?' now began to ask questions such as 'Why do you think these two pumpkin plants have grown differently?'

Peer and self-assessment: Teachers encouraged students to take more ownership of their own learning by helping them to understand learning targets. 'Traffic-lighting', in which students assigned red, amber or green to a piece of work, according to the degree to which they did or did not understand it. They use an iPad or mobile (cell) phone to photograph their progress for a digital portfolio. And with peer assessment, students were more likely to challenge each other's judgments of their work, thereby sparking further discussion and debate and suggestions for improvements.

Marking: Feedback that identifies what the student has done well and focuses on what she or he can do to improve is central to AfL. Giving a comment rather than a mark has revealed that students pay more attention to what the teacher thinks about their work which leads to a better learning environment as all saw that learning was improving.

AfL through summative assessment: AfL has to operate in the real world of summative assessment, but Black and Wiliam showed that if students understand the criteria against which exams were marked, then exams can be used to directly improve learning. Active revision by generating their own exam questions, explaining parts of units to each other and marking peers' work again improves learning.

(See also: Black, Wiliam and Assessment for Learning, 2022).

As mentioned above, this process is being introduced in Wales where the government recommends focussing on Questioning, Feedback and In-class Assessment as seen in Table 5.3.

Using these 'assessment for learning' tools, STEM student learning can be mapped. To make this manageable, they need to be integrated naturally into lessons.

- Assessments need to be made during a project as well as at the end. The burden of assessment needs to be spread out; indeed, some important attributes can only be assessed as the work is being done. Such learning mapping opportunities need planning.
- Students can help in recording assessments by noting points in their books, project folders or design portfolios; for example, the outcome of a discussion that led to a decision, a new idea or modifications to a design.
- Internal moderation between teachers is necessary to come to a shared meaning of what mapping is required for specific levels. A collection of evidence in the school will help to establish a common agreement.
- Several STEM projects are needed to build up a view of the capability of a pupil as different projects bring different types and levels of responses from students. This is a case where 'Looking-Sideways' not only helps with the construction of the STEM Curriculum but also an understanding of the capability of the students.
- The aims and purposes of STEM project work should not be unduly affected by the assessment process. As described above, in high-stakes assessment regimes certain procedures will be strongly suggested by teachers because 'it earns marks in the exam', but the relationship between the student and the teacher and the desire of the students to take ownership of the task should not be compromised.

Table 5.3 Summary of assessment for learning principles

Questioning	Feedback	Peer and self-assessment
Improving the quality of answers	STEM project target setting	On-going lesson assessment
Peer discussion	Immediacy of feedback	Recording process, e.g. digital portfolios
Active involvement of all learners	Comments on how to improve rather than just a mark	Uses of summative assessment

Welsh Assembly Government (2010:30)

5.5 Conclusion

Teachers of STEM subjects are first and foremost teachers of young people, but they are also teachers of a subject and feel responsible for keeping up-to-date, as knowledge and processes expand exponentially. It is increasingly easy to access information—using the computing power of even a cheap mobile phone puts video, sounds and information in the hands of everyone; the ready availability of sophisticated computing power puts a library in our pocket—teachers and learners alike. But, as has been shown in the examples above, for successful STEM school projects it is the school's professional environment that facilitates different teachers sharing their enthusiasm, knowledge, ideas and resources. This is not only formally through collaborative schemes of work but also through the all-important shared coffee time conversations that make the sharing of STEM information and ideas and collaborative processes professionally permissible. I have worked in both large and small schools and found that the large schools shared a cross-subject STEM ethos less well simply due to where staff chose to meet at break time; big subject departments stuck together and chatted informally with those who taught the same curriculum area, while small department subject groups of two or three teachers went out to seek company at break time and so the informal conversations were across STEM and hence more fruitful.

A recurring theme in this chapter, whether STEM education is taught by coordination, collaboration or in an integrated way is the need to talk to colleagues and to find out what fellow teachers are doing so that curriculum knowledge can be shared, and pupil learning can be enhanced—what Banks and Barlex (2021:28) call 'Looking-Sideways'. The quickest practical way to improve STEM learning, to gain credibility as a teacher, change the attitudes of students and colleagues and address the necessary development of curriculum knowledge of the 'T in STEM' projects is through informal collaborative arrangements. All staff have clear responsibilities and an understanding of how they relate to those they work with is so important in creating a successful supportive professional environment that improves STEM teaching and, through a shared understanding, enhances learning for our students.

References

Applebee, A. (1989). The Enterprise we are a part of Learning to Teach. In B. Moon & P. Murphy (Eds.), *Developments in learning and assessment.* Hodder and Stoughton, London. ISBN: 978-0-340-51437-5

Architecture for Humanity. (2006). *Design like you give a damn: Architectural responses to humanitarian crises.* Architecture for Humanity. ISBN: 978-1-933-04525-2.

Banks, F., & Barlex, D. (2021) *Teaching STEM in the secondary school: Helping teachers meet the challenge*, (2nd ed.). Routledge, London. ISBN: 978-0-367-33046-0

Barlex, D., & Steeg, T. (2017) Big Ideas for Design & Technology available at https://dandtfordandt.wordpress.com/working-papers/big-ideas-for-dt/. Accessed 22 June 2022

Banks, F., Barlex, D., Jarvinen, E-M, O'Sullivan, G., Owen-Jackson, G., & Rutland, M. (2004) DEPTH—Developing professional thinking for technology teachers: An international study. *International Journal of Technology and Design Education, 14*(2), 141–157. Kluwer Academic Publishers, Holland. ISSN 0957-7572. https://doi.org/10.1023/B:ITDE.0000026475.55323.01

Banks, F. (2008) Learning in DEPTH: Developing a graphical tool for professional thinking for technology teachers. *International Journal of Technology and Design, 18*(3), 221–229. ISSN 0957-7572. https://doi.org/10.1007/s10798-008-9050-z

Black, P. J., & Wiliam, D. (1998). Assessment and classroom learning. *Assessment in Education: Principles, Policy & Practice, 5*(1), 7–74. https://doi.org/10.1080/0969595980050102

Black, P.J. and Wiliam, D. (2022) *Black, Wiliam and Assessment for Learning*. https://www.teachingexpertise.com/articles/black-wiliam-and-assessment-for-learning/. Accessed 25 June 2022

Chan, C. M. (2008). Using model-eliciting activities for primary mathematics classroom. *The Mathematics Educator 11(1),* 47–66. http://hdl.handle.net/10497/135

Coe, R., Searle, J., Barmby, P., Jones, K., & Higgins, S. (2008) *Relative difficulty of examinations in different subjects*, Durham University, Durham UK. http://epc.ac.uk/wp-content/uploads/2012/10/SCORE-A-levels-relative-difficulty.pdf. Accessed 6 July 2022

ES. (2022). Education Scotland *STEM Central* https://education.gov.scot/improvement/learning-resources/stem-central/. Accessed 4 July 2022

Harlen, W. (2010). *Principles and big ideas of science education*. Association for Science Education. https://www.ase.org.uk/bigideas. Accessed 6 July 2022

Hwb. (2022). https://hwb.gov.wales/curriculum-for-wales/. Accessed 15 January 2022

Just One Africa. (2022). https://www.justoneafrica.org/clean-water/. Accessed 23 June 2022

McNamara, D. (1991). Subject Knowledge and its Application: Problems and possibilities for teacher educators. *Journal of Education for Teaching, 17*(2), 113–128. https://doi.org/10.1080/0260747910170201

Norström, P. (2013). Engineers' non-scientific models in technology education. *Intentional Journal of Technology & Design Education, 23*, 377–390. https://doi.org/10.1007/s10798-011-9184-2

Pilloton, E., & Chochinov, A. (2009). *Design revolution: 100 products that are changing people's lives*. Thames and Hudson ISBN: 978-0-500-28840-5

Practical Action. (2022a). *Pumpkin Against Poverty*. https://practicalaction.org/our-work/projects/pumpkins-against-poverty/. Accessed 25 June 2022

Practical Action (2022b) *Pumpkin Against Poverty: School resources* https://practicalaction.org/schools/pumpkins-against-poverty/. Accessed 25 June 2022

Rosenberg, N. (1982). *Inside the black box: Technology and economics*. Cambridge University Press. https://doi.org/10.1017/CBO9780511611940

Shulman, L. S. (1986) Those who understand: knowledge growth in teaching. *Educational Researcher, 57*(2), 4–14. http://www.jstor.org/stable/1175860

Smith, C. (2008). *Design for the other 90%*. Smithsonian Cooper-Hewitt National Design Museum.https://doi.org/10.1080/18626033.2012.749603

UNESCO. (2022). https://en.unesco.org/sustainabledevelopmentgoals. Accessed 22 June 2022

Winn, D. (2021). Developing links with other subjects. In A. Hardy (Ed.), *Learning to teach design and technology in the secondary school: A companion to school experience*, Routledge, Abingdon. ISBN: 978-0-367-33681-3.

WJEC (Welsh Joint Education Committee). (2022). https://www.wjec.co.uk/qualifications/applied-science-gcse-double-award/#tab_overview. Accessed 15 January 2022

Welsh Assembly Government. (2010). *How to develop thinking and assessment for learning in the classroom*, WAG, Cardiff. https://hwb.gov.wales/api/storage/663d00fd-9c3b-4003-9a1f-0012bad39fad/how-to-develop-thinking-and-assessment-for-learning-in-the-classroom.pdf. Accessed 6 July 2022

Frank Banks is Emeritus Professor of Teacher Education at The Open University in the United Kingdom, where he was Director of Initial Teacher Education also with responsibility for the University's strategic direction of teacher-education projects in sub-Saharan Africa, Bangladesh and India. Frank worked for 11 years as a secondary (High) school teacher of Science, Technology, Engineering and Mathematics in different schools in England and in Wales, including head of department; and as a primary (elementary) school advisory teacher. He has authored or edited 32 books/handbooks for teachers, over 100 academic papers, and has acted as a consultant in the professional development of teachers to Egyptian, South African and Argentinean government agencies, UNESCO and the World Bank.

Chapter 6
STEM Literacy in Technology Education

Virginia R. Jones and Thomas Roberts

Abstract The development of STEM literacy is commonly utilized as the outcome of STEM projects; however, despite its common usage, there is little conceptual agreement in the notion of STEM literacy. This chapter reviews disciplinary literacies and integrated STEM literacy before positioning technological literacy as the necessary first step in achieving STEM literacy.

Keywords STEM literacy · Integrated STEM · Authenticity · Technology education

6.1 Introduction

The purpose of this chapter is to critique the current use of the term STEM literacy and suggest a way forward that recognizes the importance of technological literacy in developing STEM literacy. Governments and policymakers consistently advocate for more integrated STEM education in countries ranging from the USA (National Research Council, 2014), the UK (Council for Science & Technology, 2013), Indonesia (Blackley et al., 2018), Saudi Arabia (El-Deghaidy et al., 2017), and Australia (Office of the Chief Scientist, 2014). Despite the expressed desire for increased STEM education, there are competing names for STEM (e.g., integrated STEM, STEM integration, and interdisciplinary STEM) and competing perspectives of STEM based on grade level (e.g., Baker & Galanti, 2017; Berland, 2013; Thibaut et al., 2018). There is no common global definition of integrated STEM (English, 2016; Moore et al., 2020). The authors recommend a concerted effort to identify the *basics* that define integrated STEM. Although some argue there is strength in diversity and danger in uniformity and others argue conversely, a broad-based definition

V. R. Jones (✉)
Old Dominion University, Norfolk, VA, USA
e-mail: vrjones@icloud.com

T. Roberts
Bowling Green State University, Bowling Green, OH, USA

W. Fox-Turnbull and P. J. Williams (eds.), *Locating Technology Education in STEM Teaching and Learning*, Contemporary Issues in Technology Education,
https://doi.org/10.1007/978-981-97-1995-2_6

would allow *flexibility* in the definition so that each educational system can adapt the standard to their own specific needs.

6.2 Disciplinary Literacies

Scientific literacy was described by the American Association for the Advancement of Science (1989) as "being familiar with the natural world and respecting its unity; being aware of some of the important ways in which mathematics, technology, and the sciences depend upon one another… and being able to use scientific knowledge and ways of thinking for personal and social purposes" (p. 20). The National Research Council (2012) described scientific literacy as "possessing sufficient knowledge of science and engineering to engage in public discussions on related issues; [being] careful consumers of scientific and technology information related to their everyday" (p. 1). Just as definitions of scientific literacy have evolved over the years, mathematics literacy has been defined in varied ways depending on the goals of mathematics education at the time (Tout, 2000). The International Bureau of Education, UNESCO (2009) defines mathematical literacy as not only one's ability to understand and identify role of mathematics but also the ability to "make well-founded judgements and to use and engage with mathematics in ways the meets the needs of individual's life as a constructive, concerned and reflective citizen" (para. 1). Engineering literacy is defined as "systematic and often iterative approach to designing objects, process, and systems to meet human needs and wants" (National Assessment Governing Board [NAGB], 2018, p. 4). Reviewing these individual literacy definitions, literacy is inherently an active, social process as societies evolve through their language and cultural ideologies; it can no longer be thought of as an autonomous set of knowledge, but a set of practices woven into the culture, ideologies, and rapid expansion of global technologies in our world. The definitions share a common thread of ensuring learners can be cognizant of the influence of their ways of thinking for personal and societal consequences. Technological literacy in this context, should "place value and importance on ethno-technologies and cultures, scaffoldings, social justice, language and dialects, design, and the Internet of things" (Hilgarth, 2020, abstract). DeVries et al. (2018) supported a definition of technological literacy as

> The ability and willingness, on the basis of a technology-oriented concept, everyday related and evaluation knowledge, to successfully execute typical technical forms of actions in different application contexts and to be able to estimate the consequences for themselves and for society" (p. 35).

The authors stated this definition was not based on any specific technological or philosophical theories or literature. The NAGB defined technological and engineering literacy as, "the capacity to use, understand, and evaluate technology as well as to understand technological principles and strategies needed to develop solutions and achieve goals" (p. 4). Yet, there is a strong connection between the literacy definitions for each discipline in STEM to promote general standards for good stewardship of resources and understanding of societal influences and purposes.

Literacy is an important construct in most disciplines. Literacy is commonly used in English language societies to describe the practice of one's ability to read and write or as a way to distinguish one who is literate about certain subjects from those who are not (Bloome et al., 2018). Societies commonly accept that all citizens need a certain level of literacy to be productive, contributing members of society. Hilgarth (2020) stated, "The task of literacy is to be able to develop a knowledge which then can be recalled and applied to a particular situation" (p. 2). Hilgarth (2020) further asserted that "literacy enables extraction of knowledge related to or extrapolation of knowns into the region where an understanding of a particular aspect of technology or engineering is at issue" (p. 2). In all STEM disciplines, literacy was defined in a range of ways; nonetheless, all promote an element of citizenry and societal needs.

Given the increased prevalence of and emphasis on integrated science, technology engineering, and mathematics (STEM) in today's world and in education (Ehlert & Roberts, 2021; Moore et al., 2020), STEM literacy is a newer form of literacy. Hallström and Ankiewicz (2019) indicated the difficulty of an accepted definition of STEM literacy was predicated on the fact that each of the distinct disciplines in STEM is not necessarily connected either through "content nor pedagogy" (p. 188). Integrated STEM is important because it promotes skills such as critical thinking, problem-solving, creativity, collaboration, and communication (Mohr-Schroeder et al., 2020). However, at times, integrated STEM curricula or coursework overemphasizes concepts from science and mathematics. The same holds true for STEM literacy definitions (International Technology and Engineering Educators Association [ITEEA], 2020). Thus, integrated STEM literacy must emphasize the importance of each discipline. Tang and Williams (2019) after noting "similarities found in several languages and thought processes of each discipline" (p. 675) promoted an inclusive yet distinctive definition of STEM literacy as follows:

> A unitary STEM literacy that reflects the shared general capabilities required in all the STEM disciplines. At the same time, there are also substantial differences that support the retention of the existing literacy constructs (i.e. S.T.E.M literacies) to reflect the specific linguistic, cognitive and epistemic requirements found in each disciplinary area. This distinction from the singular STEM literacy is necessary to highlight the skills and practices that are unique to each particular discipline, and therefore not applicable in all the other disciplines (p.675).

The conundrum of defining STEM literacy and especially the "T" continues to confound practitioners, so the authors suggest a broader basis in understanding literacy as a concept, be integral to our definition.

6.3 Operationalizing STEM Literacy

As the prior section discussed, integrated STEM education literature is full of competing interests, competing epistemologies, and competing perspectives of literacy. Standardization of teaching and learning should not be the goal in integrated STEM education. The engineering habits of mind, Standards for Technological and Engineering Literacy (ITEEA, 2020), and other frameworks (e.g., Roberts et al., 2022) emphasize the importance of creativity and authentic problem-solving. This cannot be achieved through standardization of teaching and learning. However, as integrated STEM continues to develop as a field, the community must operationalize terms. Common definitions are needed to ensure when teachers, policy-makers, researchers, and other stakeholders are working within integrated STEM, there is a common understanding.

When terms are not properly operationalized, goals are less clear because there is no common understanding. For example, many view STEM as any curriculum, instruction, and activities that use any of the four disciplines (English, 2016). There is growing evidence that STEM should be treated as an integrated or interdisciplinary field that is different from the individual subjects (Honey et al., 2014; Holmlund et al., 2018; Gao et al., 2020). Many definitions of STEM do not include a goal of STEM education (Roberts & Roberts, 2022). This lack of a common definition could be caused by a misalignment between STEM initiatives and the siloed nature of educational standards (Bybee, 2018). Although different frameworks and approaches to integrated STEM are important, a consensus is needed to work toward common goals and to reliably measure effective teaching and learning within integrated STEM (Roberts & Roberts, 2022). To do this, the authors advocate for situating technological literacy as central to integrated STEM literacy.

6.3.1 Understanding Literacy

Historically, literacy was associated with a skills-based approach with a focus on the application of these skills in meaningful ways. The focus in the mid-1960s through 1970s was functional literacy, including "teaching literacy as a set of skills thought to be universal and applicable anywhere, …one literacy which everyone should learn in the same way" (Fransman, 2005, p. 11). Fransman further asserts that a newer recognition of literacy must include an ideological approach, understanding that literacy is not a predetermined set of technical or neutral skills, but a social practice rooted in "conceptions of knowledge, identity and being" (p. 13). The current global and social contexts of society require this broader definition of literacy, one which recognizes the importance of being critical consumers, users, and designers of technology (Jackson et al., 2021; ITEEA, 2020). In the *Standards for Technological and Engineering Literacy* (STEL; ITEEA, 2020) the definition of being literate is considered "a fluid construct, meaning that knowledge, skills, and abilities in a given field will change over time" (ITEEA, 2020, p. 2). This expanded understanding of literacy demonstrates the need for enhanced literacy in process and action and designing and making, which are hallmarks of technology education (ITEEA, 2020). Avsec and Jamšek (2018) stated technological literacy is a multilayered goal of technology education specifically for cognitive, practical, and strategic abilities.

Avsec and Jamšek (2018) emphasized, "Despite the critical importance of TE [technology education] a mismatch between the knowledge and skills delivered by education and training systems required by the labour market has been reported in several countries" (p. 146). An emphasis on technology and engineering practices, defined as "abilities and dispositions that are…fundamental to student success" (ITEEA, 2020, p. 70), is one way to build technological literacy and correct the curricular mismatch described by Avsec and Jamšek (2018). These practices focus on ideas such as creativity, optimism, ethics, making and doing, collaboration, and critical thinking.

The need to acknowledge the importance of each discipline along with their commonalities is critical to developing a definition of STEM literacy. By promoting STEM literacy as an outcome of these practices, this chapter offers a strong reconceptualization of STEM literacy situated within the other STEM disciplines as the first step in ensuring all learners achieve this needed literacy. One definition that achieves these goals defines integrated STEM literacy as.

> the dynamic process and ability to apply, question, collaborate, appreciate, engage, persist, and understand the utility of STEM concepts and skills to provide solutions of STEM-related personal, societal, and global challenges that cannot be solved using a single discipline (Mohr-Schroeder et al., 2020, p. 33).

6.3.2 Developing Authentic STEM Literacy Through Technological Literacy

Authenticity in STEM is an often-desired goal, but is not always achieved (Roberts & Chapman, 2017). Reeves et al. (2002) suggest authentic activities have a combination of ten design characteristics: real-world relevance, ill-defined problems, complex tasks requiring ongoing investigation, multiple perspectives, collaboration, reflection, interdisciplinary connections, integrated assessment, polished products, and multiple interpretations and outcomes. The focus in authentic STEM learning is "the problem is grounded in the real world but is open for interpretation, complex enough to require sustained work with the help of peers, layered to require reflection, and results in a variety of tangible solutions" (Roberts & Chapman, 2017, p. 16). Authenticity should not only describe the learning experiences (e.g., real-world focused problems), but should also apply to how the learning experiences leverage connections between the STEM disciplines (ITEEA, 2020). One way to accomplish this goal is by focusing on the practices of integrated STEM (Roberts et al., 2022). These practices offer specific behaviors and habits for teachers to model and develop in students; moreover, the practices are important ways of approaching discipline-specific content in ways that are similar to how experts in each discipline approach their work (Roberts et al., 2022). Students must work together, question, apply knowledge, and use a range of ideas from different disciplines to design solutions to problems. These characteristics of authentic STEM instruction have their foundations in technology and engineering education (ITEEA, 2020). Central to technology and engineering education is "hands-on, design-based strategies... [through] authentic experiences" (ITEEA, 2020, p. 6) that often have transdisciplinary connections due to the broad nature of design.

When students engage in integrated STEM through authentic activities, not only is their STEM literacy increased but their learning is more meaningful (Maiorca & Mohr-Schroeder, 2020; Roberts et al., 2018). This is true both for individual disciplines (e.g., Kelley & Knowles, 2016) and for integrated STEM. In particular, technology and engineering education's emphasis on hands-on, design-based activities can create authentic learning experiences for students (ITEEA, 2020). Roberts and Roberts (2022) noted students typically perceive STEM learning experiences as opportunities to apply content to the real world, as opportunities to see career connections, as opportunities to engage in hands-on activities grounded in STEM practices, and as opportunities to engage in authentic STEM learning. The students' perceptions align with the hallmarks of technology education (ITEEA, 2020). The perception of STEM being connected to the real world is particularly defining as it provides context and purpose for the learning that occurs in content-specific classrooms (Densen et al., 2015; Jackson et al., 2021; Roberts et al., 2018, 2019). The real-world connections not only make learning more meaningful, but authentic STEM learning activities can also address higher-order thinking skills. Students recognize authentic STEM learning as a place to apply knowledge from specific disciplines. For example, robotics is a common way students make connections to mathematics (Coxon et al., 2018). Students identify how they actively use logic, measurement, and decimals when programming robots (Roberts et al., 2019). This is a markedly different context than completing worksheets about decimals and decimal operations; instead, students must understand decimals and then apply that understanding to program the robots. Using robotics to facilitate problem-based learning makes the content meaningful. Using problem-based and design-based learning is also ways to promote authentic learning (ITEEA, 2020).

Many of the benefits afforded by integrated STEM education developed from important principles in technology and engineering education. A person's "ability to understand, use, create, and assess the human-designed environment in increasingly sophisticated ways over time" (ITEEA, 2020, p. 161) requires scientific and mathematical literacy. The ability to use multiple disciplines to solve authentic problems centers on design. Design is central to technological literacy due to the emphasis on iterative decision-making to create solutions that meet needs or wants (ITEEA, 2020). Focusing on technological design builds technological literacy, which in turn forms a foundation for developing integrated STEM literacy.

6.4 Technological Literacy and Social-Emotional Learning

In addition to promoting authentic learning, integrated STEM can also impact Social-Emotional Learning. Social-Emotional Learning (SEL) and executive control skills are reputed to be stronger indicators of students' academic and personal success than general intelligence (Immordino-Yang, 2016). SEL and executive capacities according to Immordino-Yang depend on students' understanding of the context of the task and their abilities to interpret a task as "possible for them to accomplish, important enough to persist on, and relevant to their personal goals" (p. 212). Social emotional learning should not be limited in scope or depth; not a "one-size-fits-all approach… [but] an approach to SEL that is developmental, flexible, and responsive…focused on strategies for teachers to implement as appropriate" (Bailey et al., 2019, p. 54). Bailey et al. asserted that SEL development is vitally important for low-income students, specifically those with lower socioeconomic status who are lagging in their academic or behavioral adaptations in formal education. Effective SEL involves the use of experiences that are both responsive to students' needs and related to their lived experiences in school, home, or communities (Bailey et al.). Technology education supports students' sense of agency and "how they feel about themselves in relation to the task, and how they believe others think about them" (Immordino-Yang, p. 212) through the integration of SEL and technological literacy.

Hsieh-Hua et al. (2012) asserted that limited emphasis has been placed on the intersection of technological literacy and social factors. Social factors, also viewed as societal inputs, are key components of one's ability to develop social capital. Stolle-McAllister (2011) defined social capital as a person's acquaintances and social networks; the more prestigious and well-developed the more capital one has (as cited in Bourdieu, 1986). Social capital is a learned behavior. Social capital gives learners a toolkit of technological literacy skills to understand, manage, or resolve concerns encountered while learning. Technological literacy skills assist in preparing students for the societal workplace requirements for collaboration, communication, and information management (Jones, 2016b). Strong social capital helps "students realize they are not in competition with each other but are working together to improve themselves as a group as well as individuals" (Stolle-McAllister, 2011, p. 16). Using integrated STEM projects assists learners in using collaboration, design thinking, and group work "making …connections with other people who have the same ideas as you, or some that even have different ideas from you" (Morganson, et al., p. 355). Social capital development equips learners to be immersive problem solvers building their technological literacy skills for inquiry, questioning, and analyzing problems (Jones, 2016b). Cropley (2018) states that STEM pedagogy should promote "psychological factors to foster adaptability to change, willingness to try something new, the ability to see unexpected solutions, the 'joy of discovery, self-belief, and resistance to social pressure to conform'" (p. 91).

Technological literacy necessitates cognitive (ability to know, understand, and evaluate), practical (use), and strategic (responding to challenges) dimensions (Avsec & Jamšek, 2018). Research in brain-based learning reinforces that effective learning involves engagement, strategies, and principles but also needs a "kicker" to ensure its relevance (Jensen & McConchie, 2020). Relevance, according to Jensen and McConchie, involves the classroom triggers provided in the learning environment. The STEL supplies a framework with three organizers which include relevant core elements of learning, the practices for 21st-century learning skills, and contexts for teaching the standards (ITEEA, 2020). Inclusivity is at the heart of these organizers as their design is focused on multidisciplinary education and stresses the importance of whole-child learning with a strong integration in design, reading, writing, arts, physical education, and history. Technological literacy reinforces the integration of all subjects with interwoven activities stressing that changes in any discipline affect all of society. By developing strong integrations in all educational subjects, technological literacy infuses abilities to use lateral thinking and divergent options and to evaluate all viable solutions in a project (Jones, 2016a).

6.5 Looking Forward

The authors recommend a global adoption of a broad-based definition of technological literacy that includes flexibility for meeting the needs of diverse groups of learners yet still provides a strong emphasis on Social-Emotional Learning, design thinking, and fluidity of technological concepts. Mohr-Schroder et al. (2020) emphasized the need for social-emotional skills (apply, collaborate, question, engage, persist) as hallmarks of integrated STEM literacy to meet the challenges of STEM-related personal, societal, and global events. Each discipline of the acronym STEM must share equally in the problem-solving arena, not competing but each contributing in an integrated way to the desired conclusion. Also, other learning opportunities such as the arts and history are key factors in technological literacy as supported by the core Standards for Technological and Engineering Literacy, "Influence of Society on Technological Development, Influence of Technology on Human Progress" as published by ITEEA (2020, p. 11). Learners must be grounded in the impacts of historical decisions, design, and societal input to understand current global issues of technological literacy. According to ITEEA (2020) adopting these eight core standards "defines a level of literacy that is expected of all learners across the Pre-k-12 spectrum, much as we expect all students to achieve a certain level of language literacy, scientific literacy, and mathematical numeracy" (p. 6). Models of discipline learning that keep each in a silo must be redesigned to encourage learners to develop concepts by applying them to a problem. This design thinking approach encourages learners to question, think about interrelationships of science, technology, engineering, and mathematics, develop a concept or idea, and problem-solve to achieve the answer. As stated previously, this is a marked difference from learning about discipline content through lectures and worksheets; it encourages learners to use trial and error to solve a problem.

Integrated STEM is inherently a "changing" definition as all disciplines of the acronym continue to evolve and expand their knowledge bases. While knowledge evolves, an emphasis on problem-solving remains. As noted above, key components of integrated STEM education (e.g., authenticity, hands-on, real-world connections) have their roots in technology education. Thus, design is a central component of integrated STEM education. Technological design is foundational to technological literacy as it provides a structured, iterative process to solve a problem to meet human needs and wants (ITEEA, 2020). Therefore, the authors deem any operationalization of integrated STEM literacy must center on technological and engineering literacy in its definition.

References

American Association for the Advancement of Science. (1989). Science for all Americans: A project 2061 report on literacy goals in science, mathematics, and technology.

Avsec, S., & Jamšek, J. (2018). A path model of factors affecting secondary school students' technological literacy. *International Journal of Technology & Design Education, 28*(1), 145–168. https://doi.org/10.1007/s10798-016-9382-z

Bailey, R., Stickle, L., Brion-Meisels, G., & Jones, S. M. (2019). February) Re-imagining social-emotional learning: Findings from a strategy-based approach. *Phi Delta Kappan, 100*(5), 53–58. https://doi.org/10.1177/0031721719827549

Baker, C. K., & Galanti, T. M. (2017). Integrating STEM in elementary classrooms using model-eliciting activities: Responsive professional development for mathematics coaches and teachers. *International Journal of STEM Education, 4*(1), 1–15.

Berland, L. K. (2013). Designing for STEM integration. *Journal of Pre-College Engineering Education Research (J-PEER), 3*(1), 3.

Blackley, S., Sheffield, R., & Koul, R. (2018). Using a Makerspace approach to engage Indonesian primary students with STEM. *Issues in Educational Research, 28*(1), 18–42.

Bloome, D., Castanheira, M., Leung, C., & Rowsell, J. (2018). *Re-Theorizing literacy practices*. Routledge.

Bourdieu, P. (1986). The forms of capital. In J. G. Richards (Ed.), *Handbook of theory and research for the sociology of education* (pp. 241–258). NY. Greenwood Press.

Bybee, R. W. (2018). *STEM education now more than ever*. Arlington, VA: National Science Teachers Association.

Council for Science and Technology. (2013). *Science, technology, engineering and mathematics education: Update*. https://assets.publishing.service.gov.uk/government/uploads/system/uploads/attachment_date/file/230509/13-1131-stem-education.pdf

Coxon, S. V., Dohrman, R. L., & Nadler, D. R. (2018). Children using robotics for engineering, science, technology, and math (CREST-M): The development and evaluation of an engaging math curriculum. *Roeper Review, 40*(2), 86–96. https://doi.org/10.1080/02783193.2018.1434711

Cropley, A. (2018). Creativity of doom: Integrating STEM education to forestall disaster. *The International Journal of Creativity and Problem Solving, 28*(1), 89–100.

Densen, C. D., Stallworth, C. A., Hailey, C., & Householder, D. L. (2015). Benefits of informal learning environments: A focused examination of STEM-based program environments. *Journal of STEM Education, 16*, 11–15.

De Vries, M. J., Fletcher, S., Kruse, S., Labudde, P., Lang, M., Mammes, I., Max, C., Münk, D., Nicholl, B., Strobel, J., & Winterbottom, M. (Eds.) (2018). The impact of technology education: International Approaches. Münster & New York: Waxmann

Ehlert, J., & Roberts, T. (2021). Exploring public schools' approaches to STEM: A pilot study. In: Proceedings of the 120th annual convention of the school science and mathematics association, vol 8, pp 22–29.
El-Deghaidy, H., Mansour, N., Alzaghibi, M., & Alhammad, K. (2017). Context of STEM integration in schools: Views from in-service science teachers. *EURASIA Journal of Mathematics, Science, & Technology Education, 13*(6), 2459–2484. https://doi.org/10.12973/eurasia.2017.01235a
English, L. D. (2016). STEM education K-12: perspectives on integration. *International Journal of STEM Education, 3*, 3. https://doi.org/10.1186/s4059%204-016-0036-1
Fransman, J. (2005). Understanding literacy: A concept paper. *The EFA Global Monitoring Report, Literacy for Life, 31.*
Gao, X., Li, P., Shen, J., & Sun, H. (2020). Reviewing assessment of student learning in interdisciplinary STEM education. *International Journal of STEM Education, 7*(24), 1–14. https://doi.org/10.1186/s40594-020-00225-4
Hallström, J., & Ankiewicz, P. (2019). Laying down the "T" and "E" in STEM education: Design as the basis of an integrated STEM philosophy. In S. Pulé & M. J. de Vries (Eds.), *Proceedings PATT 37: Developing a knowledge economy through technology and engineering education* (pp. 187–194). Msida: University of Malta. https://www.researchgate.net/profile/Jonas-Hallstroem/publication/333681977_Laying_down_the_T_and_E_in_STEM_education_Design_as_the_basis_of_an_integrated_STEM_philosophy/links/5cff75b34585157d15a216a6/Laying-down-the-T-and-E-in-STEM-education-Design-as-the-basis-of-an-integrated-STEM-philosophy.pdf
Hilgarth, C. O. (2020, June). Should we consider transforming the definition of technological and engineering literacy. In *2020 ASEE Virtual Annual Conference Content Access.*
Holmlund, T. D., Lesseig, K., & Slavit, D. (2018). Making sense of 'STEM education' in K-12 contexts. *International Journal of STEM Education, 5*(32). https://doi.org/10.1186/s40594-018-0127-2
Honey, M., Pearson, G., & Schweingruber, A. (2014). *STEM integration in K-12 education: Status, prospects, and an agenda for research.* National Academies Press.
Hsieh-Hua, Y., Huang, F. F., Yi-Horng Lai, Hung-Jen, Y., & Jui-Chen, Y. (2012). Social capital and technological literacy in Taiwan. *The Scientific World Journal*, 2012.https://doi.org/10.1100/2012/149132
International Bureau of Education, UNESCO. (2009). *Mathematical literacy.* http://www.ibe.unesco.org/en/glossary-curriculum-terminology/m/mathematical-literacy
Immordino-Yang (2016, June 29). Emotion, sociality, and the brain's default mode network: Insights for educational practice and policy. *Policy Insights from the Behavioral and Brain Sciences, 3*(2), 211–219. https://doi.org/10.1177/2372732216656869
International Technology and Engineering Educators Association. (2020). Standards for Technological and Engineering Literacy. ITEEA.
Jackson, C., Mohr-Schroeder, M. J., Bush, S. B., Maiorca, C., Roberts, T., Yost, C., & Fowler, A. (2021). Equity-oriented conceptual framework for STEM literacy. International Journal of STEM Education.
Jensen, E. P., & McConchie, L. (2020). *Brain-based learning: Teaching the way students really learn.* Rev. 3rd ed. Corwin.
Jones, V. R. (2016a). The nature of technology: Unraveling technological literacy. *Children's Technology and Engineering, 20*(1), 22–24.
Jones, V. R. (2016b). Social and cultural capital: The heart of STEM success. *Children's Technology and Engineering, 20*(3), 16–19.
Kelley, T. R., & Knowledge, J. G (2016) A conceptual framework for integrated STEM education. *International Journal of STEM Education, 3*(11). https://doi.org/10.1186/s40594-016-0046-z
Maiorca, C., & Mohr-Schroeder, M. J. (2020). Elementary preservice teachers' integration of engineering into STEM lesson plans. *School Science and Mathematics, 120*(7), 402–412.

Mohr-Schroeder, M., Bush, S. B., Maiorca, C., & Nickels, M. (2020). Moving toward an equity-based approach for STEM literacy. In C. Johnson, M. J. Mohr-Schroeder, T. Moore, & L. English (Eds.), *Handbook of research on STEM education* (pp. 29–38). Routledge.

Moore, T. J., Johnston, A. C., & Glancy, A. W. (2020). STEM integration: A synthesis of conceptual frameworks and definitions. In C. Johnson, M. J. Mohr-Schroeder, T. Moore, & L. English (Eds.), *Handbook of research on STEM education* (pp. 3–16). Routledge.

Morganson, V. J., Major, D. A., Streets, V. N., Litano, M. L., & Myers, D. P. (2015). Using embeddedness theory to understand and promote persistence in STEM majors. *Career Development Quarterly, 63*(4), 348–362. https://doi.org/10.1002/cdq.12033

National Assessment Governing Board. (2018). Abridged technology & engineering literacy framework for the national assessment of educational progress. https://www.nagb.gov/content/dam/nagb/en/documents/publications/frameworks/technology/2018-technology-framework-abridged.pdf

National Research Council. (2012). A framework for K-12 science education: practices, crosscutting concepts, and core ideas. National Academies Press.

National Research Council. (2014). *Developing assessments for the next generation science standards*. The National Academies Press.

Office of the Chief Scientist. (2014). Science, technology, engineering and mathematics: Australia's future. Canberra: Commonwealth of Australia.

Reeves, T. C., Herrington, J., & Oliver, R. (2002). Authentic activities and online learning. Annual Conference Proceedings of Higher Education Research and Development Society of Australasia. Perth, Australia.

Roberts, T., & Chapman, P. (2017). Authentically engaging elementary students in the designed world. *Children's Technology & Engineering, 21*(4), 15–17.

Roberts, T., Jackson, C., Mohr-Schroeder, M. J., Bush, S. B., Maiorca, C., & Delaney, A. (2019). Exploring applications of school mathematics: Students' perceptions of informal learning experiences. In Otten, S., Candela, A. G., de Araujo, Z., Haines, C., & Munter, C. (Eds.), Proceedings of the forty-first annual meeting of the North American American chapter of the international group for the psychology of mathematics education (pp. 1515–1520). St. Louis, MO: University of Missouri.

Roberts, T., Jackson, C., Mohr-Schroeder, M. J., Bush, S. B., Maiorca, C., Cavalcanti, M., Schroeder, D. C., Delaney, A., Putnam, L., & Cremeans, C. (2018). 'Students' perceptions of STEM learning after participating in a summer informal learning experience,' *International Journal of STEM Education, 5*(35). https://doi.org/10.1186/s40594-018-0133-4

Roberts, T., Maiorca, C., Jackson, C., & Mohr-Schroeder, M. J. (2022). Integrated STEM as problem-solving practices. *Investigations in Mathematics Learning, 14*(1), 1–13. https://doi.org/10.1080/19477503.2021.2024721

Roberts, T., & Roberts, A. C. (2023). Promises and perils of STEM education: Synthesizing teacher, student, and research perceptions. In R. J. Tierney, F., Rizvi, and K. Erkican (Eds.), International Encyclopedia of Education, vol. 11 (4th ed., pp. 262-269). Elsevier.

Stolle-McAllister, K. (2011). The case for summer bridge: Building social and cultural capital for talented black STEM students. *Science Educator, 20*(2), 12–22.

Tang, K.-S., & Williams, P. J. (2019, October). STEM literacy or literacies? Examining the empirical basis of these constructs. *Review of Education, 7*(3). https://doi.org/10.1002/rev3.3162

Thibaut, L., Ceuppens, S., De Loof, H., De Meester, J., Goovaerts, L., Struyf, A., Boeve-dePauw, J., Dehaene, W., Deprez, J., de Cock, M., Hellinckx, L., Knipprath, H., Langie, G., Sturyven, K., vande Velde, D., van Petegem, P., & Depaepe, F. (2018). Integrated STEM education: A systematic review of instructional practices in secondary education. *European Journal of STEM Education, 3*(1), 2.

Tout, D. (2000). Numeracy up front: Behind the international life skills survey. *ARIS Resources Bulletin, 11*(1), 1–5.

Thomas Roberts is a former elementary school teacher, is an Associate Professor and co-program coordinator of the Inclusive PreK-5 Education Program at Bowling Green State University where he teaches STEM education and mathematics education courses. He received his doctorate in Education Sciences specializing in STEM Education from the University of Kentucky. Dr. Roberts' research explores students' perceptions of STEM learning environments and ways to increase the effectiveness of teachers' instructional practices so that all students have the opportunity to participate in high-quality STEM learning. He is also the Editor-in-Chief of Technology and Engineering Education, a peer-reviewed practitioner journal for integrated STEM teachers published by the International Technology and Engineering Educators Association (ITEEA).

Chapter 7
The Role of Assessment of Technology in STEM Education

Niall Seery and Donal Canty

Abstract This chapter positions technology subjects as a major component of STEM. The contribution explores the nature of activity in technology as it features discrete subject knowledge and skills and its importance in mediating the elements of STEM to support an integrated learning experience. Considering the role of designerly learning in the transactional activities of technology education highlights its utility in drawing together STEM learning in a contextual and tangible way. The challenge of assessing the T in STEM is discussed in the context of the dependent relationship between designing and creating and the associated difficulties in capturing evidence of performance resulting from a fluid, non-determinist approach to student learning. The ability to articulate designerly decisions and actions is developed through a dichotomous representation of designing and making as the value proposition. Holistic assessment is presented as a useful construct to be considered in support of liberating the learners' approach to tackling the learning task while having the capacity to adjudicate on the summation of evidence through the teacher's professional judgement. The chapter will also address how assessment can be planned for and managed so that it can inform rather than drive teaching and learning.

Keywords Technological performance · Designerly activity · Integrated STEM assessment

---

N. Seery (✉)
Technological University of the Shannon, University Road, Athlone, Westmeath, Ireland
e-mail: Niall.Seery@tus.ie

D. Canty
University of Limerick, Castletroy, Limerick, Ireland

W. Fox-Turnbull and P. J. Williams (eds.), *Locating Technology Education in STEM Teaching and Learning*, Contemporary Issues in Technology Education,
https://doi.org/10.1007/978-981-97-1995-2_7

7.1 The Nature of the Activity

The relationship between designing and making is a central tenet of technological education. Project-based learning is a well-established practice within the discipline, with the emphasis on designing and creating being core to the definition of technological activity. This action orientation can manifest as a cyclical act of action and reflection, an iterative dialectic between the mind and hand, and an experimental disposition where value and direction become apparent through the realisation of 'what works' and by association 'what does not'. Designerly acts create and monitor a multitude of variables as ideas become more tangible. While technology education has specific knowledge that supports 'good practice' in the evolution and realisation of the idea, the subject develops core knowledge and skills by adhering to standards and conventions, standard operating procedures, health and safety, properties of materials, principles of CAD modelling, functions and applications of electronics and control systems, process technologies and mechanics. While this declarative and procedural knowledge can be easily articulated in policy and curricular documentation with an explicit vocational focus, it is the treatment and utility of this knowledge and skill that change when considered from a broader STEM perspective. The integrated nature of STEM has further highlighted the role and value of designerly activity and brought with it the need for conditional/strategic knowledge to become more central to the development of capability. With a focus on the application of knowledge within a context set by a design challenge and not just the acquisition of knowledge as an entity in itself, the importance of conditional knowledge (that is difficult to express) as it mediates designerly problem-based activity makes the epistemological definition of contemporary technological education more aloof. McGarr and Lynch (2017) describe the bounded nature of post-primary school subjects as defined areas of study. The classification of technology education as having weak boundaries as knowledge is 'borrowed' from other subjects to give meaning and context to the design problem or innovation. While the goals of technology education are often difficult to articulate, as the contextual and situational factors are integral to success, understanding the nature of its weak boundaries has significant utility as an enabler for the development of an integrated approach to STEM education.

Learning environments and activities that liberate what is imagined through the realisation of artefacts, systems, and technologies are central to a general education that embeds agency and active citizenship. Action-oriented education that builds the capacity to realise what is in the mind's eye recentres the importance of technical (knowledge and skills to realise and execute correctly, accurately, and safely) and technological (making contextual and situational decisions on how to apply, manipulate, and transfer technical capability in response to designerly activity) competency. Knowing 'how' and 'why', supports knowing 'what' to do and 'when' to do it, emphasises the value and significance of the T in an integrated STEM education. Integrated STEM extends further the epistemological challenge in the treatment of knowledge and the definition of knowing, bringing with it further assessment tension when grappling with evidence of learning in a valid and reliable way.

Technical manufacturing, commonly referred to as 'make' within the subject area, has evolved from what was once (in most countries) vocationally positioned education and training to now embracing advanced technologies and techniques. The complexity of what can be produced through digital design technologies, coupled with additive techniques affords greater scope for designing and increased feasibility to develop tangible outcomes. For example, the speed, accuracy, and ease of prototyping early conceptions of a design solution have added significant value to learning activities that encourage a dialectic of speculative and critical thinking. Increased conceptual engagement and the need to embrace function as a fundamental, requires comprehension of space, orientation, technical specification, and application as apparent outcomes of authentic performance. These are less predictable when compared to outcomes captured in technical/craft development tasks. Doyle (2020) describes the contemporary goals of technology education as the capacity to acquire knowledge and skills for application, be able to think in a technological way and be able to act in a technological way. Although this appears to frame a macroscopic view of technological education, it is useful in framing the learning intent in what are often multi-faceted, value-laden, technically complex learning activities. Doyle's work captures the transactional nature of technological activity and questions if the subject should be defined by activity and learning that supports the development of technological literacy and capability. Actively and reflectively translating knowledge and skills to new applications supports a comprehensive learning activity, as the 'transfer' of understanding and meaning to a new context becomes a central pedagogical approach. By extension, new contexts cannot always be assimilated. Therefore, a more complex transactional negotiation is required, where the causal impacts of action (sometimes speculative or even tacit) form the approach to pupil-led enquiry and knowledge and skill construction. Translations and transactions are best captured in the iterative dialectic of mind and hand described by Kelly et al. (1987); this in turn shifts the focus away from a reliance on the traditional articulation of knowledge and skill acquisition (Doyle et al., 2018). Critically, any approach that relies heavily on translations and transactions is dependent on the nature and quality of the learning activity. Depending on the specific learning intention being addressed, tasks and activities can span the continuum of prescribed projects that focus on developing foundational knowledge and skills to more permeable tasks that apply this learning in a novel and creative way, requiring a more designerly approach. Contemporary Technology education embodies designerly activity as a central tenet of what it means to be capable, and although epistemologically challenging, it formulates a foundational relationship between proof, speculation, critique, and innovation in what are definable outputs and outcomes. The purpose of this approach stems from the inception of design as an activity to educate children that they would be actively involved in shaping the future of material culture through personal decision-making and citizenship (Baynes, 2010). The speculative and critical activity not only applies to the design of products and artefacts but also encompasses the impact of technology and technological developments on consumers, society, and the environment.

7.2 The Continuum of Evidence

While technology education can be described with a specific knowledge base and associated skills, the essence of technology education is in the problem, project, and design-based challenges that give meaning to the definition of capability and literacy. Frameworks and descriptions of technological capability and literacy have been modelled by many (Barlex, 2007; Gibson, 2008; ITEEA, 2020, Kimbell et al., 1991) and largely share the foundational elements of knowing what (knowledge), knowing how (skills), knowing why (values/ethics) and knowing when (problem-solving) (Buckley et al., 2019). Yet, categorical criteria may not be sufficient in capturing the complexity of an idea and the creative processes of innovating within the physical and environmental limits of the facilities accessible to the learner. Kelly et al. (1987) based on the description of observed practice modelled how pupils move from the development of 'hazy ideas' through externalisation techniques showing evidence of a deepening understanding of the design solution through an iterative dialectic of thinking and doing. The conversation with materials, environment, the self, peers, and the teachers are often idiosyncratic, and the design journey may be scoping, exploratory, experiential, speculative, evaluative, staged, ordered, ad hoc or all the above at the same time. This makes the learning environment where the designerly activity takes place a complex one for the alignment of teaching, learning and assessment and requires careful planning of fluid activities. Kimbell (2011, p. 7) argues that students operate in a space of hunch and half-knowing, working through refining their thoughts through action. As a result, the trail of evidence that emerges must be untangled, interpreted, and categorised (declarative, procedural, conditional, iterative, evaluative, etc.) to support the value judgement that comes with assessment. Balancing the need for assessment criteria with the influence these criteria can have on the execution of the design solution, or the narrative of the process, is becoming an increasingly significant challenge when embracing designerly activity. Overly defined and explicit criteria can inhibit the very nature of the designerly activity leading to formulaic and recipe-type approaches to teaching, learning and assessment. Therefore, the purposeful integration of assessment in the teaching and learning activity is critical to achieving the intended goals of a design-based technology curriculum.

7.3 Planning for Good Evidence

Design briefs play a central role in the learning activity and guide the intention of the learning to formulae evidence of the intended outcomes and outputs. With a focus on design and creating and fostering the evolving design journey, it is not surprising that there has been much support for formative assessment in technology education. Teachers guide the students to move learning forward while facilitating their natural curiosity by exploring the technical dimensions necessary to realise their ideas. The 'need to know' can be the catalytic stimuli and confirmatory factors of the 'jigsaw' that support the actions to move towards the desired solution. While this approach may seem unstructured, the specificity of the design brief is critical in defining the direction of the enquiry, ultimately ensuring curricular coherence. The level of 'permeability' of the brief can reflect the students' stage of development and the experience of the teacher to confidently allow their students to meaningfully engage with the permeable nature of the brief. As a result, closed tasks are used to develop specific knowledge and skill and more open (permeable) tasks are implemented to test hypotheses and explore alternatives and encourage innovation and creativity as the learners grapple with how their actions may impact consumers, the environment, and society.

Project e-scape (Kimbell, 2007) in England highlighted that the 'permeability' of tasks is not the only consideration to capture robust evidence of performance. The nature of the brief can also have an impact on engagement and performance, with evidence demonstrating that levels of engagement differ depending on the thematic area of the design challenge. Some students are less likely to engage with industrial briefs, while social and environmental agendas proved more inclusive learning media. This is critical to consider when one values meaningful engagement as a significant function of success and thus will have a significant impact on how the learner navigates the learning and assessment task. Creating a design challenge or brief that may intrinsically motivate learners beyond 'doing well in the exam' must be a key consideration when planning for learning and assessment. This will provide an opportunity for better engagement, ownership of the learning, and authentic evidence of learner capability. In the context of an integrated STEM approach, the nature of evidence needs to be carefully considered, as the balance between being prescriptive enough to ensure there is evidence of the discrete competencies may direct the evidence towards the notion of a 'single' correct answer where the ownership of the design task is perceived to be held external to the learner. This approach could limit the exploration of the plurality of designerly activity that is inherent in technological problem-solving. Acknowledging that these are the extremes, the inverse is also problematic, the intention to acquire fundamental knowledge and skills that are useful can be limited, suboptimal, or missed entirely when the speculative is separated from the critical. STEM as an integrated proposition needs to balance these resulting epistemological tensions. Any lack of epistemological 'balance' between the treatment of discrete subjects and that of an integrated STEM will be amplified through assessment incoherency.

The craft-based origin of technology education is well acknowledged in most countries (Buckley 2023). However, the legacy of vocational education as a definition of contemporary provision may not be presently representative in most jurisdictions. Although definitions occasionally fall short of being general education, there is certainly an acknowledgement that the contribution of technology education is far beyond building technical competency. Contemporary views of technology education and its future position is well captured by the term *ethnotechnological literacy* coined by Daker (2022, pp. XII). His definition makes the distinction '… 'technological literacy' is often perceived in objective terms, whereas the addition of 'ethno' clarifies ethnotechnological literacy as a subjective, ergo socio-cultural concept. It is thus the study of the human-technology relationship.' With each evolution of technology education, the dominant view of technology education as vocationally orientated is less apparent. Dunbar et al. (2023) manage the relationship between general education and vocationally oriented skills and knowledge by proposing a transactional model for technology teaching and learning. This model frames the continuum from technical competency to active citizenship and from future industrial skills needs to general education. The philosophical alignment of active citizenship and the goals of a general education bring forward the need to support a more comprehensive understanding of technological developments, their application, and their impact. The teaching intention is made explicit by focusing pedagogy on technology as a critical and emancipatory activity providing learners with the opportunity to make decisions that are responsible and sustainable—considering environmental impact, energy consumption, material selection and usage, etc. mediated through a local, regional, and global context. Developing future skills needs to be required to support and assess the ongoing technological development, and aligning with labour force and societal requirements begins to frame a more balanced integration of the T in STEM. Although the model by Dunbar et al. is not intended to frame learning intentions as mutually exclusive, it is intended to clarify the focus of the learning and understand the best 'fit' pedagogical design to support this learning, framing a continuum of foundational attitudes, knowledge, and skills that can liberate designerly acts of psychosocial learning. Technology education is built on intricate connections between technology as objects, activities, knowledge, and volition (Mitcham, 1994). Although it is not appropriate to separate the process from the output, or the realisation from the design, or to separate objects, activities, knowledge, and volition into separate elements, the following sections will discuss these separately, in part, to illustrate the separate assessment philosophies, where viewing them from the extremes highlights how they are contradictory and when considered as bespoke practices they remain tense.

7.3.1 Realisation of Systems and Products

Traditional approaches to assessment in technological education often involve categorised series of technical criteria that expand in response to contemporary technologies (e.g.—advanced computer-aided equipment, materials, etc.) and techniques (skills to operationalise new technologies e.g. parametric modelling). Consideration is given to best practice, material usage and integrity, accuracy, precision, function, and finish. All of which can be articulated, comprehended, prioritised, and weighted within a rational construct of the traditional vocational/craft view of capability. Typically, assessments that focus on making skills take the shape of a skills-based exam, taken over a confined time frame under proctored conditions, with the objective to measure specific practical craft skills in a way that can capture precision, efficiency, and best practice. The prescribed nature of the assessment instrument can include key indicators that may focus on the assembly of parts or the function of the artefact as a measure of performance. Alternatively, skills can be assessed in a more developmental manner by utilising a prescribed project that is manufactured over a longer period (often a school term), that is designed to capture specific skills aligned with the subject curriculum. The realisation of a system or artefact can be easily evaluated on the quality of the make. This can focus on precision, accuracy, finish, and execution of craft skills. This measure aligns with the use of explicit criteria that can be objectively assessed with weighted priority. Over time, in response to the changing nature of activity in technology education, these criteria have expanded to capture and value the evolving subject definition. The problem that manifests is captured by Kimbell (1994) where this approach to assessment is problematic for the evaluation of designerly activity. As designerly activity is fluid and dynamic, leading learners through multiple different pathways to achieve their goals the rigidity of explicit criteria for assessment becomes problematic. Expanding the list of criteria in a quest to capture all eventualities conflicts with the uniqueness of each learner's design journey and can lead to 'box ticking' in the quest for marks rather than demonstrating individual capability through their authentic actions. While lists of expanding criteria are useful in measuring what can be articulated, the consistent dilemma in technology education is how to value the performance that cannot be neatly predicted, captured, organised, narrated, or explained, as it emerges from designerly activity.

As an example in the digital space, the ability to use CAD tools can be framed as the representation of a component or artefact, with a strong emphasis on projection systems, drawing conventions, standards, and precision, with the utility of function in the communication, prescription, and control of the realisation stage, regardless of it being manual and/or CAM related. Alternatively, the processes of developing a design, especially with the aid of CAD modelling can be adjudicated on the 'intelligence' of the model that embraces design intent within the function of the digital solution. The question is, what is the definition of capability and what is it that you must measure? Rynne et al. (2010) developed a schema for parametric CAD that captures the sophistication of modelling capability by illustrating the declarative knowledge, procedural knowledge, visualisation, model deconstruction capability, strategic knowledge and metacognitive processes that are required to develop an 'intelligent' parametric CAD model. Therefore, depending on treatment, CAD modelling can be reduced to a 'driving licence' of features to execute specific geometric relationships or the very sophisticated and complex modelling of strategic decisions and predictive design intent that supports alterations, configurations, and modifications to ensure flexibility, efficiency, and robust functionality. The complexity of articulating and communicating the latter for the purposes of assessment is challenging when confined by weighted criteria as decisions by the learners are in response to context and need to be unique to their design and should be considered as part of the 'whole' rather than discrete elements of their work. This sets out the inherent assessment challenge for Technology within an integrated STEM framework.

The CAD modelling as an example of designerly activity is useful in exemplifying the goal of developing active citizenship, where the complex multi-faceted thinking, decision-making and predictive capacities are critical functions of intelligent modelling that need to be fully valued and supported by any assessment approach. Understanding the continuum of design stages and functions can often require an alternating of assessment methods and methodologies in response to the nature of expected activity and evidence. The role of technology in an integrated STEM mandates designerly actions to give meaning to the learning activity. The assessment of design is a function of how design is treated. Seery et al. (2022) propose a model for the treatment of design in technology education as a means of making explicit the intended learning outcomes in the context of the curricular subjects. While they present treatments that include *learning about design*, *learning to design* and *learning through design,* they are not intended to be exclusive to classroom activity but instead have a primary focus. While learning *about* and *to* design primarily encapsulates declarative and procedural knowledge to develop an understanding of design philosophy and practice, learning through design creates an assessment dilemma that highlights the limitations of traditional assessment practices.

At a macro level, this dilemma can be best represented by the emerging dichotomous schools of thought that consider design assessment. Assessing evidence of the process is well supported and encourages evidence to be presented in the form of a design portfolio that elaborates on the design journey, highlighting critical decisions made as the solution evolves. Criticism of this approach includes the influence of assessment criteria in overtly prescribing the evidence that is presented and the creation of this evidence to meet the assumptions of the specified criteria. Largely the creation of the portfolio is a reflective exercise, often developed following the completion of the solution. Alternatively, Williams (2009) notes that 'The domain of knowledge as a separate entity is irrelevant; the relevance of knowledge is determined by its application to the technological issue at hand', while Norman (2010) argues that designing is an idiosyncratic act that cannot be judged by the evidence of the process, but more appropriately by the evidence of the product. The product argument acknowledges the limited capacity of the designer to articulate what cannot be easily described in natural language. It also highlights the discussions with self. This forms a critical function in the development of the idea and the refining of the (re)solution but can see a lack of alignment between the internal dialogic activity and the outputs required for assessment purposes. The former approach can lead to formulaic pedagogical methods and a recipe model for designing. The latter however can overly rely on the inference of the assessor on intent by the learner which can challenge the validity of the assessment outcome.

When considering the assessment of designerly outcomes in technology education there are two opposing schools of thought: one approach is that the assessment should focus on the evolution of the design decisions, with a strong emphasis on the processes that demonstrate a level of performance. An alternative view is that the assessment should be output focused with the final solution representing the best fit solution to the given challenge. This school of thought acknowledges that design decisions cannot be explained and emerge because of tacit experiences and idiosyncratic synthesis. Students will make decisions in relation to aspects of their design that sometimes are difficult to explicate. The capacity to articulate their journey is a complex activity, yet critical for the assessor to determine the student's performance.

7.3.2 Capturing and Communicating Good Evidence

The relationship between teaching and learning and the process of creating evidence that supports a valid and reliable assessment of performance requires the foresight to align intention, activity, evidence, and assessment instruments. Therefore, it is useful to think about assessment in the context of the continuum of development and learning as students progress towards capability. Canty et al. (2019) and Canty et al. (2021) used the work of Jay and Johnson (2002) on reflective practice and Tosey et al. (2011) on learning to express an assessment framework for lower secondary school technology education in Ireland that describes a shift from the evidence that is declarative, to evidence that is comparative, to evidence that concludes by demonstrating the capacity to critique. This developmental approach facilitates meaningful formative intervention not limited to feedback, but also creates the space for a more empowering 'feed-forward' (Orsmond et al., 2013).

The development of fundamental skills and knowledge focuses on prescriptive actions and emphasises descriptive evidence. Evidence of learning can be egocentric, with work focused on developing competency and producing accurate and precise work. The capacity to describe the known reality is set in a determinist context, with relevant reference to standards, conventions, technical details, material properties and best practice. With this intent and understanding the need for foundational skills and knowledge is best created in response to 'closed' or controlled tasks that are limited by pedagogical factors such as materials, resources, and time. Consideration is given to the idea of 'foundational' understanding techniques and values and framed in the context of future transfer and application.

Seeing design and technology as complementary dyadic problem-solving strategies, synthesises design and creation as a continuous and singular learning activity. This broader epistemological view considers the utility of design in shifting the focus of technology education from the production of a prescriptive output to the creation of a solution to a given problem, or more accurately, the resolution to a current situation, challenge, or problem. Engaging with designerly activity begins with the need to evolve the nature of the expected evidence, with learning moving to include empathy as a function of innovation. Design learning activities support an emphasis on assumptions and speculations and encourage a methodological approach of comparative and critical analysis. Comparisons between known realities and hazy conceptions, together with initial concepts and their future state create the medium for discourse and collaboration. Learning to manage multiple perspectives allows the learner to become aware of new insights and phenomena. The capacity to assimilate comparable insights and accommodate new ideas that may not be within their lived experiences requires the capacity to understand and resolve alternatives or different interpretations. Evidence of effective performance moves away from known realities, towards the identification and analysis of critical issues relevant to their (re)solution, drawing inferences to shape future actions. This approach is well captured in Kelly et al.'s iterative dialectic model that illustrates how the learner moves from 'hazy ideas' to comprehensive solutions, developed through iterations between the hand (action) and head (reflecting on action). Designerly activity brings with it the context for realisation and the need to critically engage with the technical and technological knowledge and skills together with the broader knowledge bases of Science and Maths.

Reflecting on and in action (Schön, 1983) develops evidence of ethical, moral, or democratic dimensions as they emerge through the creative or developmental process. Context is central to the evidence of the designing and creating activity, where decision-making is revealed through the student voice as they critique their own ways of thinking and working and other sources of information. This evidence is indicative of performance. Individual learning and behaviour, mediated through the social factors that give meaning and context to the design decisions and creative expression are core to how performance should be valued. It is the relationship between the technology, the individual, and the context that makes apparent the challenge in assessing designerly and technological decisions and actions.

7.4 The Importance of Communication

Learning to solve technological problems can be sub-optimal and inefficient when compared to a subject that values the mastery of knowledge and skills evidenced in a prescriptive activity. However, capturing the navigation of a multi-faceted technological problem aids in the reflection of decision 'in' action and 'on' action. These are very different skills, requiring very different knowledge types. One of the significant challenges is the nature of dialogue when engaging with the materials and techniques—conversation with the self and conversation with the artefact—the ability to concisely reflect the complexity of the journey and the variables and criteria considered in making decisions when often the task is assessment driven where assumed transactional contracts are present.

The interplay between explicit and implicit knowledge is well recognised within educational assessment and much work has been done on recognising that tacit knowledge makes a significant contribution to developing technological capability and by extension STEM performance. The iterative nature of the activity, while it supports the curiosity and ingenuity of the student, creates a difficulty in 'seeing' the story of the design journey. The student's ability to communicate what can be tacit is a complex assessment process. The ability to design and engineer a solution to a problem is different to the skill of communicating the process of designing. Furthermore, the skills in using the communication media accessible to the student may not be developed to the level required to adequately communicate their 'journey'. Seery

(2017) emphasises modelling as a form of critique and the importance of developing teachable modelling and communication skills as foundational to technological education. The alignment of these media and the assessment instrument is then critical. For example, if an assessment instrument required students to communicate their journey in a written format, independent of sketches, models and demonstrators, performance may appear poorer than if students could utilise multi-modal representations. This example of construct under-representation demonstrates the need for an acceptance of school-based outcomes beyond that of traditional assessment practice that allows for a valid assessment. We would not assess the ability to play a musical instrument by asking music students to write an essay. Technology education is a performative subject and as such requires an assessment construct and instruments that can robustly interpret the breadth of activity that is encouraged.

The work of the Assessment Performance Unit (1991)—understood that Design and Technology was a performative subject that did not fit the traditional view of assessment. Their research made a significant contribution to our understanding of D&T (in England) practice and the challenges associated with assessment. When it came to trying to elicit and capture student learning and design decision-making throughout the journey, their view was to convert the design portfolio into *'dynamic dialogue mode'* which allowed the research team to choreograph the learning activity and capture real-time evidence of student actions and decisions. With such rich evidence, the recognition of the need for holistic assessment was born.

7.5 Valuing Holism

Although it is well supported that students can be successful in technology education in multiple different ways (Kimbell, 2007), the dichotomy of valuing the well-executed solution over the evidence of creative or innovative ideas remains a dominant orientation and one not easily resolved. While student progression and standards of performance can be established through the evidence of descriptive, comparative, and critical outcomes, the nature of the activity and what is assumed to be 'good' evidence of learning can limit creative endeavour and often see the retrofitting of a narrative to neatly fit the assessment or portfolio headings in a 'ritual' type of approach to learning and assessment (Banks & Barlex, 2020). While the portfolio may include features of quality of student performance such as evidence of knowledge, skills, research, analysis, communication, reflection, etc. its dissonance with the activity reduces its utility. So, while the approach can drive teaching and learning, it is not informing it.

The previously mentioned E-scape project (Kimbell, 2007) shifted the focus of this challenge by utilising the overarching criteria of *having*, *growing*, and *proving* an idea to help support the development of the student's ideas, but also to help steer the teaching, learning, and assessment direction. The clarity of these criteria proved to be very useful in support of specific projects while remaining generalisable and robust. This concept was explored by Canty et al. (2012) and Seery et al.

(2012) to capture the intentional acts of Initial Teacher Education (ITE) students as they managed and navigated through a design challenge that required them to demonstrate their capability. Using the criteria to guide the direction of enquiry, the assessment instrument acted as an enabler by using a holistic assessment method that facilitated students collectively engaging in peer assessment. The Adaptive Comparative Judgement (ACJ) process is well documented (Kimbell, 2007, 2012; Pollitt, 2011) across a variety of subject domains, but in brief, uses pairwise comparisons as a means of gaining a collective consensus on the rank order of performance. In its simplest form judges (students or teachers) are presented with a random pair of portfolios and asked to decide which is better. Simultaneously, several judges are also making binary decisions. These decisions are then aggregated, and a rank order is recorded. Although the algorithm appears to be a simple idea, its application is a powerful means of developing and identifying a shared construct of capability. This is derived from actual decisions on authentic evidence and not interpreted or mapped from abstracted assessment criteria. This approach forces judges to develop and apply their own assessment criteria derived in response to the work presented to them. These emerging assessment criteria generated from the evidence and validated by the level of consensus, liberate the creative, risk-taking, exploratory, and speculative approach supported by the contemporary view of designerly activity in technology education. Synthesising the evidence in a holistic way to 'see' the connectivity, and connectedness of evidence, increases the validity of the judgement made. Aggregating the collective interpretation of performance increases its reliability. As such, a holistic approach to assessment can ensure high levels of validity and high levels of reliability based on the holistic interpretation of capability as presented by the learner in response to the design challenge.

Facilitating students in exploring opportunities and alternatives, shifts the nature of pedagogical practice, from one of transmission to co-creation. Guided teaching practices by overarching criteria provide direction and intent while avoiding prescriptive acts, in response to assumed didactic contracts (students must know the 'correct' answer and teachers are looking for a correct answer). The ACJ process can also be utilised for both formative assessments where the learners themselves can be the judges creating the platform for holistic peer assessment. Through this approach, the teacher works with students to develop a shared understanding of what success looks like through the judging activity. This provides opportunity for dialogue between teacher and learner and the learner and peers where the co-creation of the assessment criteria can emerge as their work evolves. The purpose of this practice is to make learners 'partners in learning' (Alexander, 2004) and help them to recognise the features of quality in work, and how to incorporate quality into their own work-therefore enhancing self-regulation skills. The success of ACJ as a formative peer assessment tool to facilitate this process can be seen in the work of Bartholomew et al. (2019); Canty et al. (2017); Seery and Canty (2017), and Jones and Alcock (2013) where the ACJ activity creates the opportunity for meaningful integration of assessment in the learning activity.

Evolving technology education from a vocational orientation to embracing a plurality of outcomes and outputs requires full alignment of the teaching, learning and assessment ecosystem. Positioning Technology within a contemporary frame of reference and defining it through the lens of assessment increases its utility when supporting an integrated STEM agenda where a holistic approach to STEM assessment offers much promise.

7.6 Concluding Comments

Technology education can be a contested construct and although its manifestations in various countries share many features, context and situation play a significant role in how it is practised, valued, and treated. Suggesting that technology education is universally understood is laudable on a categorical level but challenged epistemologically and pedagogically. How knowledge is treated has direct implications for assessment methodologies and how performance standards are derived and maintained. The validity of the thinking that technology education is evolving towards being a critical contributor to general education is becoming more apparent. This evolution brings with it a renewed emphasis on designerly activity and the creative relationship between learning to, through, and about design (Seery et al., 2022), but also an emerging challenge as to how best to assess technological performance.

While the holistic approach to assessment is arguably more liberating for both teaching and learning, it challenges the traditional atomised approach to listing and weighting assessment criteria. However, ACJ acts as a valid operational platform that supports the interpretative demands of designerly activity and possibly the breadth of demands that will emerge from a fully integrated framework for STEM education.

References

Alexander, R. (2004). Still no pedagogy? Principle, pragmatism and compliance in primary education. *Cambridge Journal of Education, 34*(1), 7–33. https://doi.org/10.1080/0305764042000183106

Banks, F., & Barlex, D. (2020). Teaching STEM in the secondary school: Helping teachers meet the challenge. Routledge.

Barlex, D. (2007). Assessing capability in design and technology: The case for a minimally invasive approach. *Design and Technology Education: An International Journal, 12*(2), 49–56.

Bartholomew, S. R., Zhang, L., Strimel, G. J., & Garcia Bravo, E. (2019). A tool for formative assessment and learning in a graphics design course: Adaptive comparative judgement. *The Design Journal*, 1–23.

Baynes, K. (2010). *Models of change: The impact of 'designerly thinking' on people's lives and the environment: Seminar 4 modelling and society*. Loughborough University.

Buckley, J. (2023). Historical and philosophical origins of technology education. In D. Gill, D. Irving-Bell, M. McLain & D. Wooff (Eds.) *The bloomsbury handbook of technology education*. Bloomsbury.

Buckley, J., O'Connor, A., Seery, N., Hyland, T., & Canty, D. (2019). Implicit theories of intelligence in STEM education: Perspectives through the lens of technology education students.

International Journal of Technology and Design Education, 29(1), 75–106. https://doi.org/10.1007/s10798-017-9438-8

Canty, D., Seery, N., Buckley, J., & Dunbar, R. (2021). A conceptual framework for assessment of learning in technology classroom based assessments. *Techne Series—Research in Sloyd Education and Craft Science A, 28*(2), 213–220.

Canty, D., Buckley, J., & Seery, N. (2019). Research paper on features of skills development in technology education. Report commissioned by the National Council for Curriculum and Assessment (NCCA) with respect to the reform of technology subjects in the Irish Junior Cycle.

Canty, D., Seery, N., Hartell, E., & Doyle, A. (2017). Integrating peer assessment in technology education through adaptive comparative judgment. PATT34 Technology & Engineering Education–Fostering the Creativity of Youth Around The Globe, Millersville University, Pennsylvania, USA, pp. 10–14.

Canty, D., Seery, N., & Phelan, P. (2012). Democratic consensus on student defined assessment criteria as a catalyst for learning in technology teacher education. PATT 2012: Technology Education in the 21st Century.

Daker, J., (2022) A Nomadic Pedagogy about Technology: Teaching the Ongoing Process of Becoming Ethnotechnologically Literate: 18 (International Technology Education Studies) ISBN 9004536981

Doyle, A. (2020). Consolidating concepts of technology education: From rhetoric towards a potential reality (Doctoral dissertation, KTH Royal Institute of Technology).

Doyle, A., Seery, N., & Gumaelius, L. (2018). Defining a curriculum through activity: Facilitating epistemic autonomy in technology education. In Proceedings of the 10th Biennial International Design and Technology Teacher's Association Research Conference (DATTArc) (pp. 29–37). Hawthorn, VIC: Design and Technology Teacher's Association Research Conference jointly with Swinburne University of Technology.

Dunbar, R., Seery, N., & Phelan, J. (2023). Industrial perspectives: Translational and transactional agendas. In D. Gill, D. Irving-Bell, M. McLain & D. Wooff (Eds.) *The bloomsbury handbook of technology education.* Bloomsbury

Gibson, K. (2008). Technology and technological knowledge: A challenge for school curricula. *Teachers and Teaching: Theory and Practice, 14*(1), 3–15.

International Technology and Engineering Educators association. (2020). Standards for technological and engineering literacy: The role of technology and engineering in STEM education. www.iteea.org/STEL.aspx

Jay, J., & Johnson, K. (2002). Capturing complexity: A typology of reflective practice for teacher education. *Teaching and Teacher Education, 18*(1), 73–85.

Jones, I., & Alcock, L. (2013). Peer assessment without assessment criteria. *Studies in Higher Education, 39*(10), 1774–1787.

Kelly, A. V., Kimbell, R., Patterson, V. J., Saxton, J., & Stables, K. (1987). *Design and Technology: A framework for assessment.* HMSO.

Kimbell, R., Stables, K., Wheeler, T., Wosniak, A., & Kelly, V. (1991). *The assessment of performance in design and technology.* School Examinations and Assessment Council/Central Office of Information.

Kimbell, R. (1994). Assessment of design and technology. *Teaching technology*, 161–172.

Kimbell, R. (2007). E-assessment in Project e-scape. *Design and Technology Education: An International Journal, 12*(2).

Kimbell, R. (2011). Wrong...but right enough. *Design and Technology Education: An International Journal*, *16*(2), 6–7.

Kimbell, R. (2012). The origins and underpinning principles of e-scape. *International Journal of Technology and Design Education, 22*, 123–134.

McGarr, O., Lynch, R. (2017). Monopolising the STEM agenda in second-level schools: exploring power relations and subject subcultures. *International Journal of Technology and Design Education, 27*, 51–62.https://doi.org/10.1007/s10798-015-9333-0

Mitcham, C. (1994). *Thinking through technology: The path between engineering and philosophy.* University of Chicago Press.

Norman, E. (2010). A Silent 'D.' *Design and Technology Education: An International Journal, 15*(2), 3–5.

Orsmond, P., Maw, S. J., Park, J. R., Gomez, S., & Crook, A. C. (2013). Moving feedback forward: Theory to practice. *Assessment & Evaluation in Higher Education, 38*(2), 240–252. https://doi.org/10.1080/02602938.2011.625472

Pollitt, A. (2011). Comparative Judgement for Assessment. *International Journal of Technology and Design Education* Published online: 14 December.

Rynne, A., Gaughran, W. F., & Seery, N. (2010) Defining the variables that contribute to developing 3D CAD modelling expertise 161–178. In Norman, E. W. L., Seery, N. (2011): IDATER online conference: graphicacy and modelling 2010. Loughborough University. Book. https://hdl.handle.net/2134/9015

Seery, N. (2017). Modelling as a form of critique. *Critique in design and technology education,* 255–273.

Seery, N., Phelan, J., Buckley, J., & Canty, D. (2022). Epistemological treatment of design in technology education. *International Journal of Technology and Design Education.* https://doi.org/10.1007/s10798-022-09781-y

Seery, N., Canty, D. (2017). Assessment and learning: The proximal and distal effects of comparative judgment. In de Vries, M. (eds) *Handbook of Technology Education.* Springer International Handbooks of Education. Springer, Cham.

Seery, N., Canty, D., & Phelan, P. (2012). The validity and value of peer assessment using adaptive comparative judgement in design driven practical education. *International Journal of Technology and Design Education, 22,* 205–226.

Schön, D. A. (1983). The reflective practitioner: How professionals think in action. New York: Basic Books (Reprinted in 1995).

Tosey, P., Visser, M., & Saunders, M. (2011). The origins and conceptualizations of "triple-loop" learning: A critical review. *Management Learning, 43*(3), 291–307.

Williams, P. J. (2009). Technological literacy: A multliteracies approach for democracy. *International Journal of Technology and Design Education, 19*(3), 237–254.

Niall Seery He in a qualified teacher with a PhD in Engineering Education. Niall spent 15 years as an academic in Teacher Education with a specialist interest in pedagogical practice. He has served as director of studies at both undergraduate and masters level. In 2010, Niall founded and continues to direct the Technology Education Research Group (www.TERG.ie), where he is still active in research development and mentorship.

Donal Canty is Lecturer in Initial Teacher Education at the University of Limerick. He is qualified as a Post-Primary Teacher of Graphics and Architectural Technology and has taught Construction Studies, Technical Drawing, Materials Technology (Wood), Technical Graphics and LCA Graphics and Construction at Post-Primary level (1996–2004). He is currently lecturing in Subjects Pedagogics at undergraduate and post-graduate levels in the University of Limerick. Donal is a director of the Technology Education Research Group (TERG). Donal has also successfully supervised both PhD and Masters students to graduation in the area of Design and Technology education.

Chapter 8
Approaches to Teaching STEM

Piet Ankiewicz

Abstract When it comes to the teaching of STEM, the literature often refers to inquiry-, discovery-, problem-, project-, design-based learning and or teaching. However, it largely mentions it in a fairly haphazard and messy way without contextualising it within a sound theoretical framework. Thus, the purpose of this chapter is to relate the abovementioned constructs to relevant learning theories, congruent teaching approaches and strategies of particular relevance to technology education. These constructs are generally linked to constructivism as a learning theory and approach to learning. This chapter provides a brief overview of cognitive, radical and social constructivism as three perspectives on constructivism. The Piagetian view, which focuses on the experiential world of the student, is related to cognitive and radical constructivism, while the Vygotskian view, which focuses on the social and cultural domain, is classified as social constructivism. According to cognitive and radical constructivism students are constantly engaged in the process of adaptation of knowledge which is based on the individual's unique experience. Social constructivists see learning as a process of "enculturisation", whereby students are inducted into the cultural practices of the societies they live in, learning from various sources of information. Subsequently, the chapter distinguishes between cooperative, inquiry- and discovery-based learning as some sub-approaches to constructivism by relating them to cognitive, radical and social constructivism. It also succinctly addresses the relationship between cooperative, inquiry- and discovery-based learning. Based on these sub-approaches, it then discusses problem-, project- and design-based teaching approaches or strategies from a technology education perspective. Lastly, it discusses the affordances of these perspectives, approaches, sub-approaches and related teaching strategies for the application of STEM education from a technology education perspective.

Keywords Constructivism · Teaching approaches and strategies · Technology education · Engineering education · Integrated STEM education

---

P. Ankiewicz (✉)
Department of Science and Technology Education, Faculty of Education, University of Johannesburg, Auckland Park, South Africa
e-mail: pieta@uj.ac.za

W. Fox-Turnbull and P. J. Williams (eds.), *Locating Technology Education in STEM Teaching and Learning*, Contemporary Issues in Technology Education,
https://doi.org/10.1007/978-981-97-1995-2_8

8.1 Introduction

When it comes to the teaching of integrated or integrative STEM in secondary schools, teachers need to build a cohesive understanding of STEM education and the process of integrating science, technology, engineering, and mathematics in authentic contexts (cf. Hallström & Schönborn, 2019). The recent STEM education literature advocates for STEM education to be offered by cooperative, inquiry-, discovery-, problem-, project-, design-based learning and teaching in authentic contexts (Carlson & Sullivan, 1999; Daugherty & Carter, 2018; Frykholm & Glasson, 2005; Han et al., 2022; Hmelo-Silver, 2004; Kelley & Knowles, 2016; Kolodner, 2006; Kolodner et al., 2003; Krajcik et al., 1998; Nadelson & Seifert, 2017; Nguyen et al., 2020). Even when STEM is used as a multidisciplinary approach to learning which integrates knowledge from the separate STEM subjects into existing science, technology and engineering, or mathematics instructional programmes (Sanders, 2009) or into inter- or transdisciplinary approaches to integrated STEM programmes (cf. English, 2016) the abovementioned constructs still apply, such as problem-, project-, and engineering design-based learning strategies (Ritz & Fan, 2015). However, the literature largely mentions these constructs in a fairly haphazard and messy way without contextualising them within a sound theoretical framework. Thus, the purpose of this chapter is to relate the abovementioned constructs to relevant learning theories, congruent teaching approaches and strategies of particular relevance to technology education. These constructs are generally linked to constructivism as a learning theory and an approach to learning. The discussion shifts now to the constructivist theory and approach to learning.

8.2 Constructivism as a Theory on Learning or Learning Approach

An approach in this context should be viewed as a broad means of tackling a job. Despite the chapter's focus on constructivism as a learning theory and an approach to learning which underlies STEM education, we also recognise behaviourism as an important learning theory and an approach to learning. Behaviourist learning theories focus on ways in which responses that students learn to make are influenced by the stimuli that precede and/or follow their responses (Biehler & Snowman, 1997). It would emphasise the immediate feedback students receive and the benefits they receive from increased practice (Killen, 2007).

Teaching based on behaviourist learning theories, for example, in technology education is appropriate when students must remember important information or to develop their skills in using and operating tools and equipment (Reddy et al., 2005). Factual knowledge (embedded in conceptual knowledge) and basic skills (associated with lower-order procedural knowledge), as part of the epistemology of technology open the door to a teaching approach or strategy which relates more to behaviourist learning theories (Ankiewicz, 2003). It is for teachers to translate behaviourist learning theories into congruent behaviourist teaching approaches or strategies, for example direct teaching and demonstration (De Swardt et al., 2005; Killen, 2007). However, behaviourism alone is insufficient for technology education (De Swardt et al., 2005; Reddy et al., 2005).

Constructivism has been defined in many ways by different scholars, including a Piagetian and Vygotskian view (Barak, 2018; De Swardt et al., 2005; Kanuka & Anderson, 1999; Muller et al., 2017). The basic premise of constructivism is that knowledge is obtained and expanded through active construction and reconstruction of theory and practice (De Swardt et al., 2005; Reddy et al., 2005), which means that learning is not a passive process (Taylor, 1999) but requires active engagement by the student (Jenkins, 2000). The construction and reconstruction of theory and practice claim to be meaningful learning in the sense that it is the active creation of knowledge structures from personal experience (e.g. concepts, rules, hypotheses, associations). This means that students build a personal view of the world from their own experiences (Biehler & Snowman, 1997). Davis et al. (1990) define constructivism as the process in which students construct their own knowledge individually and cooperatively.

The constructivist theory and learning approach view teachers as facilitators in learning, rather than them taking centre stage (Jonassen, 1994). Students are responsible for their own knowledge making and understanding (Jenkins, 2000) through interaction with the environment and collaboration with peers. This interactive nature of learning allows students to formulate knowledge through their own perspective, which constantly evolves. When students interact with their environment it provides a better experience and better prepares students for a lifetime of independence and development (Rieber, 2004).

The two theories or approaches (i.e. behaviourism and constructivism) should not be viewed as two opposites but rather as the two extremes on a continuum. Learning and subsequently teaching take place between the two extremes on the continuum at various points in time, depending on the nature of the content (conceptual and procedural knowledge) and the student tasks (Killen, 2007). Johnson (1997) asserts that if intellectual skills development is to become the larger focus of the technology education curriculum, it does not imply that behaviourist theories of learning be abandoned in favour of the more contemporary constructivist theories. In fact, some teaching approaches or strategies could be based on a combination of both theories (Ankiewicz, 2003; Hammer, 1997) where teachers blend, for example, transmissionist lecture with cooperative, team-based, constructivist pedagogy [cf. Fig. 1 in De Swardt et al. (2005) for a detailed description of how teachers can manoeuvre between behaviourist and constructivist teaching during a lesson]. It is, however, important to establish a balance between the two theories and approaches to learning in the technology classroom. Bowen (2009) alludes to the fact that content-based learning, which is associated with behaviourism, is not the antithesis of cognitive-based learning, which is associated with constructivism (Reddy et al., 2005).

Brooks and Brooks (1993) refute the notion that constructivism is a theory about teaching and define it as a theory about knowledge and learning. However, constructivist learning theories and approaches to learning can be translated by teachers into constructivist approaches or strategies to teaching in their classrooms to bring about a constructivist learning environment (Killen, 2007). The various perspectives and sub-approaches that arise from the constructivist approach (cf. Nkadimeng & Ankiewicz, 2022) will be discussed next.

8.3 Perspectives on and Sub-approaches to Constructivism

8.3.1 Perspectives on Constructivism

The perspectives representing constructivist theories of learning are *radical, cognitive and social constructivism* (Cobb, 1994; De Swardt et al., 2005; Killen, 2007; Reddy et al., 2005). Both *cognitive* and *radical constructivism* relate to the Piagetian view, which focuses on the experiential world of the individual student (De Swardt et al., 2005). The view of knowledge suggested by both *cognitive* and *radical constructivism* is that human beings are, from birth, constantly engaged in the process of adaptation of knowledge which is based on the individual's unique experience (Biehler & Snowman, 1997).

The perspective of *cognitive constructivism* is underpinned by Piaget's principle that knowledge is not passively received, but actively constructed by people (Dart, 1994; Dart et al., 1998; Muller et al., 2017). A person learns individually by the process of adaptation which creates a good match between one's conception of reality and the real-life experiences one encounters. Adaptation consists of two subprocesses, namely assimilation and accommodation. When someone adapts by interpreting an experience so that it fits an existing scheme (organised mental patterns or behaviour), it is called assimilation. When a someone adapts by changing an existing scheme to incorporate an experience it is referred to as accommodation (Biehler & Snowman, 1997). According to Abbott and Ryan (2001: 9), "(A) a person learning something new brings to that experience all of their previous knowledge and present mental patterns. Each new fact or experience is assimilated into a living web of understanding that already exists in that person's mind".

Through a meta-synthesis of Piaget's scholarly work, Von Glasersfeld established the perspective of *radical constructivism* by adding a second underlying principle to *cognitive constructivism*. This principle, which is more difficult to claim as a Piagetian view, states that the function of cognition is adaptive and serves the organisation of the experiential world and not the discovery of ontological reality. Most branches of educational constructivist theory consider this second principle as radical as it suggests that there is no independent reality as nothing can be taken as proven. Thus, the emphasis is solely on the individual's knowledge construction, focusing the analysis on the manner in which an individual constructs personal understanding (Candy,

1991; Muller et al., 2017; Reddy et al., 2005; Zietsman, 1996). Emanating from this perspective to learning, is the question of reflection and choice on the part of students (hence the freedom for alternative conceptions) implying that any person can be an autonomous student (Zietsman, 1996). This does not suggest that students should be left to their own devices, since they have the potential to perform on their own, but radical constructivism rather suggests that teachers "create situations in which students could develop and exert their innate drive towards acting independently" (Candy, 1991: 258). Radical constructivism is concerned with both the structure and the construction of meaning in learning (Doolittle & Camp, 1999), and it advocates for the independent learning of students through creative and critical thinking, decision making, problem solving and design (Reddy et al., 2005).

Although both cognitive and radical constructivism draw heavily on the work of Piaget (essentially focusing on the individual's constructions of knowledge that are developed through personal interactions with the environment), it fits well with current models of cognition that advance students' understanding (De Swardt et al., 2005; Killen, 2007). Teachers should understand the continuum where students' knowledge is either entirely individually constructed within themselves through interacting with their own thoughts and unique activities in the world, or socially constructed through interacting with other students and other students' views and experiences (Guzdial, 2018; Kanuka & Anderson, 1999; Phillips, 2000). Despite the distinction between individually and socially constructed knowledge, knowledge is always constructed within the individual.

While the experiential world of the individual student is the central focus of the Piagetian view (cognitive and radical constructivism), learning in the Vygotskian view (social constructivism) is placed firmly in the social and cultural domain (Barak, 2018; De Swardt et al., 2005; Kanuka & Anderson, 1999). *Social constructivism* places the student in a specific "cultural space" so that the student is seen to be engaged in constructing knowledge not in isolation, but in a "context that is shaped and modified by the influence of the society he/she finds himself/herself in" (Magadla, 1996: 83). Social constructivists see learning as a process of "enculturisation", whereby students are inducted into the cultural practices of the societies they live in, learning from various sources of information. This implies that the student appropriates meaning from the social and cultural environment (Barak, 2018; Magadla, 1996) before meaningful learning takes place. Social constructivism is more concerned with meaning rather than structure in learning (Doolittle & Camp, 1999; Kanuka & Anderson, 1999). Teachers act as mediators in personal meaning making in social and cultural environments (Cobb, 1994; Reddy et al., 2005). It emphasises the construction of an agreed upon, socially constructed reality (Doolittle & Camp, 1999).

The implications of social constructivism for learning and teaching in technology education are that students' social and cultural environment, i.e. home and family milieu, living conditions, religious and socio-economic factors, need to be considered in teaching and learning (Reddy et al., 2005). Technology can be regarded as both minds-on (complex thinking) and hands-on (practical activities). The essence of technology is the interaction of mind and hand; thus, inside and outside the head. Ideas conceived in the mind need to be expressed in concrete form (artefacts/products) (Ankiewicz, 2015, 2016; APU, 1994; McCormick & Davidson, 1996). Both the minds-on activities in technology—which consist of complex thinking processes (creative, critical, decision making, problem solving and design thinking)—and the hands-on activities have been associated with technological procedural knowledge (Ankiewicz, 2016; Ankiewicz et al., 2006; Jakovljevic & Ankiewicz, 2016). The nature of technological activities lends itself to social constructivist theories of learning where students engage in practical activities and interact with one another when solving problems and designing in cooperative learning environments as part of projects (Barak, 2018; Grobler & Ankiewicz, 2022; Johnson & Thomas, 1992; Reddy et al., 2005). Learning and teaching that characterise these complex thinking skills flow naturally from constructivist theories of learning (Grobler & Ankiewicz, 2022; Reddy et al., 2005).

Constructivism parallels the idea of constructionism of which the latter is a theory that was invented by Piaget's student, Papert. Constructionism holds that students learn best when they actively design and construct (Barak, 2018; Guzdial, 2018). The computer programme, Logo was originally advocated as the primary vehicle for constructionism. The idea was that learning to draw geometric objects using Logo was a way to promote mathematical learning (Aydin, 2005). It was mentioned earlier that constructivism holds that students are perceived as active constructors of knowledge. Constructivism relates to the mental (minds-on activities) and constructionism to the physical level (hands-on activities). Thus, following from designing (minds-on activities) and making (hands-on activities) physical artefacts/products in technology education, both constructivism and constructionism are relevant to technology education.

Table 8.1 provides a summary of the various perspectives on constructivism (column 1) with each one's specific focus (column 2) and its relationship with a specific scholarly view (column 3). The arrow indicates the Piagetian link between cognitive and radical constructivism.

Arising from the constructivist approach to learning are various sub-approaches, for example experiential, cooperative, inquiry- and discovery-based learning (Black & Ammon, 1992; De Swardt et al., 2005; Reddy et al., 2005). The deliberations shift to these sub-approaches now.

8.3.2 *Sub-approaches to Constructivism*

8.3.2.1 Experiential Learning

It was mentioned earlier that experience is a key tenet of constructivist learning theories and approaches to learning. Thus, experiential learning is the overarching sub-approach to constructivism. Experiential learning emphasises the importance of learning from experience that is goal driven, activity-based and problem centred (Johnson, 1997). One reason why student centredness emphasises problem solving

Table 8.1 Summary of the perspectives on constructivism

Perspectives on constructivism	Focus	Related scholarly view
Cognitive	Knowledge is not passively received, but actively constructed through personal interactions with the environment The experiential world of the individual student is the central focus Human beings are constantly engaged in the process of adaptation of knowledge which is based on the individual's unique experience	Piaget
Radical	Similar to cognitive constructivism with the addition that there is no independent reality as nothing can be taken as proven The emphasis is solely on the individual's construction knowledge, focusing the analysis on the manner in which an individual constructs his or her personal understanding	Piaget (as expanded by Von Glasersfeld)
Social	Learning is a process of "enculturization", whereby students are inducted into the cultural practices of the societies they live in, learning from various sources of information. Social and cultural interactions are critical to cognitive functions.	Vygotsky

as opposed to direct teaching is that learning from experience is a more active and valid form of learning (Down, 1996). Henak (1992) contends that meaning is added when students are actively involved in learning experiences. Meaning can come from activities that are relevant to real-life problems (Reddy et al., 2005).

Current teaching approaches or strategies that emphasise experiential learning include among others, activity-, action-, project-, inquiry- and discovery-based learning. Technology education specifically and STEM education in general focus on real life, authentic problems. Experiential learning is seen as a powerful and empowering tool to make technology education more meaningful and relevant to the lives of students. The activity-based nature of technology education provides ample opportunities for students to engage in experiential learning (Reddy et al., 2005).

8.3.2.2 Cooperative Learning

Cooperative learning is a sub-approach to constructivism and relates to the social constructivist perspective. It shifts the focus of intellectual development from the solitary individual to the individual in a highly interactive social milieu (Black &

Ammon, 1992). Rogoff and Lave (1984; cited by Johnson, 1997) state that all cognitive activity is socially defined. Cooperative learning is a learning approach in which students work together in groups to achieve a particular learning outcome (Slavin, 1980). Cooperative learning encourages positive interdependence, individual accountability, shared leadership, and improved social skills (Cavalier et al., 1995; Johnson & Johnson, 1989; King, 1984).

Regarding learning, social constructivism can be implemented through group discussions or collaborations on projects, which allows students to generalise and transfer their knowledge (Rezniskaya et al., 2007). These discussions allow the students the opportunity to exercise self-regulation and perseverance while improving communication and problem-solving skills (Corden, 2001; Nystrand, 2006). Several scholars (Cavalier et al., 1995; Johnson & Johnson, 1989; King, 1984) found that students' group interaction behaviour is related to improved performance in a cooperative learning environment. Through the exchange of ideas, students develop shared meanings that allow group members to communicate effectively with each other (Wheatley, 1991).

As mentioned earlier, the nature of technological activities lends itself to social constructivist theories of learning where students engage in practical activities and interact with one another when solving problems and designing in cooperative learning environments as part of projects (Barak, 2018; Grobler & Ankiewicz, 2022; Johnson & Thomas, 1992; Reddy et al., 2005). By interacting with others in the technology classroom, students can learn from one another, share knowledge, engage in co-operation and competition and negotiate meaning. Through the social activity of learning, therefore, teachers can develop communities of students (Grobler & Ankiewicz, 2022; Johnson, 1997; Reddy et al., 2005). Cooperative learning is probably the most important skill students acquire, especially if one considers the technological work environment which entails collaboration and teamwork. It is important that students undertake technology practice in classrooms as close as possible to technologists' practice in the work environment (Ankiewicz, 2013).

8.3.2.3 Inquiry Learning

Inquiry learning is whereby the student questions, investigates and seeks out information, thereby assuming responsibility for his/her own learning, with the teacher playing a facilitative role. Inquiry learning is an investigative learning process which is often equated to experimentation, problem solving and discovery (Schwaller, 1995). Duke (1990) claims that it is more important that students become familiar with the inquiry process than to obtain the correct answers. Orlich et al. (1994) advocate that the inquiry process must be learnt, demonstrated and assimilated into learning styles (Reddy et al., 2005).

The problem-solving nature of the technological tasks provides ample opportunities for students to engage in inquiry learning (Best, 2018; Reddy et al., 2005). Investigation and research, which are important for technology education, are rooted in inquiry learning. Involving technology students in inquiry learning is one of the most effective ways to help them develop their higher order critical thinking skills (Eggan & Kauchak, 1996; Reddy et al., 2005). In accordance with the Piagetian view, inquiry learning advocates for independent learning through problem solving and discovery (Friesen & Scott, 2013).

8.3.2.4 Discovery Learning

Discovery learning is a sub-approach of inquiry learning (Bruner, 1961). Discovery learning involves the discovery by students of what they are capable of thinking and doing for themselves (Bruner, 1966). In order to achieve discovery, the student must incorporate and integrate information with what is already known so that new relationships are formed (Biehler & Snowman, 1997). Discovery does not necessarily mean coming up with knowledge that is unknown to someone else but coming up with knowledge by oneself (Mwamwenda, 1995). An important component of discovery learning is the discovery of relationships between parameters (Biehler & Snowman, 1997). Discovery learning involves rearranging or transforming information to obtain new insights.

Given the problem-solving nature of technology education, the discovery approach to learning is a valid form of knowledge construction, whereby students take responsibility for their own learning (Reddy et al., 2005). Innovation—the process of improving and refining technological solutions—in technology education has its roots in discovery learning as it is a product of discovery (Reddy et al., 2005).

Table 8.2 provides a summary of the sub-approaches associated with constructivism with the main principle underlying each sub-approach, its focus and implications for technology education.

As learning is neither entirely individual, nor entirely cooperative, but a mix of both, Fig. 8.1 depicts the Piagetian (cognitive and radical constructivism)—on the left—and the Vygotskyan view (social constructivism) as part of constructivism—on the right. The sub-approaches associated with constructivism have been superimposed onto the three perspectives. Generally, students out of school may learn entirely individually by inquiry-based learning and discovery-based learning (Band 1). However, when learning in school they also present as individuals who might learn in a similar way (Band 2) than out of school (Band 1). Furthermore, when learning together with their peers in school (Band 3), they may also learn cooperatively, including inquiry- and discovery-based learning. The fact that students interpret and construct meaning from existing socially constructed systems does not diminish the importance of personal construction of knowledge (Guzdial, 2018; Phillips, 2000). It was mentioned earlier that in reality knowledge is always constructed within the individual.

The discussion shifts now from the perspectives on and the sub-approaches to constructivism to translating the constructivist approaches into congruent teaching approaches or strategies (De Swardt et al., 2005; Killen, 2007).

Table 8.2 Summary of the sub-approaches associated with constructivism

Sub-approach	Main principle	Focus	Implications for technology education
Experiential learning/teaching	Students are guided by the teacher to learn from specific real-life experiences	Help students to grasp real-life situations Learning through doing	The technology teacher formulates a project brief with an implicit problem to be identified and solved. It must also relate to both an authentic, real-life problem and the current value system of the students, taking psychological and sociological aspects of the students' situation into consideration (Ankiewicz, 2019)
Cooperative learning/teaching	Students work together to accomplish a learning task	Individual responsibility, communal sharing, joint responsibility, ethos of collaboration	Technology students interact with one another in a co-operative learning environment (e.g. groupwork) by sharing knowledge, constructing and negotiating meaning about the problem to be solved (De Swardt et al., 2005)
Inquiry learning/teaching	Students investigate and research real-life situations which encourage thinking and problem-solving skills under supervision and guidance of the teacher/facilitator	Data gathering allows students to work outside the school context	Technology students investigate/research, collect, record and report information/data in relation to the problem to be solved (De Swardt et al., 2005)
Discovery learning/teaching	Students discover how ideas relate to each other and to existing knowledge by solving authentic problems under supervision and guidance of the teacher/facilitator	Teach students how to find and process relevant information themselves	Technology students discover the relationships between parameters relating to the problem to be solved (Reddy et al., 2005), e.g. the relationship between speed or torque and gear ratio of a gear train

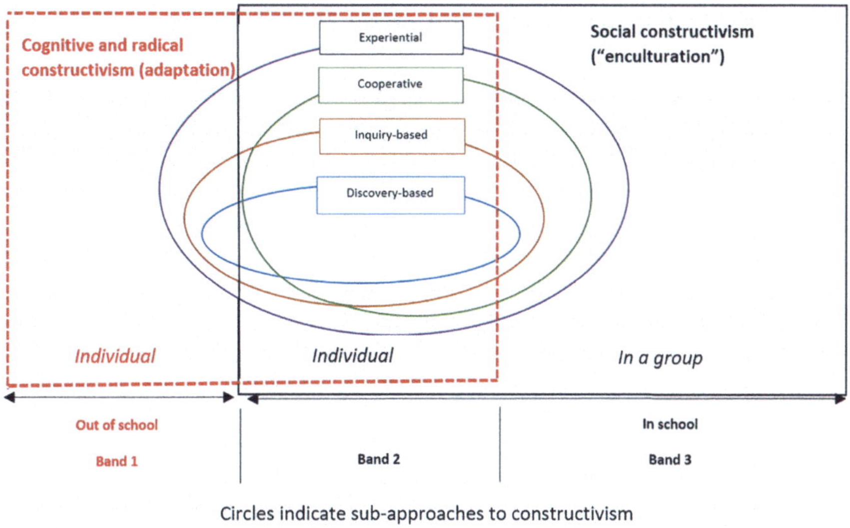

Fig. 8.1 Superposition of the sub-approaches associated with constructivism onto the three perspectives on constructivism (Nkadimeng & Ankiewicz, 2022)

8.4 Constructivist Teaching Approaches or Strategies from a Technology Education Perspective

8.4.1 Translating Learning Approaches into Teaching Approaches or Strategies

The nature of technology education involves, inter alia, creativity, problem solving and design (Jones et al., 2013). As "… (learner) activity is that pivotal event in which knowledge and skills unite to bring artefacts into existence …", technology education should at its core also be activity-based and constructivist (Ankiewicz, 2003: 10, 2013: 5; Jones et al., 2013). Unlike constructivism as a learning theory and an approach to learning, teaching strategies are more confined/narrowed and they indicate "a plan", "…a procedure of how to go about it…". As stated earlier, technology education is best situated within constructivist theories of learning (Grobler & Ankiewicz, 2022; Reddy et al., 2005).

When applying a constructivist or behavioural teaching approach, a variety of teaching strategies (e.g. brainstorming, inquiry/investigation, discussions, case studies, activity-based practice and project work) are essential for technological problem solving (Jakovljevic & Ankiewicz, 2016). Based on constructivism, congruent teaching approaches or strategies which are common to technology education and STEM education, namely problem-, project- and design-based teaching will be briefly discussed in the next sections.

8.4.2 Problem-Based Teaching Approaches/Strategies

Problem solving is a key feature that the STEM subjects have in common (Banks & Barlex, 2014). Problem-based learning can be traced back to McMaster University in Canada in the 1960s where the lecturers developed a teaching approach/strategy that would force the medical students to use their textbook knowledge in case study situations and, in turn, lecturers could assess students' ability to work as practicing doctors. Nowadays, problem-based learning is a curriculum model constructed around authentic, real-life problems that, by their very nature, are ill-structured, open-ended and ambiguous (Banks & Barlex, 2014; Williams et al., 2008). Problem-based learning is driven by the inquiry process (Best, 2018). Problem-based learning and the congruent teaching approaches or strategies are based on the social constructivist perspective on and the cooperative and inquiry-based sub-approaches to constructivism (Allen et al., 2011; Best, 2018; Jakovljevic & Ankiewicz, 2016; Schwartz, 2013; Stage et al., 1998).

Technology, as well as technology education, is essentially problem solving (Best, 2018). Although a distinction is made in technology between design and problem-solving (Johnsey, 1995; McCormick, 1997), there is a connection between these concepts: The design process is seen as the manifestation of the problem-solving process (McCormick et al., 1994). According to De Vries (ITEA, 1997) design is a particular type of problem solving in technology (Ankiewicz, 2003, 2013; Ankiewicz et al., 2006). Many technologies and engineering textbooks introduce engineering design as a technological problem-solving process (Sung & Kelley, 2022). Therefore, essential to technology education, problem-based learning "is not about problem solving per se, but rather it uses appropriate problems to increase knowledge and understanding" (Wood, 2002: 8; cited by Best, 2018).

8.4.3 Project-Based Teaching Approaches/Strategies

Ongoing practical work and its related tasks tend to have a range of names depending to some extent on the emphasis fashionable when the name was coined; in technology and engineering examples include capability and design-and-make tasks. In science project work, the constructs of project-based and problem-based learning are used to describe activity where the focus is on discovering an aspect of the natural world for oneself. The nature of mathematical tasks can be truly characterised as problem solving. Work linked to authentic real-world tasks is the principal driver in STEM activities and for many students, doing a project—in class or extracurricular—is what it is all about (Banks & Barlex, 2014).

Project-based learning is founded on the Piagetian and the Vygotskian views which account for learning as individuals constructing knowledge in interaction with others and with the environment (Major & Govers, 2014). The importance of learning through doing, which underpins problem solving through practical work, is paramount. It goes as far back as the end of the nineteenth century when Dewey (1899: 60) claimed that as part of learning "Children simply like to do things, and watch to see what will happen". Dewey (1963) also stressed the importance of active experience with objects (hands-on activities) as a means of developing thinking (minds-on activities) (Banks & Barlex, 2014; Best, 2018; Grobler & Ankiewicz, 2022; Major & Govers, 2014).

Teachers are a resource to the students, but should act more as facilitators, requiring students to organise their own work and manage their own time and compelling them to communicate and collaborate with one another. Project-based learning engages students in complex tasks based on challenging or "driving questions" (Banks & Barlex, 2014: 137; Hawari & Noor, 2020).

Project-based learning is a teaching approach or strategy which provides a learning environment designed for students to systematically and actively solve problems like working in real life. Project-based learning involves a dynamic classroom approach, which emphasises long-term, interdisciplinary and student-centred learning (Hawari & Noor, 2020). Project-based learning, where the project is based on a real-world problem, encourages students to carry out authentic tasks based on their "real-life" experience (Banks & Barlex, 2014; Cheng et al., 2007; Hawari & Noor, 2020; Major & Govers, 2014; Svärd et al., 2017).

Student activities in a project-based classroom require different approaches to learning (e.g. experiential, cooperative, inquiry and discovery learning) (Bednar et al., 2013; Jakovljevic & Ankiewicz, 2016; Johnson, 1997; Johnson & Johnson, 1991; Rogoff & Lave, 1984). It was already mentioned that the nature of technological activities lends itself to social constructivist theories of learning where students engage in practical activities and interact with one another when solving problems and designing in cooperative learning environments as part of projects (Barak, 2018; Grobler & Ankiewicz, 2022; Johnson & Thomas, 1992; Reddy et al., 2005).

Project- and problem-based learning tend to be two sides of the same coin to such an extent that they are both represented by the acronym *PBL* in the literature. Problem-based learning focuses on students proposing solutions to a problem or a situation to increase knowledge and understanding. Project-based learning involves an authentic context for doing something—collecting data, designing a product, or doing practical work. The desired outcome of problem-based learning is new knowledge and understanding; the desired outcome of project-based learning is an artefact, system, or process manifesting that new knowledge (Banks & Barlex, 2014).

As mentioned earlier, there is a connection between problem-solving and design in technology education. Therefore, design-based teaching approaches/strategies will be discussed next.

8.4.4 Design-Based Teaching Approaches/Strategies

Design is the primary methodology of technology which is more intuitive (De Vries, 2018) than engineering design as it has an element of trial and error to it (Williams, 2011). Design also features in engineering which is part of the broader field of technology (De Vries, 2018). Engineering is a sub-set of the broad area of technology (Williams, 2011). However, design as a characteristic of engineering is different from design in technology as it is infused with the more structured elements of modelling, quantification and the use of concepts (De Vries, 2018; Hallström & Ankiewicz, 2019, 2023).

From a philosophical perspective, Hallström and Ankiewicz (2019, 2023) view technological design as the integrator of the STEM subjects. De Vries (2018, 2021) suggests the use of particular technological design challenges in which engineering principles, scientific concepts and mathematical ways of thinking are essential for finding solutions to the challenge. Thus, De Vries (2018, 2021) advocates an approach where there is knowledge development in all STEM subjects simultaneously, and not merely the application of existing knowledge related to them—similar to how it happens in practice. In such design challenges, students will "design" and do experiments to understand scientific phenomena. Design in science may also be more experimentally oriented—such as, for example, in synthetic chemistry—whereas design in technology and engineering are not really experiments in the scientific sense but rather an epistemic practice of its own that is open ended, iterative and produces its own knowledge (Ammon, 2017). Design in technology and engineering—and even sometimes in mathematics, according to Bishop (1988)—will typically lead to the making of an object, or system (cf. Vermaas et al., 2011). The students will also "design" mathematical expressions or functions and do calculations to optimise their design by modelling to test their optimisations. The steps of mathematical modelling resemble, to a large extent, the stages of technological design. Such challenges will fit the nature of design as a process, and in the philosophical sense a methodology, in which both new knowledge is developed (about designing itself, but also science and engineering), and existing knowledge (previously learnt in science, technology and/or mathematics) is applied (De Vries, 2018, 2021; Hallström & Ankiewicz, 2019, 2023; Vossen et al., 2019).

With engineering as a subset of technology (Williams, 2011), Kelley and Knowles (2016) view engineering design in contemporary technology education as the integrator to connect technology with science, mathematics, and engineering (Hacker, 2018; Han et al., 2022; Sung & Kelley, 2022). When considering integrating STEM content from a pedagogical perspective, engineering design can become the situated context and the platform for STEM learning. Using engineering design as a catalyst to STEM learning is vital to bring all four STEM disciplines on an equal platform. The very nature of engineering design provides students with a systematic approach to solving problems that often occur naturally in all the STEM fields. Engineering design provides the opportunity to locate the intersections and build connections among the STEM disciplines, which has been identified as key to subject integration

(Barnett & Hodson, 2001; Daugherty & Carter, 2018; Frykholm & Glasson, 2005; Han et al., 2022; Kelley & Knowles, 2016; Sung & Kelley, 2022).

Engineers' inquiry is comparable with scientific inquiry (Han et al., 2022; Sung & Kelley, 2022). Science education can be enhanced by infusing an engineering design approach because it creates opportunity to apply science knowledge and inquiry as well as provides an authentic context for learning mathematical reasoning for informed decisions during the design process. The analytical element of the engineering design process allows students to use mathematics and science inquiry to create and conduct experiments that will inform the student about the function and performance of potential design solutions before a final prototype is constructed. This approach to engineering design allows students to build upon their own experiences and provide opportunities to construct new science and mathematics knowledge through design analysis and scientific investigation (Han et al., 2022; Kelley & Knowles, 2016; Sung & Kelley, 2022).

Although there is no clear evidence which approach is dominantly excellent or bad, technology and engineering educators need to consider balanced problem- and solution-driven design strategies depending on students' prior knowledge and skills in engineering design (Sung & Kelley, 2022). In design-based learning, students need to seek out information in service of reaching a solution to a problem that has become important to them. Learner engagement fosters deep, probative, just-in-time learning. The discussion shifts now to the affordances of the aforementioned approaches, sub-approaches, perspectives and related teaching strategies for STEM education.

8.5 The Affordances of These Perspectives, Approaches, Sub-approaches and Related Teaching Strategies for the Application of STEM Education

8.5.1 Orientation

It applies generally to all the discussions below that once the students are grouped together in STEM classrooms, the learning environment is framed by the social constructivist perspective and the cooperative teaching sub-approach (refer to Fig. 8.1). It still does not prevent the students from learning entirely individually occasionally while they are supposed to be learning (also individually within themselves) as part of cooperative learning groups (Guzdial, 2018; Phillips, 2000), thus also framing it within the radical and cognitive constructivist perspectives. Furthermore, all of these are contained within the experiential learning sub-approach.

8.5.2 Inquiry-Based Teaching

The problem-solving nature of the technological tasks provides ample opportunities for students to engage in inquiry-based learning (Reddy et al., 2005). Furthermore, Kelley and co-workers focus on the role that engineering inquiry, which is similar to scientific inquiry, can play to bring the STEM subjects on the same platform (Kelley & Knowles, 2016; Kennedy & Odell, 2014; Sung & Kelley, 2022). Thus, the inquiry-based sub-approach to constructivism, which includes the discovery-based approach (refer to Fig. 8.1), is crucial in STEM education.

According to the literature, students enhance design thinking, which is required for a competent STEM workforce in the future, while engaging in inquiry-based, real-world problems (Dare et al., 2018; Han et al., 2022; Jonassen et al., 2006; Kelley & Knowles, 2016; Moore et al., 2014). One should also remember that inquiry-based learning often translates into discovery-based learning as the latter is part of the former. When adopting the engineering design approach in schools, appropriate teaching about the engineering design process must build these practical problem-solving abilities (Sung & Kelley, 2022), and the same applies to the technological process.

8.5.3 Problem-Based and Project-Based Teaching

As problem-based and project-based learning tend to be two-sides of the same coin where the latter is more extensive than the former, the discussion rather focuses on project-based teaching. As stated earlier, project-based learning engages students in complex tasks based on challenging or "driving questions". According to Banks and Barlex (2014) an example of project-based learning in the area of technology and engineering is offered by the Design and Technology Association (DATA). The significant driving question in this case is: What can nature offer architectural design? The issue of significance is sustainability and the project-based learning unit of work is focused on education for sustainable development. There are some suggested "big tasks" (or projects) and some contributory "small tasks" from across the STEM subjects that might be useful in undertaking the big task.

The projects at the centre of this unit of work pose considerable challenges for students. To support the students in meeting these challenges a range of small tasks may be provided to structure opportunities for learning that will be useful in tackling the big tasks. There are three sets of tasks, one set for each of the contributing STEM subjects. For technology (and engineering) education, the small tasks, for example, are concerned with architecture informed by nature; for science reducing heat loss; and for mathematics scale drawings, plans and elevations (Banks & Barlex, 2014). However, projects in technology and engineering education usually entail design activities with an understanding of the impact and influencing factors of proposed designs.

8.5.4 Design-Based Teaching

Kelley and Knowles (2016), and Hallström and Ankiewicz (2019, 2023) assert that both engineering and technological design can respectively be used as an integrator of the four STEM subjects. Hence, it makes design-based teaching essential in STEM education. We have also stated that De Vries (2018, 2021) advocates an approach where there is knowledge development in all STEM subjects simultaneously, and not merely the application of existing knowledge related to them—like how it happens in practice. Therefore, he suggests the use of particular technological design challenges in which engineering principles, scientific concepts, and mathematical ways of thinking are essential for finding solutions to the challenge.

When technology students' intuitive ideas are challenged, they will do experiments to understand unknown or misunderstood scientific phenomenon. They will do calculations to optimise their design, and they will do modelling to test this. This activity will result in a rich design experience, a better understanding of science concepts and new experiences in using mathematics (De Vries, 2018, 2021). This design-based teaching can be guided by the team teaching of a science teacher, technology/engineering teacher and a mathematics teacher. Such an activity does justice to the nature of design as a process in which both new knowledge is developed (about designing itself, but also science and engineering), and existing knowledge (previously learnt in science, technology and/or mathematics) is applied.

De Vries (2018) presents another example in real engineering. Knowledge (i.e. aerodynamics) is developed in the context of designing. The design of airplanes is still today based on tinkering with the prototypes and investigating the effect of systematically changing the design of the flying behaviour of the airplane by putting models in wind tunnels. The process of simultaneously developing knowledge about natural phenomenon and improving the design can easily be simulated in class by using paper airplane designing and testing.

8.6 Conclusion

Table 8.3 provides a simple outline of the aspects and constructs which were discussed in the chapter. The dotted lines between the aspects and constructs indicate that they are related or part of one another. The shaded aspects and constructs are particularly related to the role of the T in integrated STEM education.

Aligned with the purpose of the chapter to provide a sound theoretical framework for the teaching approaches and strategies in STEM education of particular relevance to technology education, it, firstly, described the tenets of the constructivist theory and learning approach. It also emphasised that technology education is prone to constructivism. Secondly, the chapter discussed various perspectives on constructivism, namely *radical, cognitive and social constructivism.* Thirdly, arising from the constructivist theory on learning or approach to learning, it deliberated

Table 8.3 An outline of the aspects and constructs particularly in relation to role of the T in integrated STEM education

Aspects	Constructs		
Learning theories or approaches	Constructivism		
Perspectives on constructivism	Cognitive	Radical	Social
Sub-approaches to constructivism	Experiential		
	Cooperative	Inquiry	Discovery
Teaching approaches or strategies	Problem-based	Project-based	Design-based

on the various sub-approaches for example experiential, cooperative, inquiry and discovery learning. It asserted that the nature of technological activities lends itself to social constructivist perspectives of learning where students engage in practical activities and interact with one another when solving problems and designing in cooperative learning environments. Fourthly, based on constructivism, the chapter reviewed congruent teaching approaches or strategies which are common to technology education and STEM education, namely problem-, project- and design-based teaching. Fifthly, it indicated that when both technological (Hallström and Ankiewicz) and engineering design (Kelley and co-workers) are seen as the integrator of the STEM subjects then inquiry-, problem, project, and specifically design-based teaching strategies are relevant for STEM education.

It can be concluded that, with both technological (Hallström and Ankiewicz) and engineering design (Kelley and co-workers) being the integrator of the STEM subjects, the design-based teaching strategy holds the most important affordances for STEM education. It builds on inquiry-, problem- and project-based teaching strategies which are rooted in experiential, cooperative, inquiry and discovery learning as sub-approaches to constructivist learning and teaching approaches. Furthermore, design-based teaching strategies relate to the social constructivist perspective when students learn cooperatively, and to the cognitive and radical constructivist perspective when students learn individually.

Acknowledgements Parts of the theoretical framework that underpins this chapter have been published elsewhere in a different format and genre, such as appropriate learning theories for technology education (De Swardt et al. 2005); a conceptual framework for learning and teaching in technology education (Reddy et al., 2005), and design as the basis of an integrated STEM philosophy (Hallström & Ankiewicz, 2019, 2023).

References

Abbott, J., & Ryan, T. (2001). Constructing knowledge and shaping brains. *HOW Journal, 9*(1), 9–13.

Allen, D. E., Donham, R. S., & Bernhardt, S. A. (2011). Problem-based learning. *New Directions for Teaching and Learning, 2011*(128), 21–29.

Ammon, S. (2017). Why designing is not experimenting: Design methods, epistemic praxis and strategies of knowledge acquisition in architecture. *Philosophy & Technology, 30*, 495–520. https://doi.org/10.1007/s13347-017-0256-4

Ankiewicz, P. (2003). *Technology education at school: Illusion or reality? Inaugural address, 15 October 2003.* Rand Afrikaans University.

Ankiewicz, P. J. (2013). 'n Teoretiese besinning oor die implikasies van die filosofie van tegnologie vir klaskamerpraktyk/A theoretical reflection on the implications of the philosophy of technology for classroom pedagogy. *Suid-Afrikaanse Tydskrif vir Natuurwetenskap en Tegnologie, 32*(1), Art.#386, 9. https://doi.org/10.4102/satnt.v32i1.386

Ankiewicz, P. J. (2015). The implications of the philosophy of technology for the academic majors of technology student teachers. In M. Chatoney (Ed.), *Plurality and complementarity of approaches in design and technology education. Proceedings of the PATT 29 conference* (pp. 13–25). Marseille, France.

Ankiewicz, P. (2016). The relevance of indigenous technology knowledge systems (ITKS) for the 21st century classroom. In M. J. de Vries, A. Bekker-Holtland, & G. van Dijk (Eds.), *Technology education for 21st century skills. Proceedings of the PATT 32 conference* (pp. 22–34). Utrecht, The Netherlands.

Ankiewicz, P. (2019). Andrew Feenberg: Implications of critical theory for technology education. In J. R. Dakers, J. Hallström, & M. J. de Vries (Eds.), *Reflections on technology for educational practitioners: Philosophers of technology inspiring technology education.* Brill Sense.

Ankiewicz, P. J., De Swardt, A. E., & De Vries, M. (2006). Some implications of the philosophy of technology for science, technology and society (STS) studies. *International Journal of Technology and Design Education, 16*(2), 117–141. https://doi.org/10.1007/s10798-005-3595-x

Assessment of Performance Unit (APU). (1994). Learning through design and technology (pp. 59–67). In F. Banks (Ed.), *Teaching technology.* Routledge.

Aydin, E. (2005). The use of computers in mathematics education: A paradigm shift from "computer assisted instruction" towards "student programming". *The Turkish Online Journal of Educational Technology, 4*(2).

Banks, F., & Barlex, D. (2014). *Teaching STEM in the secondary school: Helping teachers meet the challenge.* Routledge.

Barak, M. (2018). Teaching electronics: From building circuits to systems thinking and programming. In M. J. de Vries (Ed.), *Handbook of Technology Education* (pp. 337–360). Springer.

Barnett, J., & Hodson, D. (2001). Pedagogical context knowledge: Toward a fuller understanding of what good science teachers know. *Science Education, 85*(4), 426–453.

Bednar, A. K., Cunningham, D., Duffy, T. M., & Perry, J. D. (2013). Theory into practice: How do we link? In T. M. Duffy & D. H. Jonassen (Eds.), *Constructivism and the technology of instruction: A conversation*. Routledge.

Best, M. (2018). Problem-based learning in technology education. In M. J. de Vries (Ed.), *Handbook of technology education* (pp. 489–503). Springer.

Biehler, R. F., & Snowman, J. (1997). *Psychology applied to teaching* (8th ed.). Houghton Mifflin.

Bishop, A. J. (1988). Mathematics education in its cultural context. *Educational Studies in Mathematics, 19*(2), 179–192.

Black, A., & Ammon, P. (1992). A developmental-constructivist approach to teacher education. *Journal of Teacher Education, 43*(5), 323–336.

Bowen, R. (2009). Quality teaching in primary school design and technology. *Journal of Design & Technology Education, 1*(1), 12–23.

Brooks, J. G. G., & Brooks, M. G. (1993). *The case for constructivist classroom*. Association for Supervision and Curriculum Development.

Bruner, J. S. (1961). The act of discovery. *Harvard educational review*.

Bruner, J. S. (1966). *Towards a theory of instruction*. Harvard University Press.

Candy, P. (1991). *Self direction for lifelong learning*. Jossey Bass.

Carlson, L. E., & Sullivan, J. F. (1999). Hands-on engineering: Learning by doing in the integrated teaching and learning program. *The International Journal of Engineering Education, 15*(1), 20–31.

Cavalier, J. C., Klein, J. D., & Cavalier, F. F. (1995). Effects of cooperative learning on performance, attitude and group behaviours in a technical team environment. *Educational Technology Research and Development 43*(3), 61–71.

Cheng Y. Y., Liu K. S., & Chang, C. M. (2007). The effect of creative problem solving instruction on elementary schools science lessons. *Journal of Science Education, 15*(5), 565–591. https://doi.org/10.6173/CJSE.2007.1505.04

Cobb, P. (1994). Where is the mind? Constructivist and sociocultural perspectives on mathematical development. *Educational Researcher, 23*(7), 13–20.

Corden, R. (2001). Group discussion and the importance of a shared perspective: Learning from collaborative research. *Qualitative Research, 1*(3), 347–367. https://doi.org/10.1177/146879410100100305

Dare, E. A., Ellis, J. A., & Roehrig, G. H. (2018). Understanding science teachers' implementations of integrated STEM curricular units through a phenomenological multiple case study. *International Journal of STEM Education, 5*(1), 4.

Dart, B. (1994). Measuring constructivist learning environments in tertiary education. In *A paper presented at the annual conference of the Australian association for research in education, Newcastle*.

Dart, B. C., Boulton-Lewis, G. M., Brownlee, J. M., & McCrindle, A. R. (1998). Change in knowledge of learning and teaching through journal writing. *Research Papers in Education, 13*(3), 291–318.

Davis, R. B., Maher, C. A., & Noddings, N. (1990). Suggestions for the improvement of mathematics education. In R. B. Davis (Ed.), *Constructivist views on the teaching and learning of mathematics. Journal for Research in Mathematics Education: Monograph No. 4*. National Council of Teachers of Mathematics, 1906 Association Drive, Reston, VA 22091.

Daugherty, M. K., & Carter, V. (2018). The nature of interdisciplinary STEM education. In M. J. de Vries (Ed.), *Handbook of technology education* (pp. 159–171). Springer.

De Swardt, A. E., Ankiewicz, P. J., & Engelbrecht, W. (2005). Technology education in South Africa since 1998: A shift from traditional teaching to outcomes-based education. In *Proceedings of the PATT 15 conference* (pp. 1−23). Haarlem, The Netherlands. https://www.iteea.org/File.aspx?id=86630&v=4dc66db2

De Vries, M. J. (2018). The T and E in STEM: From promise to practice. In M. J. de Vries, S. Fletcher, S. Kruse, P. Labudee, M. Lang, I. Mammes, C. Max, D. Münk, B. Nicholl, J. Strobel, & M. Winterbottom (Eds.), *Research in technology education* (pp. 11–19). Waxmann.

De Vries, M. J. (2021). Design-based learning in science and technology as integrated STEM. In M. J. de Vries & I. Henze (Eds.), *Design-based concept learning in science and technology education* (pp. 14–24). Brill.
Dewey, J. (1899). *The school and society: Being three lectures*. University of Chicago Press.
Dewey, J. (1963). *The child and the curriculum*. University of Chicago Press.
Doolittle, P. E., & Camp, W. G. (1999). Constructivism: The career and technical education perspective. *Journal of Vocational and Technical Education, 16*(1), 23–46. https://files.eric.ed.gov/fulltext/EJ598590.pdf
Down, B. K. (1996). Problem-solving, CDT and child-centredness. In A. Cross & R. McCormick (Eds.), *Technology in schools* (pp. 228–237). Open University Press.
Duke, D. L. (1990). *Teaching: An introduction*. McGraw Hill Publishing Company.
Eggen, P. D., & Kauchak, D. P. (1996). *Strategies for teachers: Teaching content and thinking skills*. Allyn & Bacon.
English, L. D. (2016). STEM education K-12: Perspectives on integration. *International Journal of STEM Education, 3*(1), 1–8.
Friesen, S., & Scott, D. (2013). *Inquiry-based learning: A review of the research literature*. Paper prepared for the Alberta Ministry of Education.
Frykholm, J., & Glasson, G. (2005). Connecting science and mathematics instruction: Pedagogical context knowledge for teachers. *School Science and Mathematics, 105*(3), 127–141.
Grobler, R., & Ankiewicz, P. (2022). The viability of diverting from a linear to a parallel approach to the development of PCK in technology teacher education. *International Journal of Technology and Design Education, 32*(2), 1001–1021. https://doi.org/10.1007/s10798-020-09644-4
Guzdial, M. (2018). Constructivism versus constructionism. https://computinged.wordpress.com/2018/03/19/constructivism-vs-constructivism-vs-constructionism/
Hacker, M. (2018). Engineering and technology concepts: Key ideas that students should understand. In M. J. de Vries (Ed.), *Handbook of technology education* (pp. 173–191). Springer.
Hallström, J., & Ankiewicz, P. (2019). Laying down the "T" and "E" in STEM education: Design as the basis of an integrated STEM philosophy. In S. Pulé, & M. J. de Vries (Eds.), *Developing a knowledge economy through technology and engineering education, Proceedings of the PATT 37 conference* (pp. 187–194). Msida, Malta.
Hallström, J., & Ankiewicz, P. (2023) Design as the basis for integrated STEM education: A philosophical framework. *Frontiers in Education, 8*, 1078313. https://doi.org/10.3389/feduc.2023.1078313.
Hallström, J., & Schönborn, K. J. (2019). Models and modelling for authentic STEM education: Reinforcing the argument. *International Journal of STEM Education, 6*(1), 1–10.
Hammer, D. (1997). Discovery learning and discovery teaching. *Cognition and Instruction, 15*(4), 485–529.
Han, J., Kelley, T., & Knowles, J. G. (2022). Building a sustainable model of integrated STEM education: Investigating secondary school STEM classes after an integrated STEM project. *International Journal of Technology and Design Education*, 1–25.
Hawari, A. D. M., & Noor, A. I. M. (2020). Project based learning pedagogical design in STEAM Art education. *Asian Journal of University Education, 16*(3), 102–111.
Henak, R. (1992). Enhancing motivation with experiential learning. In G. A. Edminson, & A. E. Schwaller (Eds.), *Delivery systems: Teaching strategies for technology education* (pp. 14–17). International Technology Education Association (ITEA).
Hmelo-Silver, C. E. (2004). Problem-based learning: What and how do students learn? *Educational Psychology Review, 16*(3), 235–266.
International Technology Education Association (ITEA). (1997). *Standards for technology education. First draft*. Virginia Tech.
Jakovljevic, M., & Ankiewicz, P. (2016). Project-based pedagogy for the facilitation of webpage design. *International Journal of Technology and Design Education, 26*(2), 225–242. https://doi.org/10.1007/s10798-015-9312-5

Jenkins, E. W. (2000). Constructivism in school science education: Powerful model or the most dangerous intellectual tendency? *Science & Education, 9*(6), 599–610. https://doi.org/10.1023/A:1008778120803
Johnsey, R. (1995). The design process—Does it exist? *International Journal of Technology and Design Education, 5*(3), 199–217.
Johnson, S. D. (1997). Learning technological concepts and developing intellectual skills. *International Journal of Technology and Design Education, 7*, 161–180.
Johnson, D., & Johnson, F. (1991). *Learning together and learning alone* (3rd ed.). Prentice-Hall.
Johnson, D. W., & Johnson, R. T. (1989). *Cooperation and competition: Theory and research.* Interaction Book Company.
Johnson, S. D., & Thomas, R. (1992). Technology education and the cognitive revolution. *The Technology Teacher, 51*(4), 7–12.
Jonassen, D. H. (1994). Thinking technology: Toward a constructivist design model. *Educational Technology, 34*(4), 34–37. https://www.learntechlib.org/p/171050/
Jonassen, D., Strobel, J., & Lee, C. B. (2006). Everyday problem solving in engineering: Lessons for engineering educators. *Journal of Engineering Education, 95*(2), 139–151.
Jones, A., Buntting, C., & De Vries, M. J. (2013). The developing field of technology education: A review to look forward. *International Journal of Technology and Design Education, 23*(2), 191–212.
Kanuka, H., & Anderson, T. (1999). Using constructivism in technology–mediated learning: Constructing order out of the chaos in the literature. *Radical Pedagogy, 2*(1).
Kelley, T. R., & Knowles, J. G. (2016). A conceptual framework for integrated STEM education. *International Journal of STEM Education, 3*(1), 1–11.
Kennedy, T. J., & Odell, M. R. (2014). Engaging students in STEM education. *Science Education International, 25*(3), 246–258.
Killen, R. (2007). *Teaching strategies for outcomes–based education.* Juta and Company Ltd.
King, A. (1984). Verbal interaction and problem solving within computer assisted cooperative learning groups. *Journal of Educational Computing Research 5*(1), 1–15.
Kolodner, J. L. (2006). Case-based reasoning. In K. L. Sawyer (Ed.), *The Cambridge handbook of learning sciences* (pp. 225–242). Cambridge University Press.
Kolodner, J. L., Crismond, D., Fasse, B., Gray, J., Holbrook, J., & Puntembakar, S. (2003). Putting a student-centered *Learning by Design*™ curriculum into practice: Lessons learned. *Journal of the Learning Sciences, 12*(4), 495–548.
Krajcik, J., Blumenfeld, P. C., Marx, R. W., Bass, K. M., & Fredricks, J. (1998). Inquiry in project-based science classrooms: Initial attempts by middle school students. *The Journal of the Learning Sciences, 7*(3–4), 313–350.
Magadla, L. (1996). Constructivism: A practitioner's perspective. *South African Journal of Higher Education, 10*(1), 83–88.
Major, S., & Govers, E. (2014). Project-based learning in visual arts and design: What makes it work. https://ako.ac.nz/assets/Knowledge-centre/RHPF-c52-Project-based-learning-in-visual-arts/RESEARCH-REPORT-Project-Based-Learning-in-Visual-Arts-and-Design.pdf
McCormick, R. (1997). Conceptual and procedural knowledge. *International Journal of Technology and Design Education, 7*(1), 141–159.
McCormick, R., & Davidson, M. (1996). Problem solving and the tyranny of product outcomes. *The Journal of Design and Technology Education, 1*(3), 230–241.
McCormick, R., Murphy, P., & Hennessy, S. (1994). Problem-solving processes in technology education: A pilot study. *International Journal of Technology and Design Education, 4*(1), 5–34.
Moore, T., Stohlmann, M., Wang, H., Tank, K., Glancy, A., & Roehrig, G. (2014). Implementation and integration of engineering in K-12 STEM education. In S. Purzer, J. Strobel, & M. Cardella (Eds.), *Engineering in pre-college settings: Synthesizing research, policy, and practices* (pp. 35–60). Purdue University Press.

Muller, M., Buchheister, K. E., & Boutte, G. (2017). Multiple perspectives on cognitive development: Radical constructivism, cognitive constructivism, sociocultural theory, and critical theory. *Faculty Publications, Department of Child, Youth, and Family Studies*. 261. https://digitalcommons.unl.edu/famconfacpub/261/?utm_source=digitalcommons.unl.edu%2Ffamconfacpub%2F261&utm_medium=PDF&utm_campaign=PDFCoverPages
Mwamwenda, T. S. (1995). *Educational psychology: An African perspective*. Butterworth.
Nadelson, L., & Seifert, A. (2017). Integrated STEM defined: Contexts, challenges, and the future. *Journal of Educational Research, 110*(3), 221–223.
Nguyen, T. P. L., Nguyen, T. H., & Tran, T. K. (2020). STEM Education in secondary schools: Teachers' perspective towards sustainable development. *Sustainability, 12*(21), 8865. https://doi.org/10.3390/su12218865
Nkadimeng, M., & Ankiewicz, P. (2022). The affordances of Minecraft Education as a game-based learning tool for atomic structure in junior high school science education. *Journal of Science Education and Technology, 31*(5), 605–620. https://doi.org/10.1007/s10956-022-09981-0
Nystrand, M. (2006). Research on the role of classroom discourse as it affects reading comprehension. *Research in the Teaching of English, 40*(4), 392–412. https://www.jstor.org/stable/40171709
Orlich, D. C., Harder, R. J., Callahan, R. C., Kauchak, D. P., & Gibson, H. W. (1994). *Teaching strategies. A guide to better instruction*. D.C. Heath & Company.
Phillips, D. C. (2000). *Constructivism in education: Opinions and second opinions on controversial issues. Ninety-ninth yearbook of the national society for the study of education*. University of Chicago Press.
Reddy, V., Ankiewicz, P., & De Swardt, E., (2005). Learning theories: A conceptual framework for learning and instruction in technology education. *South African Journal of Higher Education, 19*(3), 423–443. https://hdl.handle.net/10520/EJC37155
Reznitskaya, A., Anderson, R. C., & Kuo, L. (2007). Teaching and learning argumentation. *The Elementary School Journal, 107*, 449–472. https://doi.org/10.1086/518623
Rieber, L. P. (2004). Microworlds. In D. H. Jonassen (Ed.), *Handbook of research on educational communications and technology* (pp. 583–603). Lawrence Erlbaum Associates Publishers.
Ritz, J. M., & Fan, S. C. (2015). STEM and technology education: International state-of-the-art. *International Journal of Technology and Design Education, 25*(4), 429–451.
Rogoff, B., & Lave, J. (1984). *Everyday cognition: Its development in social context*. Harvard University Press.
Sanders, M. (2009). STEM, STEM education, STEMmania. *The Technology Teacher, 68*(4), 20–26.
Schwaller, A. E. (1995). Instructional strategies for technology education. In G. E. Martin (Ed.), *Foundations of technology education. 44th yearbook of the council on technology teacher education* (pp. 29–43). Glencoe/McGraw Hill.
Schwartz, P. (2013). *Problem-based learning*. Routledge.
Slavin, R. E. (1980). Cooperative learning. *Review of Educational Research, 50*(2), 315–342. https://doi.org/10.3102/00346543050002315
Stage, F. K., Muller, P. A., Kinzie, J., & Simmonds, A. (1998). Creating learning centred classrooms. What does learning theory have to say? *Ashe-Eric Higher Education Report, 26*(4), 98–104.
Sung, E., & Kelley, T. R. (2022). Using engineering design in technology education. In P. J. Williams, & B. Von Mengersen (Eds.), *Applications of research in technology education. Contemporary issues in technology education*. Springer. https://doi.org/10.1007/978-981-16-7885-1_9
Svärd, J., Schönborn, K., & Hallström, J. (2017). Design of an authentic innovation project in Swedish upper secondary technology education. *Australasian Journal of Technology Education, 4*, 1–15. https://doi.org/10.15663/ajte.v4i1.48
Taylor, N. (1999). Operationalising Curriculum 2005. *Proceedings of the Western Cape Education Department's Making OBE work? Conference, 13–15 December 1999,* (pp. 98–107). Cape Town, South Africa.

Vermaas, P., Kroes, P., Van de Poel, I., Franssen, M., & Houkes, W. (2011). *A philosophy of technology: From technical artefacts to sociotechnical systems*. Morgan & Claypool Publishers.

Vossen, T. E., Henze, I., De Vries, M. J., & Van Driel, J. H. (2019). Finding the connection between research and design: The knowledge development of STEM teachers in a professional learning community. *International Journal of Technology and Design Education*, 1–26.

Wheatley, G. H. (1991). Constructivist perspectives on science and mathematics learning. *Science Education, 75*(1), 9–21.

Williams, P. J. (2011). STEM education: Proceed with caution. *Design and Technology Education: An International Journal, 16*(1), 26–35.

Williams, P. J., Iglesias, J., & Barak, M. (2008). Problem based learning: Application to technology education in three countries. *International Journal of Technology and Design Education, 18*, 319–335.

Wood, D. (2002). Problem based learning. In C. Cantillon, L. Hutchinson, & D. Wood (Eds.), *ABC of teaching and learning medicine*. BMJ

Zietsman, A. (1996). Constructivism: Super theory for all educational ills? *South African Journal of Higher Education, 10*(1), 70–75.

Piet Ankiewicz is Emeritus Professor of Technology Education in the Faculty of Education at the University of Johannesburg. He considers himself as a theoretical educationalist who specializes in studying and developing theoretical models and philosophical frameworks to understand and explain the tenets of STEM—science, technology, engineering and mathematics—education. His research interests include the affordances of the philosophy of technology for technology classroom pedagogy, teacher education, indigenous technology knowledge systems, and STEM education. He also has an interest in students' attitudes towards technology. He has been rated by the National Research Foundation as an established researcher with international recognition. He is a member of the Editorial Board of the *International Journal of Technology and Design Education*.

Chapter 9
Technology and Engineering in STEM Education

Michael Hacker

Abstract What is deemed to be important for people to learn varies over time and evolves in relation to societal waves of change. Due to the essential roles technology and engineering (T&E) play in addressing global and environmental challenges, support for school-based T&E programs has markedly increased over the last decade. In addition to the workforce and economic imperatives, T&E are recognized contributors to sustainable development and transformative improvement in quality of life. This chapter examines commonalities and differences between technology and engineering as disciplinary domains and practices. As T&E alter and influence virtually every aspect of life, all students should learn about their impacts and affordances as part of their fundamental education. Because complex problems facing people globally require interdisciplinary solutions, STEM education has grown in popularity as an educational paradigm that draws upon knowledge from different domains as needed. Within STEM education, T&E experiences can be uniquely valuable for students—not only in providing an effective way to integrate and contextualize learning but also in mobilizing engineering and design thinking as a way for people to approach problems of all kinds. As there are striking similarities between the requirements for undergraduate engineering majors and undergraduate technology education majors, a case is made that with focused preservice or in-service professional development, T&E teachers can be well-qualified to teach engineering-oriented curricula in school-based programs. Attention is paid to curricular and instructional starting points (e.g., content standards, overarching thematic ideas, and design challenges in authentic learning contexts).

Keywords Design pedagogy · Technology and engineering education · STEM education · Content standards · Thematic instruction

M. Hacker (✉)
Hofstra University Center for STEM Research, Hempstead, NY, USA
e-mail: Michael.hacker@hofstra.edu

W. Fox-Turnbull and P. J. Williams (eds.), *Locating Technology Education in STEM Teaching and Learning*, Contemporary Issues in Technology Education,
https://doi.org/10.1007/978-981-97-1995-2_9

9.1 Introduction

The need to educate all people to confront technological problems and issues has been made all the more daunting by the exponential rate of technological change and the impact of globalization on our culture and economy. Education plays the most fundamental role in society in preparing people to reap benefits from technological knowledge, to promote technologies that support well-being, and impose limits on malevolent applications.

9.2 Defining Technology

Defining technology is akin to defining "art" or "religion". The term is so broad and subsumes so many concepts and ideas that a single definition cannot accommodate the nuance needed to impart understanding. Drucker (1959) cautions, "All human disciplines rest on arbitrary distinctions, and technologists ought to be conscious of the artificiality of their definition lest it become a barrier, rather than a help to knowledge and understanding".

Schatzberg (2006) writes that the present meaning of *technology* evolved in the early twentieth century. Earlier, its meaning derived from German discourse about Technik and referred to methods and tools used to create the material culture. According to the U.S. National Assessment Governing Board, technology is "any modification of the natural world done to fulfil human needs or desires" (NAGB, 2018). De Vries (2016) describes technology as "the human activity that transforms the natural environment to make it fit better with human needs". Dictionary definitions reflect the commonly held but inadequate view that technology is applied science, i.e., "the application of scientific knowledge to the practical aims of human life and to the change and manipulation of the human environment" (Britannica, 2021); or "the study and knowledge of the practical, especially industrial, use of scientific discoveries" (Cambridge, 2022) .

In the authoritative National Academy of Engineering report *Technically Speaking*, the definition of technology is more inclusive: "the process by which humans modify nature to meet their needs and wants, including all the infrastructure necessary for the design, manufacture, operation, and repair of technological artefacts, from corporate headquarters and engineering schools to manufacturing plants and maintenance facilities" (NRC, 2002).

9.2.1 Is Engineering a Subset of Technology?

Using NAE's broad definition, one might conclude that engineering is a subset of technology. The American Association for the Advancement of Science refers to engineering as "the *component of technology* most closely allied to scientific inquiry and to mathematical modelling" (AAAS, 1993). However, Swyt (1989) argues (as discussed later in this chapter) that because technology and engineering have different goals, they are fundamentally different.

Some would maintain that technology is a subset of science. In the U.S., the National *Science* Foundation funds engineering and technology programs; the *Science* Advisor to the President advises on science *and* technology matters; and the U.S. Library of Congress lists *Technology* as a Division in its Science Reference Guide (LOC.gov, 2017). But in the nineteenth century, technology was seen as a subset of *anthropology* because technology includes artifacts of the material culture (Schatzberg, 2018).

9.2.2 Technology Referents

A Google search shows 12,770,000,000 results for "What is Technology". In common parlance, *technology* is a word used to mean artifacts (e.g., medications, chairs), tools (e.g., a computer), technical means/processes (e.g., cooking), or systems (e.g., transportation). As shown in Table 9.1, the term has numerous referents, explaining in part, its ambiguity.

9.2.3 Technological Practice

For centuries, technological activity was a craft engaged in by artisans with little scientific education but a great deal of mechanical genius (Kranzberg & Pursell, 1967, p. 11). What was considered to be "technology" was mainly craft-based invention; being an inventor characterized the nineteenth century technologist (e.g., Samuel Colt, Samuel Morse, Werner von Siemens, Eli Whitney). Von Siemens recollected that he "created, from cigar boxes, tin, a few pieces of iron and some insulated copper wire, a few automated telegraphs that came together and functioned" (Siemens, 2022). In 1939, Charles Goodyear discovered a way to cure rubber while working on his kitchen stove.

Work was generally done by individuals with assistants or small teams. Edison worked in Menlo Park with a team of 14 "muckers" (as he called them). Edison said, "I find out what the world needs, then I go ahead and try to invent it. To invent, you need a good imagination and a pile of junk" (Edison Innovation Foundation, 2020). Edison's work could hardly be called science-based. He tried out over a thousand different materials in his search for a long-lasting filament wire for the incandescent bulb. Even sophisticated inventions like the aeroplane initially benefited little from purely theoretical scientific work. The Wright brothers built a wind tunnel and tested over fifty different models to determine how lift and drag are affected by various design parameters (NASA, 2021).

Table 9.1 Examples of referents found in a searching for "Technology"

As artifacts tools, and specific designs	As techniques for achieving goals of any sort	As technical means to expand human capability	As knowledge about why and how things are designed, work, and are made	As large systems comprising both technical and non-technical subsystems	As a disciplinary domain to be studied to inform understanding and decision making
Smartphones	The technology of negotiation	Fuel-saving technology	Control systems technology	Sociotechnological systems	Society for the history of technology
Bridges	The technology of nonviolence	mRNA technology	HVAC technology	Smart cities as large technological systems	*On technology*, essays by Peter Drucker (2020)
Scissors	The technology of social change	Emerging technologies	Materials handling technology	Global communication technology	*Technology and Western Civilization*, by Kranzberg and Pursell (1967)
Vaccines	The technology of teaching	Artificial Intelligence	Resistance welding technology	Manufacturing technology	National Institute of Standards and Technology (NIST)

Mathematics, particularly informal knowledge passed from craftsperson to craftsperson, has underpinned technological practice since times immemorial. From the geometric forms found in early weaving to the use of ratios and proportions in mixing materials, mathematics was fundamental. Ancient Egyptian instructions on how to produce the hardest kind of bronze showed an optimum ratio of 88% of copper to 12% of tin (Malina, 1983).

Although innovations do occur in small companies and garages, they are now more often realized by large, well-funded interdisciplinary teams of scientists, engineers, and technicians in research laboratories aiming to solve particular problems. In the 1960s, Bell Labs employed about 50,000 people. Today, in the U.S., over 13,000 people work at Los Alamos national laboratory working on projects in nanotechnology, supercomputing, space exploration, and nuclear fusion.

9.2.4 *Changes in Technology: Due to Connections and the Convergence of Systems*

A key characteristic of technological change is how systems converge to create innovation and exponential growth. In the landmark BBC television documentary series and companion book "Connections," James Burke maintained that the increase in connections between events over time is what has caused the rate of innovation to accelerate (Burke, 1978). When we consider connections between the Wright Brothers' *Flyer* that flew for 59 s and today's wide-bodied intercontinental aircraft, we see the convergence of mathematics (the Kutta-Joukowski theorem relating to calculation of lifti on an airfoil); physics (aerodynamic theories); and numerous prior technological developments (wing and fuselage construction techniques, retractable takeoff and landing gears, wind tunnel technology, and jet engine development) (WIPO, 2015).

Mass production is a confluence of technologies and systems that long pre-existed Ford's Model T assembly line: Eli Whitney's use of interchangeable parts in assembling muskets in 1798; assembly lines with the division of labor used in Cincinnati's slaughterhouses in the 1860s (Gordon, 1990); and the system of conveyor belts and chutes to bring items from separate departments together designed by Otto Doering for Sears and Roebuck at the turn of the twentieth century (Deutsch, 2012).

Artificial intelligence is made possible by the convergence of advances in software, semiconductor design, computer processing, and data analysis. The astounding little smartphone draws upon the convergence of myriad complex technological systems including ICs, receivers/transmitters, batteries, cameras, software, touch displays, sensors, and loudspeaker technology. As cumulative knowledge grows and enables more systems to converge, the rate of change will continue to rise exponentially and new and more powerful technologies will be available for innovators to draw upon, inexorably leading to new capabilities beyond the scope of our present imagination (think about ChatGPT and AI-powered video generation tools like OpenAI's Sora).

9.3 Defining Engineering

Engineering is classically defined as the profession in which knowledge of the mathematical and natural sciences gained by study, experience, and practice is applied with judgment to develop ways to utilize, economically, the materials and forces of nature for the benefit of mankind (ECPD, 1979 as cited in NRC, 1986, p. 74). The U.S. Department of Labor describes engineers as those who "apply the theories and principles of science and mathematics to research and develop economical solutions to technical problems" (LOC.gov, 2017). The definition advanced by ABET (formerly the Accreditation Board for Engineering and Technology) in the U.S. is that "engineering is the creative application of scientific principles to design or develop structures, machines, apparatus, or manufacturing processes, or works utilizing them singly or in combination" (ABET, 2019).

Erich Bloch, the former Director of the National Science Foundation (NSF), wrote that "engineering is the process of investigating how to solve problems that leads to a body of knowledge consisting of concepts, methods, databases, and physical expressions of results" (Bloch, 1986). De Vries distinguishes engineering from technology when professionals called "engineers" do human activity (de Vries, 2016).

9.3.1 Engineering Practice

Engineering grew from craft roots and emerged as a profession during the 1300 s when military engineers used mathematics to design roads, bridges, canals, and fortifications. Then, engineers were those who designed and made military engines (devices such as catapults). Leonardo Da Vinci (1452–1519), in applying for a job, wrote that:

> In case of a siege, I know how to dry up the water of the moats and how to construct an infinite number of bridges, covered ways, scaling ladders, and other machines for this type of enterprise. (Sniderman, 2012)

There are many early examples of systems that were engineered, notable among them Roman aqueducts (fourth century BCE) and multimasted sailing ships (fifteenth century). In the middle ages (fifth to fifteenth centuries), harnessing of wind and water power fueled engineered systems like water mills and contributed to a revolution in agriculture. Programmable machines (a music sequencer, a drum machine, and an astronomical clock) emerged during the middle ages. During the Renaissance, the steam engine (in 1712) gave rise to the industrial revolution in Europe allowing for the beginnings of mass production (Wikipedia contributors, 2022).

The term *civil engineering*, the first engineering discipline to be named, was first used in the eighteenth century to differentiate between the design and construction of military engines and the design and construction of civilian structures.

9.3.2 Changes in Engineering: Due to Design, Science, Computation, and Mathematical Modeling

Mathematics use increased with time. The great cathedrals of the Middle Ages relied upon stonemasons with practical mathematics knowledge to produce sophisticated geometric designs in walls and ceilings. In the mid-seventeenth century, manuscripts with mathematical analyses of structures were published (Heyman, 2014), and calculus was applied to engineering problems for the first time (Hansson, 2020). During this period, empirical formulas and rules of thumb were replaced by design and mathematical modeling (Watson, 2021). Contemporary cutting-edge engineered systems (autonomous vehicles, aviation/space exploration, pharmaceuticals, quantum computing) are deeply rooted in science and are highly dependent upon computation and mathematical modeling.

Specialized branches of engineering evolved, and along with them, came training programs that included mathematics and science instruction. Major branches are bio/biomedical, chemical, civil, computer, electrical, industrial, and mechanical engineering, and are divided into numerous subdisciplines including aerospace, ceramic, environmental, mining, nuclear, petroleum, software, and textile engineering. ABET has identified ~ 40 engineering curricular areas (ABET, 2020).

Today, the world relies upon engineers to address the most significant problems of the day. The US National Academy of Engineering (NAE) has identified 14 *Grand Challenges for Engineering* "that have the potential to help people and the planet thrive" (NAE, 2017). These include such ambitious endeavors as providing ubiquitous access to clean water, reengineering better medicines, and advancing health informatics.

9.4 Differences Between Technology and Engineering

For many years, the accepted goal of research in *science* has been to build knowledge and understanding regardless of potential applications (Understanding Science, 2022). Scientists, until the end of the nineteenth century, with rare exceptions, concerned themselves little with the application of new scientific knowledge (Kranzberg & Pursell, 1967, p. 15). Neither Clerk Maxwell nor Heinrich Hertz were concerned primarily about the utility of their work—which ultimately led to Marconi's invention of the wireless radio (Flexner, 1939).

Dennis Swyt, when Deputy Director of the Center for Manufacturing Engineering at the National Institute of Science and Technology (NIST), made a similar case for technological research having non-specific applicability. He maintained that although related, technology and engineering as human endeavors are fundamentally different because they have different goals. He stated:

> **Technology** is the professional endeavor directed at the creation of new ways to manipulate and control the material world to satisfy unspecified human needs as an end in itself without immediate use or practicality being limiting considerations. This goal is apparent in Robert Goddard's work to develop rockets....for the sheer technological capability to fly in space.
>
> **Engineering** is the professional endeavor directed at the creation of new combinations of proven technologies to satisfy a user-specified need within user-specified constraints. The goal of engineering research is to find ways to optimally implement a practical device, process or system for a useful purpose. Frederick Taylor's mechanical engineering work was to improve efficiency by redesigning the shape and capacity of shovels used in a factory: not to come up with a shovel. (Swyt, 1989)

Compared to engineering, technological research (in the past) has been more open-ended, less goal-directed, less bounded by constraints, and may or may not have relied upon math or science knowledge. Engineering design is highly reliant upon the use of computation, science, and mathematics (simulation, optimization, control theory, and statistical analysis) for solutions that satisfy measurable performance criteria, are bounded by externally imposed constraints (cost, time, environmental impact, durability, disposability, etc.), and require making tradeoffs to reach optimized solutions. Burghardt writes:

> Engineers create design solutions that are intentional. They use knowledge of science to understand what is happening physically, use mathematics to create models that may be analyzed, and use prior technological solutions so they can innovate. The use of modeling, with its inherent predictive analysis, is a significant difference between engineering and technology. (Burghardt, 2013)

Whereas technological research historically was oriented toward invention, engineering research is oriented toward solving particular problems under constraints and reflects innovation more than invention. However, because new technologies have become enmeshed with and reliant upon science and mathematics, the distinction to be made between outcomes of technological research and those of engineering research becomes more and more an academic exercise.

9.5 Why Study Technology and Engineering?

Education in and about technology and engineering empowers the achievement of personal goals like economic self-sufficiency, prepares people to actively contribute to a participatory democracy through informed decision-making, and instils a sense of competence and proficiency when confronted with technologically related problems and decisions. Because technology and engineering alter and influence virtually every aspect of our lives, there are compelling reasons why all students should learn about them as a component of fundamental education.

9.5.1 To Increase Prosperity, Productivity, and Beneficent Societal Growth

For a nation to remain safe, prosperous, and globally competitive its human capital must be able to make use of technological affordances, and engineer innovations in areas of environmental preservation, national security, infrastructure, industrial productivity, health care, and education. If society is to progress in ways that enable individual fulfillment and humanitarian communal growth, its populace must be educated to be able to contribute to and benefit from technological change.

9.5.2 *To Understand How People, the Environment, History, and Culture Have Been Affected by and Can Affect Technological Change*

All corners of culture (industry, the military, politics, international relations, war, finance, the environment, education, and security) have been deeply impacted by developments in technology and engineering—and cultural demands and norms promote or constrain those developments. There is a bilateral relationship between technology and society. Social media have fueled movements for social change, but limits have been placed upon its use based upon perceived risks of incitement to violence. Military ambitions have spawned astonishingly powerful weapons, but societies' recognition of the threat of annihilation after World War II has at least until now, constrained the use of nuclear weapons to serve as a deterrent.

9.5.3 *To Contribute to and Thereby Enable Participatory Democracy to Function*

Along with the emergence of powerful technologies and complex engineered systems, comes responsibility for capably and wisely consuming their outputs and governing their use. An educated citizenry must be able to knowledgeably assess, question, and critique technological development, to question misinformation and propaganda; and to mitigate undesirable consequences. As Michael Young stated in The *Rise of the Meritocracy* (1959), "*Authority cannot be humbled unless ordinary people.....assert themselves against the mighty*".

9.5.4 *To Be Exposed to Productive Personal Career Opportunities*

Education, technology, and engineering are complementary paths to prosperity. For students, as future members of the workforce, learning about technology and engineering in school opens up a world of fascinating career possibilities in diverse areas of endeavor. Those attaining higher levels of T&E education will doubtless find gainful, productive, and interesting careers.

9.5.4.1 STEM

STEM is an acronym for *Science, Technology, Engineering and Mathematics.* The term was first suggested by Judith Ramaley in 2001 as Director of the National Science Foundation's Education and Human Resources division. Prior to then, the term used by the NSF was "SMET," but Ramaley thought it sounded too much like "smut," and changed the acronym! She said:

> In STEM, science and math are critical to a basic understanding of the universe., while engineering and technology are means for people to interact with the universe (Christenson, 2011)

Outside of education, the universality of the acronym must be questioned as STEM fields include an encyclopedic list of divergent disciplines engaged in research, discovery, development, and social welfare. The NSF lists 11 STEM categories: Chemistry, Computer and Information Science and Engineering, Engineering, Geosciences, Life Sciences, Materials Research, Mathematical Sciences, Physics and Astronomy, Psychology, Social Sciences, and STEM Education and Learning Research. These are broken down further into 190 specific fields like astronomy, bioengineering, chemical catalysis, cognitive psychology, and linguistics, physiology, polymers, quantum information science, robotics, and tectonics (NSF, 2021).

So, of what use is the acronym? For funding purposes, "STEM" becomes delimiting and determinative. However, in primary and secondary education, the acronym serves a very different purpose—as an aspirational vision for problem-based, engaging, integrative teaching and learning.

9.6 STEM Education

The paradigm is most useful in pre-college education, where it serves to enable students to apply just-in-time knowledge to the solution of challenging, interdisciplinary problems not bounded by the limits of traditional disciplinary study. Complex global problems such as food insecurity, lack of clean water availability, cybersecurity, and climate change require interdisciplinary solutions. STEM education is an international movement that supports a transition from four component subjects in individual "silos" to a more interconnected way of teaching and learning.

Three goals for secondary STEM education are espoused by the U.S. National Research Council: (1) to increase advanced training and careers in STEM fields; (2) to expand the STEM-capable workforce; and (3) to increase scientific literacy among the general public (NRC, 2011). There are several ways in which STEM education is represented in schools. Different models are emerging; they differ with respect to how well the four component disciplines are integrated (see Fig. 9.1).

In **model A**, the school department/program is referred to as STEM, but the four disciplines are still in their silos. Math is King, Science is Queen, and T and/or E are taught, but are not part of the royalty.

In **model B**, STEM programs include only science and mathematics, and they are taught as independent subjects.

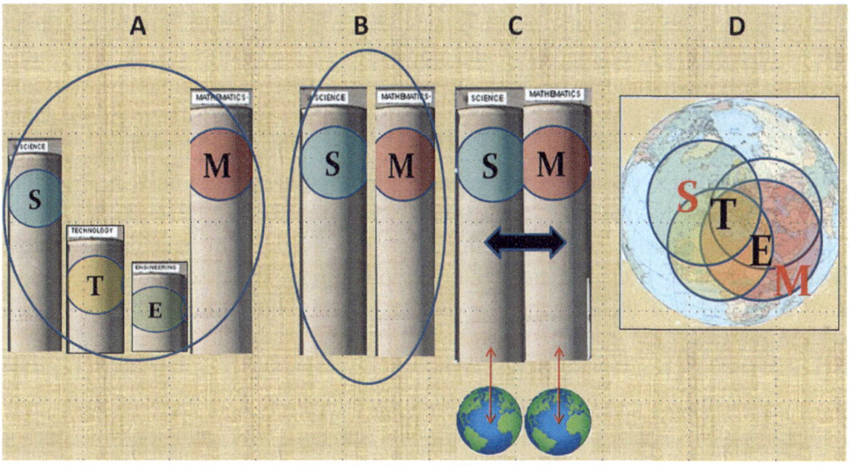

Fig. 9.1 Models of STEM education

In **model C**, science and mathematics are related in a STEM department or program, with connections between them and some attempts made at connecting instruction to real-world contexts.

In **model D,** collaboration and true interplay are reflected between and among the disciplines with T & E providing connections to contextual learning in real-world settings. What is notable in this model is how engineering design thinking draws upon disciplinary knowledge from each of the STEM domains and beyond (the arts, language, social science) as needed to inform solutions to authentic problems. Thus, design pedagogy serves as an integrating mechanism.

9.7 The Role of Technology and Engineering in STEM Education

Technology and engineering serve to establish connections among the disciplines and provide context. In this chapter, in addition to referring to technology and engineering as component STEM disciplines, T&E is used as a single term to describe school programs where technology and engineering are not instructionally separated.

Design problems are engaging contexts within which T&E students investigate, propose, model, test, modify, present, and critique alternative solutions. Curriculum materials prompt students to bring to bear knowledge and skills from other STEM disciplines and beyond *as needed*, to research, understand, realize, and analyze the relative performance of these alternatives.

T&E education becomes even more appealing to many students when instruction focuses on designing solutions to problems related to human needs—potable water, sanitation, and food scarcity. In school, T&E can leverage students' interests in social equity and their desire to make a difference in the world; and build awareness about how engineering addresses global and community-based concerns. A reasonable expectation is that students develop dispositions to forge a sustainable future and learn that further T&E study is a route to engage in rewarding, socially significant work.

T&E education is pedagogically valuable for *all* students, not only in providing an effective way to integrate, contextualize, and reinforce STEM skills, but by facilitating engineering and design thinking—associated with balancing multiple constraints, making tradeoffs, and creatively pursuing optimal solutions. By mobilizing this model of thinking, young people will be better prepared to approach problems of all kinds.

In the US, T&E teachers traditionally have been trained in university preservice programs leading to degrees in technology education. The focus of their study is not to prepare teachers to teach vocational subjects. The philosophy is one of general education where prospective teachers learn about technological resources, processes, and systems and their impacts on people, society, and the environment, and use engineering design as an *iterative process* through which potential solutions to meaningful problems are developed, modified, implemented, and evaluated.

9.8 What to Teach: T&E Content is Based on Standards and Thematic Ideas

T&E curricular content is mainly derived from content standards and crosscutting thematic ideas.

9.8.1 Standards

The development of content standards gained momentum in the late 1980s and were developed by virtually all school disciplines. Standards (and national curricula in some countries) set attainment targets for students. Developed by teams of experienced professionals in a discipline, they describe what students should know and be able to do at various stages of their primary and secondary education. However, standards have been criticized as vague, repetitive, and poorly coordinated. A caveat is that too many performance expectations lead to curriculum overload (NRC, 2008).

9.8.2 *Thematic Ideas*

When students encounter important ideas in multiple contexts, learning transfer is promoted and deeper, more holistic understanding is facilitated (Bransford et al. 2000; de Vries, 2016). Along with the development of content standards, in STEM disciplines there has also been a move toward identifying crosscutting thematic ideas.

In the *Next Generation Science Standards* (NGSS, 2013), thematic ideas are referred to as *Practices* and *Crosscutting Concepts. Standards for Technological and Engineering Literacy (STEL)* (ITEEA, 2020) has also identified *Practices* such as *systems thinking* and *attention to ethics*; as well as a set of *Core Concepts of Technology and Engineering* that "should be integrated into T&E classes at every opportunity" (ITEEA, 2020, p. 28).

Several research studies (Custer et al. 2009; ECCP, 1971; Hacker & Barak, 2017; NRC, 2010) established a consensus of expert opinion about the most important T&E competencies students should attain. A noteworthy international Delphi study was conducted in 2010 by the Delft University of Technology in collaboration with Hofstra University. Five overarching themes emerged (see Table 9.2). The study recommended their use "to rethink T&E frameworks and teaching methods and bring more conceptual coherence to the extensive list of Standards that was developed" (Rossouw et al. 2010, p.15).

9.9 How to Teach: Use of Design Pedagogy

Design thinking encourages different perspectives and approaches in viewing and solving problems. Design is the core problem-solving process used by engineers. In NRC's publication *Engineering in K-12 Education* (Katehi et al. 2009), in 574 pages "design" is mentioned 2672 times. T&E programs situate learning in the context of engineering design problems. As a pedagogical strategy, design activities:

- Engage children as active participants, giving them greater control over the learning process;
- Assist students in integrating learning from the humanities, the arts, mathematics, and science;

Table 9.2 CCETE overarching themes and sub-concepts

Themes	Sub-concepts
Design	Optimization and tradeoffs; criteria and constraints; iteration
Modeling	Representational, explanatory, predictive
Systems	Systems/subsystems; input-process-output; feedback and control
Resources	Materials, energy, information, time, tools, humans, and capital
Human values	Sustainability; technological assessment; creativity/innovation; ethical decisions

- Encourage pluralistic thinking, avoiding a right/wrong dichotomy and suggesting instead that multiple solutions are possible;
- Provide children opportunities to reflect on, revise, and extend mental models of the world;
- Encourage children to put themselves in the minds of others as they think about how their designs will be understood and used (Resnick, 1998).

Design pedagogy portrays setbacks and initial failures as sources of learning and growth and as a means to improve performance; thus, it can foster a growth mindset in learners where initial setbacks and challenges motivate persistence. Both NGSS and STEL strongly emphasize engineering design, attesting to educational policymakers' affirmation of the effectiveness of design pedagogy.

9.9.1 Studies of Design-Based Instruction

Design pedagogy has been shown to foster high student engagement and significant gains in STEM content knowledge (Burghardt et al. 2010; Haile, 1999; Hecht, 2018; Hecht et al. 2009; Satchwell & Loepp, 2002;). Studies have shown that design-based instruction is especially effective in supporting underrepresented groups such as Black, Hispanic, and special needs students (Purzer et al. 2014). In one large-scale NSF-funded project, researchers indicated that after learning to use design pedagogy, "teachers have been involved in a fundamental redefinition of what it means to teach" (Daly-Lewis, 1998).

9.10 Can Technology Teachers Teach Engineering?

There are striking similarities between what ABET requires of engineering majors (ABET, 2019) and what the National Council for Accreditation of Teacher Education (NCATE, 2003) requires of Technology Education majors (see Table 9.3).

At the undergraduate level, both engineers and technology teachers are well prepared in areas of design and problem-solving, team functioning, ethical and professional responsibility, communication skills, social and cultural impacts, and professional growth. A clear difference is how engineers are rigorously prepared in mathematics, the sciences, and computer science. With focused pre- or in-service training, technology teachers can be well-qualified to teach pre-college engineering programs. Increasingly, computer science is becoming more of a topic of study for T&E teachers (NSF, 2021; NSF, 2023).

Table 9.3 Comparison of professional competencies for engineers and technology teachers

Competency	Engineers (ABET, 2019)	Technology teachers NCATE, 2003)
Professional practice	• Prepares graduates to enter the professional practice of engineering	• Prepares graduates to design, create, and manage T&E learning environments
Design and problem solving	• Apply engineering design to produce solutions that meet specified needs considering health, safety, and welfare; and global, cultural, social, environmental, and economic factors by applying principles of engineering, science, and mathematics	• Develop an understanding of design and select design problems, including establishing criteria and constraints • Apply the processes of troubleshooting, research and development, invention and innovation, and experimentation to develop a solution to a design problem
Team functioning	• Function effectively on a team that creates a collaborative and inclusive environment, establishes goals, plans tasks and meets objectives	• Manage technological activities in both individual and group settings
Ethical and professional responsibility	• Recognize ethical and professional responsibilities in engineering situations	• Apply multicultural and global perspectives, values, and ethics as they relate to the study of technology
Communication skills	• Communicate effectively with a range of audiences	• Promote and articulate technology education to public audiences
Socio-cultural impacts	• Consider the impact of engineering in global, economic, environmental, and societal contexts	• Investigate the impacts of products and systems on people, society, and the environment
Content knowledge	• College-level mathematics and science • Engineering, computer science, and engineering design • A culminating design engineering experience reflects multiple constraints	• Learn about the designed world • Analyze medical, bio, energy and power, information and communication, transportation, manufacturing, and construction technologies

9.11 Conclusion

There are several questions that this chapter attempts to raise for readers to consider. How would you critique Swyt's distinction between technology and engineering: that they are fundamentally different endeavors because they have different goals? That is, technology is directed toward creating new ways of doing things as an end in itself, and engineering is directed at creating new combinations of proven technologies to satisfy a user-specified need within user-specified constraints. Having read this chapter pleases consider the following discussion questions:

- Do you agree that technology has been more about invention while engineering is more about using scientific knowledge and mathematical modeling to create innovation?
- Do you believe that engineering is a subset of technology?
- What model of STEM education characterizes practice in the schools that you are familiar with? If you resonate with the vision of STEM education as a model for disciplinary integration, how can it be advanced within your locales and spheres of influence?
- How would you see using standards and overarching themes to guide and focus instruction without straitjacketing teachers?
- What would be your starting points for curriculum and instruction?
- Based on the similarity of content preparation of undergraduate engineers and undergraduate preservice T&E teachers, what type of professional development do you believe is appropriate for technology teachers to enable them to become expert technology & and engineering teachers?

References

ABET. (2019). *Criteria for accrediting engineering programs.* ABET. Baltimore, MD. Retrieved January 13, 2022 from https://www.abet.org/wp-content/uploads/2020/09/EAC-Criteria-2020-2021.pdf

ABET. (2020). *ABET at a glance. 2019 ABET impact report.* See https://www.abet.org/wp-content/uploads/2021/04/2020-ABET-Accreditation-Statistics.pdf

American Association for the Advancement of Science (AAAS). (1993). *Project 2061: benchmarks for science literacy.* Oxford University Press. Retrieved January 27, 2022 from Chap. 3: The Nature of Technology (project2061.org)

Bloch, E. (1986). *Science and engineering: A continuum. In the new engineering research centers: Purposes, goals, and expectations* (p. 28). National Research Council. National Academy Press.

Bransford, J. D., Brown, A. L., & Cocking, R. R. (2000). *How people learn: Brain, mind, experience, and school.* National Academy Press.

Britannica, T., Editors of Encyclopaedia. (2021, April 15). *Technology.* Encyclopedia Britannica. https://www.britannica.com/technology/technology

Burghardt, M. D. (2013). *Interconnecting STEM with informed engineering design pedagogy.* ASEE 2013 Annual Conference.

Burghardt, M. D., Hecht, D., Lauckhardt, J., & Hacker, M. (2010). A study of mathematics infusion in middle school technology education classes. *Journal of Technology Education., 22*(1), 1–11.

Burke, J. (1978). *Connections.* Little, Brown.

Cambridge. (2022). *Technology.* In Cambridge.com. Retrieved January 22, 2022 from https://dictionary.cambridge.org/us/dictionary/english/technology

Christenson, J. (2011, November 13). *Ramaley coined STEM term now used nationwide.* Retrieved May 6, 2014 from http://www.winonadailynews.com/news/local/article_457afe3e-0db3-11e1-abe0-001cc4c03286.html

Custer, R. L., Daugherty, J. L., & Meyer, J. P. (2010). Formulating a concept base for secondary level engineering: a review and synthesis. *Journal of Technology Education, 22*(1), 4–21.

Daly-Lewis. J. (1998, December). MSTe leadership team interviews. Eastern Suffolk BOCES.

De Vries, M. J. (2016). *Teaching about technology: An introduction to the philosophy of technology for non-philosophers*. 2nd Edition. Science and technology education library. Vol. 27. Springer. https://doi.org/10.1007/1-4020-3410-5

Deutsch, S. (2012). Julius Rosenwald. Immigrant entrepreneurship. Retrieved January 22, 2022 from http://www.immigrantentrepreneurship.org/entries/julius-rosenwald/

Drucker, P. F. (1959). Work and tools. *Technology and Culture*, . *1*(1), 28–37. https://doi.org/10.2307/3100785x1. https://www.jstor.org/stable/3100785

Drucker, P. (2020). *On technology*. papers published posthumously. Harvard Business Press. Edison Innovation Foundation. (2020). *Edison Mucker's Blog*. Retrieved January 16, 2020 from https://www.edisonmuckers.org/thomas-edison-quotes

Engineering Concepts Curriculum Project (ECCP). (1971). *The man-made world*. Polytechnic Institute of Brooklyn: McGraw-Hill Book Company

Engineers Council for Professional Development (ECPD). (1979). *The engineering team*.

Flexner, A. (1939, June/November). *The usefulness of useless knowledge. Harpers*, issue 179. Retrieved January 27, 2022 from https://www.ias.edu/ideas/usefulness-useless-knowledge

Gordon, S. C. (1990). From slaughterhouse to soap-boiler: Cincinnati's meat packing industry, changing technologies, and the rise of mass production, 1825–1870. *The Journal of the Society for Industrial Archeology, 16*(1), 55–67. Society for Industrial Archeology.

Hacker, M., & Barak, M. (2017). Important engineering and technology concepts and skills for all high school students in the US: comparing perceptions of university engineering educators and high school technology teachers. *Journal of Technology Education, 28*(2), 31–52.

Haile, P. (1999). *MSTe Evaluator's Report*.

Hansson, S. O. Technology and Mathematics. (2020). *Philos. Technol.* 33, 117–139. Retrieved January 2, 2022 from https://doi.org/10.1007/s13347-019-00348-9

Hecht, D. (2018). Engineering for all (EfA) final evaluation report.

Hecht, D., Flugman, B., Russo, M., La Fata, C., & Lauckhardt, J. (2009). *Mathematics, Science, Technology Partnership (MSTP) Final Program Evaluation*. September 2003–August 2009. Center for Advanced Study in Education. CUNY Graduate Center.

Heyman, J. (2014). Geometry, mechanics, and analysis in architecture. In A. Gerbino (Ed.), *Geometrical Objects: Architecture and the Mathematical Sciences 1400–1800, Archimedes 38* (pp. 193–201). Springer.

International Technology and Engineering Educators Association (2020). *Standards for Technological and Engineering Literacy*. ITEEA.

Katehi, L., Pearson, G., & Feder, M. (Eds.). (2009). *Engineering in K-12 Education: Understanding the Status and Improving the Prospects*. Committee on K-12 Engineering Education; National Academy of Engineering and National Research Council. http://www.nap.edu/catalog.php?record_id=12635#description.

Kranzberg, M., & In Pursell, C. W. (1967). *Technology in Western civilization*. P.11. Oxford University Press.

Library of Congress (2017). *Science Reference Services. Science Reference Guides*. Retrieved January 27, 2022 from https://www.loc.gov/rr/scitech/SciRefGuides/eng-disciplines.html

Malina, J. (1983). Archaeology and experiment. *Norwegian Archaeological Review, 16*(2), 69–78. https://doi.org/10.1080/00293652.1983.9965385. See https://pdf.zlibcdn.com/dtoken/ca72689d703190db6014c6410b2094ff/00293652.1983.9965385.pdf

National Aeronautics and Space Administration, NASA. (2021). *Drag equation of the early 1900s*. Retrieved April 2, 2022 from https://www.grc.nasa.gov/www/k-12/airplane/wrights/dragold.html

National Academy of Engineering. (2017). *NAE Grand Challenges for Engineering*. The National Academies of Sciences, Engineering, Medicine. Washington, D.C.

National Research Council. (1986). *Engineering Education and Practice in the United States: Engineering Infrastructure Diagramming and Modeling*. Panel on Engineering Infrastructure

Diagramming and Modeling. Division on Engineering and Physical Sciences. Washington, DC: National Academy Press.
National Research Council. (2008). *Common Standards for K-12 Education? Considering the Evidence.* Center for Education, Division of Behavioral and Social Science and Education. Washington, DC: The National Academies Press.
National Research Council. (2010). *Standards for K-12 Engineering Education?* Committee on Standards for K-12 Engineering Education. National Academies Press.
National Research Council. (2011). *Successful K-12 STEM Education: Identifying Effective Approaches in Science, Technology, Engineering, and Mathematics.* Washington, DC: The National Academies Press. https://doi.org/10.17226/13158.
National Assessment Governing Board (NAGB) (2018). Technology and engineering literacy framework for the 2014 NAEP. Chapter One: Overview. The Framework Development Process. https://www.nagb.gov/naep-frameworks/technology-and-engineering-literacy/2014-technology-framework.html
National Science Foundation. (2021). *Graduate Research Fellowship Program.* Program Solicitation NSF 21–602. Appendix. https://www.nsf.gov/pubs/2021/nsf21602/nsf21602.htm#appendix
National Science Foundation (2023). Exploring Computation Integrated into Technology and Engineering II. Award Number 2318343. Retrieved April 13, 2024 from https://www.nsf.gov/awardsearch/showAward?AWD_ID=2318343&HistoricalAwards=false
National Science Foundation (2024). Computer Science through Engineering Design in New York. Award Number 2341962. Retrieved April 13, 2024 from https://www.nsf.gov/awardsearch/showAward?AWD_ID=2341962&HistoricalAwards=false
National Research Council. (2002). *Technically speaking: Why all Americans need to know more about technology*. The National Academies Press.
NCATE/ITEA/CTTE. (2003). *Program Standards. Programs for the Preparation of Technology Education Teachers.* The National Council for Accreditation of Teacher Education (NCATE). Washington, DC.
NGSS Lead States. (2013). *Next generation science standards: For states, by states.* The National Academies Press.
Purzer, S., Moore, T., Baker, D., & Berland, L. (2014). *Supporting the implementation of the next generation science standards (NGSS) through research: Engineering.* Retrieved from https://www.narst.org/blog/ngss-engineering
Resnick, M. (1998). *Technologies for lifelong kindergarten. educational technology research and development*, Vol. 46, no. 4.
Rossouw, A., Hacker, M., & de Vries, M. J. (2010). Concepts and contexts in engineering and technology education. *International Journal of Technology and Design Education. 21*(4), pp. 409–423. https://doi.org/10.1007/s10798-010-9129-1. Accessible from https://link.springer.com/article/10.1007%2Fs10798-010-9129-1
Satchwell, R. E., & Loepp, F. L. (2002). Designing and implementing an integrated mathematics, science, and technology curriculum for the middle school. *Journal of Industrial Teacher Education, 39*(3), 41–66.
Schatzberg, E. (2006). Technik comes to America: changing meanings of technology before 1930. *Technology and Culture., 47*, 486–512.
Schatzberg, E. (2018). Why is there no discipline of technology in the social sciences? OpenEdition Journals. *Artefact: Techniques, Histoire et Sciences Humaines*. 8, 193–213. Retrieved January 3, 2022 from https://journals.openedition.org/artefact/2150?lang=en. https://doi.org/10.4000/artefact.2150
Siemens. (2022). *Verner von Siemens: A dynamic, visionary entrepreneur*. Retrieved January 28, 2022 from https://new.siemens.com/global/en/company/about/history/stories/werner-von-siemens.html
Sniderman, D. (2012). Leonardo Da Vinci. America Society of Mechanical Engineers. NYC, NY. See https://www.asme.org/topics-resources/content/leonardo-da-vinci

Swyt, D. A. (1989). *Knowledge, Innovation, and the Administration of Publicly-Funded Research in Technology*. Unpublished Manuscript.

Understanding Science. (2022). *Science aims to Explain and Understand*. University of California Museum of Paleontology. Retrieved January 20, 2022 from https://undsci.berkeley.edu/article/0_0_0/whatisscience_04

Watson, J. G. (2021, March 25). *Civil Engineering*. Encyclopedia Britannica. https://www.britannica.com/technology/civil-engineering

Wikipedia Contributors. (2022, February 22). *History of Engineering*. Wikipedia, The Free Encyclopedia. Retrieved May 2, 2022, from https://en.wikipedia.org/w/index.php?title=History_of_engineering&oldid=1073294083

World Intellectual Property Organization (WIPO). (2015). *World Intellectual Property Report: Breakthrough Innovation and Economic Growth*. Retrieved January 25, 2022 from https://www.wipo.int/edocs/pubdocs/en/wipo_pub_944_2015-chapter2.pdf

Young, M. D. (1959). *The Rise of the Meritocracy, 1870–2033: The new elite of our social revolution*. Random House.

Michael Hacker co-directs the Center for STEM Research at Hofstra University in New York. Since 1997, he has been a PI or Co-PI on 15 large-scale National Science Foundation projects focused on improving teaching and learning in K-16 STEM education. He previously served as a New York State Education Department supervisor for Technology Education, and was a secondary school teacher, department supervisor, and university teacher educator. Michael's research interests are focused on Technology and Engineering education reform, improving engineering design pedagogy, investigating the mathematics truly needed by manufacturing technicians, and using authentic social contexts in STEM education.

Chapter 10
Food Technology-Centred Approaches to STEM Learning

Elizabeth Reinsfield

Abstract Food Technology provides an important, diverse, and authentic focus for Science, Technology, Engineering, and Mathematics (STEM) approaches to learning, to support students in applying their understanding of scientific, engineering, and mathematical principles, as an integral part of the design and development of products in *innovative* and *future-focused* ways. This chapter explores Food Education, the nature of Food Technology in New Zealand, and positions the subject to enable STEM-based curriculum learning. New Zealand highly values innovation, particularly when the technological process of generating, developing, and realising new ideas can be aligned with economic advantage (e.g., the export of milk products). Technological thinking and the ability to innovate through problem-solving are imperative. However, when approaches of this nature are translated into a secondary Food Technology classroom context, enactment can be limited by teachers' perceptions of the subject, reflecting the tension between the pedagogical intent of Home Economics and Food Technology. Both have an important role to play in food education. The focus of this chapter is to present Food Technology *in interdisciplinary ways* and to promote technological thinking and problem-solving. Emerging approaches in secondary schools, such as "I Am" projects are used to exemplify how STEM subjects can support learners' construction of their *future* or *possible* self, which is particularly pertinent because of impacts resulting from the COVID climate. Recommendations are made to ensure that curriculum learning remains true to the "T" in STEM so that technology education is positioned to connect learning *across* subjects.

Keywords Food education · Interdisciplinary · Technology education · STEM education

---

E. Reinsfield (✉)
Hamilton, New Zealand

W. Fox-Turnbull and P. J. Williams (eds.), *Locating Technology Education in STEM Teaching and Learning*, Contemporary Issues in Technology Education,
https://doi.org/10.1007/978-981-97-1995-2_10

10.1 Introduction

STEM approaches to learning present opportunities for curriculum to be enacted in culturally appropriate, innovative ways. However, any change in professional practice implies potential challenges for teachers—particularly if there is a need for them to revisit their beliefs about the nature of their practice. This chapter addresses historical issues of professional learning in New Zealand and presents Food Technology as a means to enable a STEM approach, through personalised and *future-oriented* ways of learning (Ministry of Education (MoE), n.d.; Reinsfield et al., 2021). The term *future-focused* can be used to represent technology education's evolution as a subject and include innovative and culturally responsive ways of thinking. Secondary school teachers in New Zealand are encouraged by the government to adopt a future-focused approach to education, which is sufficiently flexible to accommodate the diversity of students' learning needs in their classrooms. The reality is, however, that in New Zealand, Food Technology has a shortage of teachers who are confident enough to commit to its enactment. This can lead to students being taught using pedagogical approaches derived from a Home Economics lens (e.g., the provision of a recipe, which students then make to develop life skills) (Reinsfield, 2018; Reinsfield & Lee, 2021). In New Zealand, Home Economics has been situated in the Health and Physical Education learning area of the curriculum and focuses on improving the quality and resource management of everyday life for individuals, families, and households. A subsequent lack of confidence in the technology curriculum implementation has led to some challenges in the enactment of Food Technology as a future-focused subject.

Future-focused approaches to learning are influenced by the school context and government policy. Policy documents, like the New Zealand School Property Strategy (MoE, 2011) imply the encouragement of new approaches to learning and suggest that Innovative Learning Environments were promoted as a means to develop "a world-leading education system [able to provide] all New Zealanders with the knowledge, skills, and values to be successful citizens in the 21st Century" (p. 2). Curriculum learning approaches, such as STEM, present opportunities yet there are various factors fundamental to successful implementation and enactment. In some cases, teachers are likely to have to review their professional beliefs about their role and working practices as a teacher.

The emergence of Innovative Learning Environments (ILEs) in New Zealand has seen an increased emphasis on the establishment of school contexts capable of evolving and adapting as educational practices change, to support strengths-based teaching and learning and afford students and teachers flexibility, agency, ubiquity, and connectedness. There have been opportunities to reconceptualise secondary education, teach in interdisciplinary ways, and re-position [food] technology education as a subject that is future-focused in nature and can prepare learners for their engagement in a technologically mediated world. In ILE secondary schools, curriculum models can be integrated (where the focus is on the synthesis and reinforcement of knowledge and learning experiences), or interdisciplinary (which connects interdependent knowledge and skills to explore a problem) (Organisation for Economic Co-operation Development, 2012; Reinsfield, 2021). The benefits of an integrated approach to curriculum can include a more *coherent education*, where students can be encouraged to make connections *between* different subjects and learning can be negotiated to increase active involvement (Etim, 2005; Fraser, 2000). In such cases, teachers are expected to collaborate *across* curriculum areas when planning and responding to student needs (Beane, 1997; Virtue et al., 2009).

To facilitate learner-centred pedagogies of this nature and in the secondary school context, teachers are likely to be required to engage with *new and unique ways of planning and collaborating* that are like those required for the facilitation of STEM learning. The nature of this collaboration can be between teachers and teachers, as well as teachers and learners—in a purposeful and timely manner (Reinsfield, 2019). To model the pedagogies required to be successful in a contemporary approach to STEM education, teachers need to be motivated to explore and reflect upon learning in a *variety of contexts*. A correlation between the ILE model and improved student outcomes has been made in the literature, as the result of such collaborative approaches to teaching, but *only* where there was sustained professional learning and supportive school structures (Blackmore et al, 2011; Osborne, 2016).

10.2 Professional Learning and Development

Educational change is a complex process where curriculum implementation might occur through sufferance, and teachers can be suspicious of reform (Handal & Herrington, 2003). In this context, when practitioners come to enact the curriculum in their classrooms, they are likely to rely more heavily on their own beliefs or experiences than on currently advocated pedagogical approaches. Despite its importance in developing teachers' understandings, many professional development opportunities in New Zealand have historically been fragmented, detached from the curriculum, and disconnected from teachers' learning needs (Reinsfield & Fox-Turnbull, 2020, 2021). In the technology education community in New Zealand, there has been a tendency for passive professional development approaches where teachers rely on others to inform them of how their pedagogy should manifest. However, Reinsfield (2018) proposed a Professional Learning and Development (PLD) model that allowed teachers to reflect on their own professional needs, to interpret, engage with, and enact the curriculum.

Reinsfield's (2018) PLD model proposed a structure to provide choice for the teacher, offer a balance between external and school-based learning, and a personalised and responsive approach to address identified gaps between teachers' theory and practice. Her work illustrated how organisational challenges could be navigated according to a teacher's school context, to build sustained teacher capability in technology education. A partnership between academics, teachers and (as relevant) members of industry was proposed to establish a climate, that fostered transformative and future-focused practice; in the sense that the subsequent professional learning would lead to pedagogical practice that encouraged students to address technological needs, as identified *within* authentic contexts for learning. This was distinct from other professional development models being offered to technology teachers, which generally exposed practitioners to best practice approaches and relied on teachers practising at a level where they could *make the connections* between others' practice and their own.

Teachers need to be the agentic party in their own professional learning and development so that they feel *empowered and able* to make changes in schools, particularly when making meaning of the curriculum for implementation. Importantly, this empowerment can be furthered through the accommodation of student voice, which can, in turn, lead to the re-conceptualisation of a school curriculum, with teachers and learners collaborating to realise this (McNae, 2017). To do so, there is an expectation that teachers also work collaboratively, focus on making meaningful learning connections, and consider integrated and purposeful approaches to the curriculum as a way forward in their subject. Such approaches can be facilitated by focusing the learning on technological and critical thinking.

10.2.1 Technological and Critical Thinking

Technological thinking is not solely motivated by the development and/or replication of a product. It is also about the exploration of *why* decisions need to be made in a particular context for learning, underpinned by knowledge of the development of systems, processes, and techniques (Reinsfield, 2018; Reinsfield & Williams, 2017). The connection between technological thinking and the ability to innovate *through* critical problem-solving can be explained by an individual's ability to *make-meaning* of complex issues, such as when they develop a technological response to a solution (Reinsfield, 2016, 2018; Wertsch et al., 1995). For example, when students explore an authentic or real-world context for learning, they engage in culturally and socially mediated activity, requiring them to make sense of internal and mental processes within a socio-cultural-historical context (Galperin, 1969, 1989). New Zealand has a reputation for "Kiwi ingenuity", leading to innovations like the number 8 fencing wire, which was borne out of a motivation to be ingenious with limited resources, in creative ways. As part of the technology education learning area of the curriculum, there is an expectation that Food Technology teachers advocate for such types of innovation. The next section describes the importance and evolution of food education.

10.3 Technology Education, Food Education, and STEM Education

The technology curriculum was redeveloped in 2007 (MoE, 2007, 2017) and saw the emergence of two new strands, *Technological Knowledge* (TK) (focused on the processes and properties of materials that can be used during product development) and the *Nature of Technology* (NoT). To address the Nature of Technology strand teachers are required to foster students' critical thinking and encourage discussion about past and future technological responses, with a view to supporting them to become informed consumers who can think "outside of the box". In 2017, a revision to the technological areas saw an increased emphasis on digital technologies.

The 2017 curriculum's increased focus on digital technologies aligned with the technology learning area, but also emphasised that such knowledge should be applied to a *range of disciplines,* as well as inform the production of technological outcomes or systems. These curriculum changes more explicitly acknowledged the relationship between technological development and society, in the present day and the future (MoE, 2007). The Nature of Technology strand provides opportunities to enable STEM approaches to learning - as ways for students to think about local and global technological issues in interdisciplinary, student-centred, critical, innovative, and creative ways.

In New Zealand, agricultural and horticultural sectors are imperative for the economy, with an estimated 85–90% of our dairy, meat, fruit, and vegetables being exported (Ministry for Primary Industries, 2022). The COVID climate presents an urgent need to enhance our national food and nutrition security (e.g., an individual's ability to access enough healthy food), to ensure individuals' well-being. Health issues, like obesity and dental decay, signal the importance of food education in schools (Baker et al, 2020; Truman & Elliott, 2019; Wijayaratne et al, 2018). Healthy food advocates, like the British chef Jamie Oliver, have brought attention to the notion of *food literacy* (e.g., Vidgen & Gallegos, 2014). Food literacy emphasises learning about changing societal attitudes related to growing, preparing, and consuming healthy food—personally, as part of a community, and with a view to considering the impact of food choice on the environment or economy.

Food Technology is situated within the "Designing and developing processed outcomes" area of the technology curriculum, where it is presented to encourage learners to explore economic and cultural influences on *product development*, as well as learn about systems, processes, or techniques when designing and manufacturing quality products (MoE, 2007, 2017). New Zealand has a (food) technology education teacher shortage, and curriculum change (with limited professional learning support) has left existing food teachers feeling confused about their place in the learning area. Many are not confident enough to commit to its enactment and students continue to be taught using content and pedagogical approaches derived from the Health and Physical Education (Home Economics) curriculum (Reinsfield, 2018)—under the guise of Food Technology.

Historically, children's experience with food preparation was mostly in their homes. Changing lifestyles meant that over time, Home Economics teachers took increasing responsibility for food education, particularly to develop students' life skills (e.g., Pendergast & Dewhurst, 2012). Later, Food Technology evolved as a subject, adopting a manufacturing and commercial lens, with a view to exploring the selection, preservation, processing, packaging, distribution, and use of safe food by connecting learning to biotechnology, chemistry, engineering, and nutrition. Given its importance at both an individual and societal level, it might be assumed that food education is assured in any education system. However, according to Stitt (1996),

> The experiences in Britain, New Zealand, Ireland, Canada, and other nations suggest that this is not the case… curriculum decisions illustrate that those in positions of power, particularly economic power, will attempt to define what is taken as knowledge and how accessible any knowledge is to different groups… (pp. 27–28)

According to Rutland and Owen-Jackson (2015), *Food Technology* remained in the English curriculum to focus on the development of girls' practical skills—to prepare them for domestic employment or to be a housewife. Apparently, the subject has not evolved at the same rate as other subjects and can often still be associated with the development of life skills. STEM approaches to learning are increasingly being used to expose students to pertinent food-related issues, such as sustainability. A study in Vietnam (participants, $n = 211$) by Nguyen et al. (2020), determined that Food Technology provided a naturally occurring curriculum focus because STEM education:

- is interdisciplinary in nature, and students can solve real-world problems of interest
- encourages students to problem-solve, thereby developing soft skills (e.g., critical thinking)
- encourages students to inquire and reason, leading to the generation of a solution
- encourages the application of [food] engineering processes.

Such exciting opportunities for STEM education align well with the recent changes to New Zealand's senior secondary assessment framework, which aims to focus more on the nature of learning (rather than teaching to the assessment). Such learning is positioned to acknowledge that

> Materials and Processing Technology by its very nature, has connections with other learning areas. Technology could be considered to be at the heart of learning as what is learnt in the technology classroom can be fed *from or into* other subjects. (Emphasis added, para, Connections, Ministry of Education, 2020)

Connecting to other subjects is not as straightforward as it first appears, and there are implications for teachers' professional learning and development, as well as the pedagogical approaches that privilege technological and critical thinking. This is potentially even more challenging for those practitioners who might be experiencing difficulty in conceptualising how Food Technology might *look and be taught* in their school context.

The *draft* big ideas for Food Technology at the senior secondary level (Years 16–18) will require teachers to design learning where:

1. Sustainability through kaitiakitanga (guardianship & protection) underpins ethical intervention by design
2. Outcomes are developed to respond to peoples' needs
3. Purposeful outcome development encourages auahatanga (innovation)
4. Authentic contexts are accessed to inspire fit-for-purpose outcomes through manaakitanga (care for the user).

One of the strengths of technology education is that learning contexts can *lean into* current and important events in society. The global COVID pandemic has perpetuated socioeconomic inequity and vulnerability (Murphy et al, 2021; Ndumbe-Eyoh et al, 2021). Gupta and Jawanda (2020) cite several negative impacts on school children as the result of the COVID pandemic, including an increased learning gap between low and high socioeconomic groups, anxiety and frustration about the future, nutritional deficiency, behavioural changes (including anger and aggression), loss of enthusiasm and interest, and deprival of educational, nutritional, and healthcare services. Whilst children tend to be more adaptive than most adults, the unprecedented uncertainty has taken its toll on their well-being, particularly when thinking about the future. It is more important than ever for children to look ahead to see their "possible selves"—to be positive about and *foresee* their future and what they would like to do or become (Cross & Marcus, 1991). Developing a learning climate that encourages innovation, and incorporating a local curriculum where teachers are responsive to learners' needs, identity, language, and culture, has seen an emerging trend in the development of *"I am" projects.*

10.3.1 I Am Projects

"I am" *projects* are based on the premise that if students are engaged in their learning, they are more likely to be self-regulating. Self-regulated learners are confident, diligent, and resourceful; they know what they can do and are proactive to seek support, as they require it (Zimmerman, 1990; Zimmerman et al., 1992). Zimmerman argued that self-regulated learners can problem-solve, take responsibility for their learning, plan, set goals, reflect, and action the need to change thinking. Self-regulation can be compared to Freire's construct of conscientisation in which an individual deeply understands their place in the world (Nevin & Cardelle-Elawar, 2003). The next section describes the findings from research, to explore how teachers can manage the expectation for such approaches in an ILE, to provide insight for practitioners who are keen to apply such strategies through STEM education.

10.3.2 TENZile Project

In 2019, Reinsfield explored the nature of collaborative teaching in an ILE. Her Technology Education New Zealand: Secondary Innovative Learning Environment (TENZile) project was based in a newly established school in the Waikato. The research was designed to use an interpretivist framework to generate data by working closely in the field with a purposeful sample of four teachers, who co-taught the curriculum in an integrated manner. Lessons were observed and videoed as a means of eliciting and restructuring teachers' past and present thinking. Once a fortnight, teachers were sent an edited video of their observed lessons and asked to review and then reflect on their teaching (Maclean & White, 2007). Questions were provided to guide their reflection process, and teachers were asked how they accommodated students' academic or social needs within the ILE, how they used a range of resources, and how this enabled future-focused approaches to learning.

When asked to explain how the integrated approach to the curriculum was future-focused, Imogen (Food Technology specialist teacher) indicated that:

> … Ngaire and I teach together, we talk about and plan what we want our students to do and then decide what language [Te Reo Māori] and technology [curriculum concepts] we can incorporate…
>
> So [in this project] our teaching staff will be invited to lunch … What's so great about it is [that the] whole module is culturally bounded. It includes real-world experience - that is future-focused for me (Interview One, Lines 65–71).

Ngaire (Te Reo Māori specialist teacher) also described a need to approach learning differently when she was teaching in an integrated way, stating:

> We collaborate with both students and teachers… It is a different relationship you build for learning. We are also looking for integration. … The future is integrated; we show that life is intertwined. (Interview One, Lines 96–102)

The *Kai and Who Am I* project, which was designed by Imogen and Ngaire, was learner-centred, and exposed students to "real world" thinking (Zuga, 1989, 1994), and focused on:

> …the integration of the two subjects in the first place… By putting two subjects together students can see the inter-relationship between the two… We often focus on the 'becoming a food technologist' part in this module and sell this as a genuine career path… In terms of growing empathy and knowledge around Tikanga [customs] Māori, students become more aware of and value our indigenous culture. (Reflection Two, Lines 6–14)

Students were given a context for learning and brief, which stated:

> Kai (food) has the power to bring people together and to help us better understand each other. One of the most authentic ways to experience a culture and learn about its traditions, history, and language is through food. Despite Māori having a rich and

unique culinary heritage, you won't find any Māori fine dining restaurants anywhere. What can we do as kiwis to keep Tikanga (customs) alive?

You are to develop an indigenous menu item that fuses traditional with contemporary. You will be required to speak Te Reo when serving kai for 120 people at a Matariki lunch held at school. The menu must address the needs of stakeholders attending the lunch.

Your menu item must:

- include at least one locally sourced and foraged ingredient
- be developed within 100 min
- utilise in season vegetables or fruit
- resemble finger food, so it is easy to eat without cutlery.

Learning of this nature means that with appropriate support, students are likely to have an experience that affirms their confidence. However, the fostering of a safe and positive classroom environment depends on the professional skill of teachers to *organise and direct* the learning process, as well as being adaptive to the direction that such learning might take. What appeared to be key to participant teachers' success, were their philosophies that learning should be responsive to students' ideas and interests. Also pertinent was that both teachers negotiated the learning *with* students, whilst also using their professional experience to align learning with the curriculum and school's values. Figure 10.1 shows some of the food-based outcomes developed for the Matariki lunch held at the school.

Interdisciplinary approaches to learning (like STEM) provide opportunities for teachers to work collaboratively (e.g., across specialisms in a secondary school context) to design programmes that are socially purposeful, responsive to students' needs and interests, and support meaningful engagement. The next section presents a *potential* Food Technology learning context, which allows opportunities for diverse, integrated, and cross-disciplinary approaches, but also empowers students to take ownership of the direction that their learning takes.

Fig. 10.1 Kai and Who am I?—food outcome at Matariki lunch

10.4 Links to Science, Technology, Engineering, and Mathematics

One of the enablers for STEM education is having the ability to design learning that accommodates the knowledge and interests of the students, whilst also addressing the curriculum requirements. Subject teachers need to be confident about and identify professionally with their own subject (e.g., within technological areas such as food, or as a technology teacher) or they might prefer to maintain a siloed approach to learning. Chapter 10 identifies a project developed by Dr Lissa Carter and Dr Belinda von Mengersen at an Australian Catholic University. The project explores an ITE programme, which is designed to provide student teachers with the opportunity to develop knowledge and skills in designing and manufacturing food products. The key aim of Carter and von Mengersen's project was to challenge the historical and siloed approach to Food, Textiles, Timber, and Metal in Design and Technology in Australia. The learning was situated as innovative and interdisciplinary and used food moulding to incorporate food and product design, food testing, and prototyping—using 3D printing. Such projects present exciting and future-focused pedagogical models that can be used to facilitate STEM education.

STEM learning contexts can be presented in ways to illustrate the cross-disciplinary connections. Figure 10.2 presents potential ideas that highlight the context of *Food Security*, as explored through the lens of Food Technology education. These ideas are grouped according to the disciplines of Science (S), Technology (T), Engineering (E), and Mathematics (M) (e.g., STE). Food Technology concepts are italicised for ease of interpretation. In the example that relates to the sustainable food driver, to move away from meat protein to non-animal sources (such as algae), there might be several directions that learning could take. Sustainable food would need to be explored, to acknowledge that methods of production use processes and systems that are non-polluting and protect non-renewable energy or natural resources, with a view to supporting the needs of future generations. A focus on Food *Science* (S) could support learning about biological food materials, to develop an understanding of the behaviour of such materials in harvesting, processing, distribution, storage, and preparation. The *Engineering* (E) focus might centre on the practice underpinning the design of products and processes appropriate for material and energy efficiencies to minimise environmental impact, cut costs, minimise waste and maximise value. Other examples are provided below.

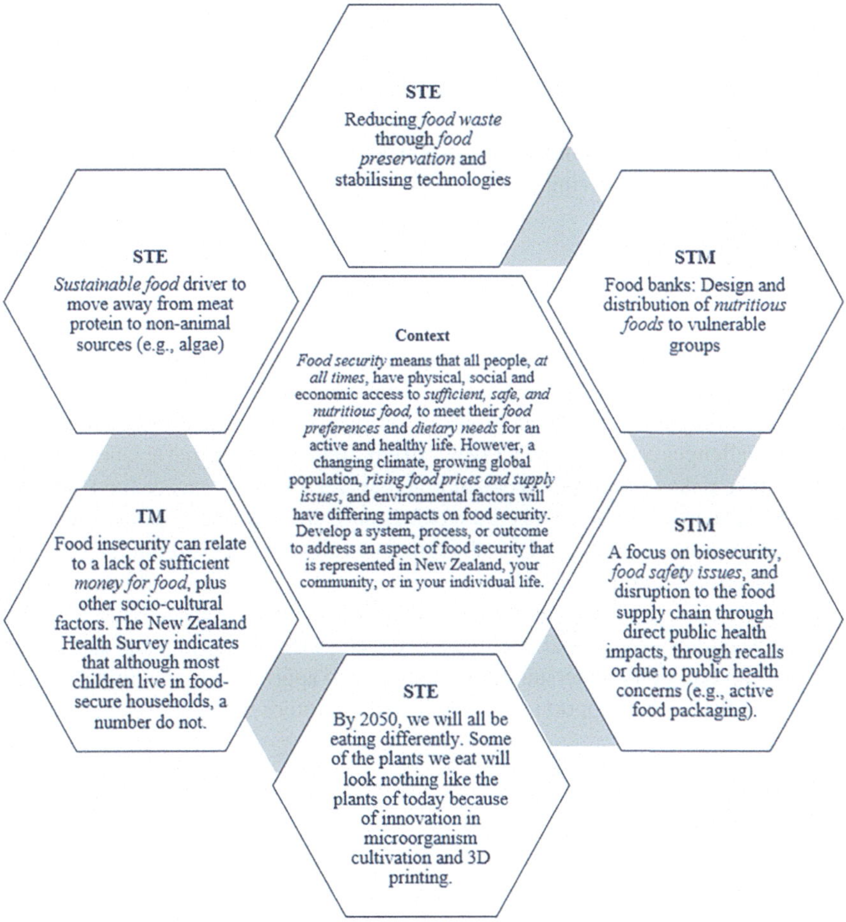

Fig. 10.2 STEM approaches to a food security context for learning

10.4.1 Pedagogical Implications

The nature of a progressive, learner-centred approach relies on the notion that students will be self-regulated and motivated by STEM education. To structure learning, teachers are required to use pedagogical approaches that support the development of a range of understandings and encourage students to take ownership of their own learning. This approach is in direct contrast to a classroom where the teacher is making all decisions about the learning and students are provided with information about the stages to replicate an outcome. Reinsfield's (2018) research suggests that teachers' practice can be significantly affected by their classroom management skills and understanding or preparedness to respond to students' learning needs. Accordingly, it is proposed here that to maximise learning for technology-centred approaches to STEM learning, students' needs to be supported to:

- utilise diverse intellectual and practical resources, to make meaning of learning contexts
- explore the dynamic relationship between technology and society when designing for a purpose
- engage in technological modelling, product, and systems development to inform innovative problem-solving.

10.4.2 Professional Development Implications

Approaches to professional learning and development in New Zealand have historically centred on traditional approaches, where teachers are introduced to new ideas, which they then must make sense of for their own learners and context. The ongoing challenges caused by the COVID-19 environment mean that teachers are less likely to take risks in their pedagogical practice, yet ongoing teacher (and student) absences mean that school leaders are seeking more efficient and responsive ways to provide curriculum learning. STEM education provides such opportunities. However, teachers need to be prepared and able to engage with new ways of thinking about learning, and open to the potential that collaborative practice may provide. There is an urgent need for professional learning and development that centres on supporting teachers' contemporary understandings of integrated approaches to planning (e.g., as modelled in ILEs), that support interdisciplinary and future-focused student learning.

10.5 Conclusion

STEM approaches to learning provide exciting opportunities to emphasise the importance of food education as a means to explore key issues facing the future of society and to design responses in engaging, innovative, and future-focused ways. This chapter explores the importance of food education, and the nature of Food Technology in New Zealand, with a particular focus on its place in the technology curriculum. A range of interdisciplinary ideas are presented, using Food Security as a context for learning. Current approaches in secondary schools, such as "I Am" projects are used to exemplify how STEM approaches can be designed for students to direct their own learning and be more likely to engage in learning that is positive and personally meaningful. The changing nature of education, particularly since the emergence of COVID-19, positions STEM learning as a means to provide efficient approaches to teaching and learning approaches and offers new ways of practising for educators. Whether such approaches can be successfully implemented, however, is likely to depend on teachers' perceptions, their confidence in enactment, as well as the professional learning support systems in place.

References

Baker, P., Machado, P., Santos, T., Sievert, K., Backholer, K., Hadjikakou, M., Russell, C., Huse, O., Bell, C., Scrinis, G., Worsley, A., Friel, & Lawrence, M. (2020). Ultra-processed foods and the nutrition transition: Global, regional and national trends, food systems transformations and political economy drivers. *Obesity Reviews, 21*(12). https://doi.org/10.1111/obr.13126

Beane, J. A. (1997). *Curriculum integration: Designing the core of democratic education.* Teachers College Press.

Blackmore, J., Bateman, D., Loughlin, J., O'Mara, J., & Aranda, G. (2011). *Research into the connection between built learning spaces and student outcomes.* Department of Education and Early Childhood Development. http://dro.deakin.edu.au/eserv/DU:30036968/blackmore-researchinto-2011.pdf

Cross, S., & Markus, H. (1991). Possible selves across the lifespan. *Human Development, 34*(4), 230–255. https://doi.org/10.1159/000277058

Etim, J. (2005). *Curriculum integration K-12: Theory and practice.* University Press of America.

Fraser, D. (2000). Curriculum integration: What it is and is not. *SET: Research Information for Teachers,* (3), 34–37.

Galperin, P. I. (1969). Stages in the development of mental acts. In M. Cole & I. Maltzman (Eds.), *A handbook of contemporary soviet psychology* (pp. 249–273). Basic Books.

Galperin, P. I. (1989). Mental actions as a basis for the formation of thoughts and images. *Soviet Psychology, 27*(3), 45–64.

Gupta, S., & Jawanda, M. K. (2020). The impacts of COVID-19 on children. *Acta Paediatrica, 109*(11), 2181–2183. https://doi.org/10.1111/apa.15484

Handal, B., & Herrington, A. (2003). Mathematics teachers' beliefs and curriculum reform. *Mathematics Education Research Journal, 15*(1), 59–69. https://doi.org/10.1007/BF03217369

Maclean, R., & White, S. (2007). Video reflection and the formation of teacher identity in a team of pre-service and experienced teachers. *Reflective Practice, 8*(1), 47–60.

McNae, R. (2017). Introduction: Partnerships with students. In R. McNae & B. Cowie (Eds.), *Realising innovative partnerships in educational research: Theories and methodologies for collaboration* (pp. 3–8). Sense.

Ministry for Primary Industries. (2022). *Situation and outlook from primary industries.* https://www.mpi.govt.nz/resources-and-forms/economic-intelligence/situation-and-outlook-for-primary-industries/

Ministry of Education. (2007). *The New Zealand curriculum.* Learning Media.

Ministry of Education. (2011). *The New Zealand school property strategy, 2011–2021*. Learning Media.

Ministry of Education. (2017). *Digital technologies: Hangarau Matihiki*. https://education.govt.nz/assets/Documents/Ministry/consultations/DT-consultation/DTCP1701-Digital-Technologies-Hangarau-Matihiko-ENG.pdf

Ministry of Education. (2020). *Materials and Processing Technology* | NCEA (education.govt.nz)

Ministry of Education. *Technology: Materials and processing technology*. https://ncea.education.govt.nz/technology/materials-and-processing-technology?view=learning

Murphy, B., Benson, T., McCloat, A., Mooney, E., Elliott, C., Dean, M., & Lavelle, F. (2021). Changes in consumers' food practices during the COVID-19 lockdown, implications for diet quality and the food system: A cross-continental comparison. *Nutrients, 13*(1), 1–14. https://doi.org/10.3390/nu13010020

Ndumbe-Eyoh, S., Muzumdar, P., Betker, C., & Oickle, D. (2021). 'Back to better': Amplifying health equity, and determinants of health perspectives during the COVID-19 pandemic. *Global Health Promotion, 28*(2), 7–16. https://doi.org/10.1177/17579759211000975

Nevin, A. I., & Cardelle-Elawar, M. (2003). Dialogic retrospection as a metacognitive research tool. *Australian Journal of Educational & Developmental Psychology, 3*, 1–14.

Nguyen, T. P. L., Nguyen, T. H., & Tran, T. K. (2020). STEM education in secondary schools: Teachers' perspective towards sustainable development. *Sustainability, 12*(21), 8865.

Organisation for Economic Co-operation Development. (2012). *Equity and Quality in Education: Supporting Disadvantaged Students and Schools*. OECD Publishing. https://www.oecd.org/education/school/50293148.pdf

Osborne, M. (2016). *Innovative learning environments*. CORE Education White paper. Retrieved from http://blog.core-ed.org/files/2017/01/Innovative-Learning-Environments-FINAL-web.pdf.

Pendergast, D., & Dewhurst, Y. (2012). Home economics and food literacy: An international investigation. *International Journal of Home Economics, 5*(2), 245–263.

Reinsfield, E. (2016). Technology education in the New Zealand context: Disparate approaches to meaning making of the curriculum and the implications for teachers' evolving knowledge for practice. *Australasian Journal of Technology Education, 3*, 1–18.

Reinsfield, E. (2019). A future-focused conception of the New Zealand curriculum: Culturally responsive approaches to technology education. *International Journal of Technology and Design Education*. https://doi.org/10.1007/s10798-019-09510-y

Reinsfield, E., & Williams, P. J. (2017). New Zealand secondary teachers' perceptions "Technological" or "Technical" thinking. *International Journal for Technology and Design Education*. https://doi.org/10.1007/s10798-017-9418-z

Reinsfield, E. & Fox-Turnbull, W. (2020). A new approach to professional learning and development for technology teachers in New Zealand: Developing networks of expertise. *Australasian Journal of Technology Education*. Retrieved from https://ajte.org/index.php/AJTE/article/view/67

Reinsfield, E., & Fox-Turnbull, W. (2021). The Mātanga Project. Developing a self-sustaining model for technology teacher education for and within the New Zealand community. In *Techne Services, 28*, 375–384). Finland.

Reinsfield, E. & Lee, K. (2021). Exploring the technology teacher shortage in New Zealand: The implications for quality teaching and learning. *International Journal of Technology and Design Education*. http://link.springer.com/article/https://doi.org/10.1007/s10798-021-09668-4

Reinsfield, E., Eames, C. & Fox-Turnbull, W. (2021). Quality STEM education. *Teachers and Curriculum, 21*, 5–8. https://www.tandc.ac.nz/tandc/article/view/387/413.

Reinsfield, E. (2018). *The potential for a future-focused curriculum in New Zealand: The perceptions and practice of six secondary school technology teachers* (Doctoral thesis). https://hdl.handle.net/10289/11939.

Reinsfield, E. (2021). Thinking about the future for learning: ILE realities and possibilities. *Pedagogies and partnerships in innovative learning environments: Case studies from New Zealand contexts*. Springer. https://link.springer.com/book/https://doi.org/10.1007/978-981-16-5711-5.

Rutland, M., & Owen-Jackson, G. (2015). Food technology on the school curriculum in England: Is it a curriculum for the twenty-first century? *International Journal of Technology and Design Education, 25*, 467–482. https://doi.org/10.1007/s10798-014-9293-9
Stitt, S. (1996). An international perspective on food and cooking skills in education. *British Food Journal.*
Truman, E., & Elliott, C. (2019). Barriers to food literacy: A conceptual model to explore factors inhibiting proficiency. *Journal of Nutrition Education and Behavior, 51*(1), 107–111.
Vidgen, H. A., & Gallegos, D. (2014). Defining food literacy and its components. *Appetite, 76*, 50–59.
Virtue, D. C., Wilson, J. L., & Ingram, N. (2009). In overcoming obstacles to curriculum integration, less can be more! *Middle School Journal, 40*(3), 4–11. https://doi.org/10.1080/00940771.2009.11495581
Wertsch, J. V., del Río, P., & Alvarez, A. (1995). *Sociocultural Studies of Mind*. Cambridge University Press.
Wijayaratne, S. P., Reid, M., Westberg, K., Worsley, A., & Mavondo, F. (2018). Food literacy, healthy eating barriers and household diet. *European Journal of Marketing., 52*(12), 2449–2477. https://doi.org/10.1108/EJM-10-2017-0760
Zimmerman, B. J. (1990). Self-regulated learning and academic achievement: An overview. *Educational Psychologist, 25*(1), 3–17. https://doi.org/10.1207/s15326985ep2501_2
Zimmerman, B. J., Bandura, A., & Martinez-Pons, M. (1992). Self-motivation for academic attainment: The role of self-efficacy beliefs and personal goal setting. *American Educational Research Journal, 29*(3), 663–676. https://doi.org/10.3102/00028312029003663
Zuga, K. F. (1989). Relating technology education goals to curriculum planning. *Journal of Technology Education, 1*(1), 34–58. https://doi.org/10.21061/jte.v1i1.a.5
Zuga, K. F. (1994). *Implementing technology education: A review and synthesis of the research literature*. Center on Education and Training for Employment.

Elizabeth Reinsfield is the Learning Area Lead for Technology Education at the Ministry of Education. Her interests include cross-generational perceptions of innovation and teachers' engagement with and enactment of curriculum. Liz's passion is working with teachers to facilitate change and build capability in technology education, leading to her achieving the Technology Education New Zealand (TENZ) Award: Outstanding Teacher Educator in Technology and Teacher Education Forum of Aotearoa New Zealand (TEFANZ): Emerging Teacher Educator Awards.

Chapter 11
Textile Technologies in STEM Education

Belinda von Mengersen

Abstract What role do Textile Technologies play in STEM education? (Spendlove and Barlex, Des Technol Educ Int J 16(1):4, 2011) concur that: "textiles as a subject remains an underrepresented area within STEM discussions". Research literature in Design and Technologies education indicates that relatively few educators are using Textiles in STEM projects or interventions apart from Electronic Textiles (e-textiles). Why is the perceived role of Textiles as a discipline in education devolving in inverse proportion to the expanding role Textile technologies play in a global economy? In 2011, Hughes et al. stated that: "it is essential that design and technology textiles education is underpinned by STEM-based teaching and learning" to "reduce the practice gap between the design and technology curriculum and the needs of the textile/design manufacturing industry". Considering these questions, this chapter aims: to explore potential applications for Textiles in STEM that can be used by educators; consider the potential value of these applications in relation to evolving conceptual frameworks for STEM; reflect upon how Textiles in STEM can foster "genuine design-based interaction" (Williams, Des Technol Educ Int J 16(1):32, 2011) and increase students self-efficacy.

Keywords Textile · Technical · Traditional · Textile science · Textile chemistry · Textile engineering · Design · Education · Technological textiles · e-textiles · Textile literacy · Bioengineering · Zero-waste · 3D knitting · 3D printing · Sustainability

11.1 Introduction

What are Textiles? As Gale and Kaur observe: "anyone who has taught textiles or has been asked 'what is textiles?' will know the impossibility of trying to explain something that can be science or art" (2002, p. ix). Textiles are a dynamic and

B. von Mengersen (✉)
Australian Catholic University, Strathfield, NSW, Australia
e-mail: Belinda.VonMengersen@acu.edu.au

W. Fox-Turnbull and P. J. Williams (eds.), *Locating Technology Education in STEM Teaching and Learning*, Contemporary Issues in Technology Education,
https://doi.org/10.1007/978-981-97-1995-2_11

multifaceted material category with a global market valued at "$577.83 billion in 2022" (TBRC, 2022). Textiles are created from a diverse range of, animal, mineral or vegetable-based source materials and "associated industries constitute one of the largest sources of employment in the world" (Gale & Kaur, 2002, p. ix), "430 million are thought to work in fashion textile and production" (CO, 2022). Unlike some materials many Textiles can be modified, or chemically engineered: both molecular and morphological structures modified to meet specific properties and performance requirements, resulting in precise product attributes (Kadolph & Marcketti, 2017). Textile products are bespoke and modified to suit specific needs of an industry or user. Modifications can occur at many points along the supply chain—in both the chemical and mechanical aspects of textiles engineering: from source matter to fibre, yarn, fabric, product design, manufacture, and finish. Textile Engineering includes chemical aspects, like: the application of stimuli-sensitive materials in smart textiles, functional finishes, protective (against UV, insect, camouflage, micro-organisms), anti-bacterial (including nano-technologies), irradiation dyeing, electroconductive textiles (conductive polymers, metals, graphene), laminated textiles for aerostat and stratospheric airships; and mechanical aspects like: textile composites, and soft body amour technology (Ul-Islam & Butola, 2018). Whilst Textiles are predominately known for fashion and clothing applications, they're also used in technical, household, and other diverse applications. The global technical textile market—including textiles for transport, sport, agriculture, construction, hygiene, and medical applications, "industrial products and components were valued at USD$190 billion in 2020 and projected to reach USD$285 billion by 2028" (VMR, 2022). High performance textiles "provide technical, functional, and performance properties" including textiles for: households, packaging, sports, comfort, medical and healthcare, personal protective, military and law enforcement personnel, industrial and filtration, geotextile and environmental protection, agriculture, and crop protection, building and construction, automotive, marine and composites, aeronautical and space, wearable, and smart responsive textiles (Paul, 2019). So, whilst Textiles are commonly perceived by many to be a traditional crafts-oriented creative artform, aspects of technical textile development, design and manufacture can be studied internationally, at postgraduate level in a Master of Science in Textiles: specializing in Technology, Chemistry, Design or Engineering, alongside programmes specifically offering Technical Clothing design, like LaSalle College in Montreal in collaboration with Techno-Espace (2022), a technological research clothing laboratory and the Outdoor Product Design and Development degree (OPDD) from Utah State University (2022) where: "students develop fleece mid-layers with technical fabric from textile companies such as 3M™ Nextel™; or test protective outerwear for high-impact sports like snowmobiling to see how it responds to wear and tear" (Tilton, 2018). Whilst textile scientists, engineers and designers are far less visible than fashion designers, or even textile artists they're changing the world in more profound ways. Some of the most revered fashion designers are the exception to the norm, running ateliers like research centres, redesigning the textiles as well as apparel. Issey Miyake (2022) is an example of a fashion designer who experimented philosophically, aesthetically, and technically: interlacing tradition and innovation. Iris van Herpen uses emerging textiles technologies with conceptual design thinking: "fusing technology with traditional haute couture craftsmanship" (HIGH, 2022). Van Herpen is credited with transforming fashion by working between fashion, design, technology, and science (HIGH, 2022). This is how Textiles inform STEM, through radical, rapidly evolving trans-disciplinary intersections in a word where textiles are ubiquitous.

Each section of this chapter responds to provocative questions in this exploration of Textiles in STEM (Science, Technology, Engineering & Mathematics) education. A question-based model research model has generated a collection of ideas, examples, and propositions, revealing complex and contradictory findings. This exploratory mode was used to keep the discussion open and generative. This set of framing questions emerged during that process and guide the chapter structure:

- How are Textiles currently used in STEM education?
- How do STEM principles apply to Textile design?
- If Textiles design and manufacture already inherently utilize all areas of STEM (historically, contemporaneously, and speculatively), why is this not broadly understood in education, industry or in society?
- Which Textile case studies could be used in STEM?
- How are Textiles used in design and industry? Isn't that just a fashion thing?
- Science v's Art: why is there opposition between Science and the Arts?
- Why aren't Textiles and Fashion designers STEM literate?
- Textile engineering? Is that a thing? Textile chemistry?
- What happened to Textile education?
- What happened to the Textiles industry?
- Can Textiles reinstate its STEM framework to develop solutions for significant design problems?

11.2 Literature Review

Textiles is an 'underrepresented' area (Spendlove & Barlex, 2011) within design and technologies education research (refer to Table 11.1). Yet, the discipline and subdiscipline areas which integrate Textiles and STEM topics in research is diverse as shown in this literature review. Textiles and STEM are used in conjunction in broad-ranging discipline research within science, engineering, humanities, medicine, economics, and business as well as education (refer to Table 11.2). An analysis of the interdisciplinary use of Textile's broadly in STEM education contexts (beyond the scope of Design and Technologies education) mirrored this diversity (refer to Table 11.2), indicating how widely Textiles are used in STEM education contexts.

A lack of research evidence (refer to Table 11.1) leads to two questions: Are Textiles being used in STEM education contexts? If so, why has this not been reported in the literature? Nor research undertaken into strategies for their inclusion? If the answer is that there is very little evidence of Textiles in STEM education because Textiles are being used in limited or unacknowledged ways, if at all—we ask why? If

Table 11.1 Literature review results: Textile* in design and technology education research

Journal	Date range	Search term/s	Results	Topic
Design & Technology Education (UK)	2010–2022	Textile*	5	1—STEM, Technological Textiles 2—e-Textile 1—Craft, Makerspace 1—Technical and textile technology
International Journal of Technology and Design Education (NL)	2010–2022	Textile*	10	1—Textile design 1—Textbook resources 2—Material knowledge in design education 1—e-Textiles 1—Smart textiles 1—Sketching 2—Collaborative design, CAD 1—Craft, Sloyd, Technical, and Textile domain
Journal of Technology Education (USA)	2010–2022	Textile*	0	
Australasian Journal of Technology Education	2014–2022	Textile*	1	1—Case study, design fictions

*Implies a search was conducted using the term: Textile as a base and included plural forms

Textiles are only infrequently used—given their potential for applications in innovative design contexts—we ask why? Especially, we consider: Why is the perceived role for Textiles in design shrinking—when the range of diverse Textile materials, technologies, and applications emerging in many fields—including Engineering, Science, and Mathematics—is rapidly evolving?

Hughes et al. (2010) have outlined the value of Textiles, especially, Technological Textiles in Design and Technologies education, and advocate for curriculum change, and inclusion within undergraduate Initial Teacher Education (ITE) courses (Wooff et al., 2011) to mirror their findings. They propose contemporary textiles education should have a technological focus (Hughes & Bell, 2011) in line with societal and industry shifts, rather than a traditional one aligned with home-economics, visual arts, or craft. Their research also demonstrates how Technological Textiles have been applied in STEM education contexts in the UK (Hughes & Bell, 2011; Wooff et al., 2011). They've established how easily Initial Teacher Education (ITE) students felt Textiles could be applied in STEM education contexts (Wooff et al., 2011), 64% indicated easily or very easily; 36% indicated not very easily; and no participants stating that wasn't possible. Overall, their research indicates that Textiles (especially technological textiles) can be applied by Design and Technologies educators within STEM projects and educational contexts.

Table 11.2 Literature review results: (a) Textile* AND STEM (21,708 peer reviewed journal articles); (b) Textile* AND STEM AND Education (6002 peer reviewed journal articles); (c) Electronic Textiles (e-Textiles) in STEM and Education research (1875 peer reviewed journal articles); date range 2010–2022

Broad discipline categories and topics	Search term results: (a) Textile* AND STEM	Search term results: (b) Textile* AND STEM AND Education	Search term results: (c) STEM AND Education AND Research AND Electronic AND Textile*
Science and Technology	13,441	0	861
Life Sciences and Biomedicine	6459	0	348
Technology	5923	953	495
Physical Sciences	4614	663	335
Material Science	4237	445	235
Material Science Interdisciplinary	0	0	0
Social Sciences	4182	2287	709
Chemistry	3480	463	247
Engineering	2267	0	227
Analysis	2257	594	206
Science and Technology	0	2234	0
Science and Technology—other topics	2249	471	213
Business	0	418	137
Business and Economics	1858	935	297
Environmental Sciences and Ecology	1781	454	0
Nanoparticles	1599	0	0
Environmental Sciences	1568	341	0
Studies	1538	601	214
Materials Science, Textiles	1513	0	0
Polymer Science	1466	0	0
Chemistry and Materials Science	1430	0	0
Mechanical Properties	1420	0	0
Economics	0	534	174

(continued)

Table 11.2 (continued)

Broad discipline categories and topics	Search term results: (a) Textile* AND STEM	Search term results: (b) Textile* AND STEM AND Education	Search term results: (c) STEM AND Education AND Research AND Electronic AND Textile*
Arts and Humanities	0	485	0
Education	0	441	185
Education and Educational Research	0	367	168
Management	0	430	154
Physics	0	0	151

*Implies a search was conducted using the term: Textile as a base and included plural forms

This review revealed the most common Textile applications used in STEM education to be electronic textiles or e-Textiles (Table 11.2). Overall, the diversity of disciplines in which Textiles and e-Textiles are used was striking, demonstrating the authenticity of the Textiles discipline in both STEM research and STEM education contexts. These findings indicate that Textiles offer considerable scope for researchers and educators in STEM.

11.3 How Are Textiles Currently Used in STEM Education?

Textiles in STEM education are synonymous with 'e-Textiles' or electronic textiles. Literature reviews reveal that Textiles in STEM education predominately focus on electronic or e-Textiles. Studies demonstrate how e-Textiles have developed "students' STEM/technological self-efficacy, teach debugging, develop computational thinking, experiment with aesthetics, and create culturally relevant artefacts" (Lovell et al., 2022, p. 2). Educators working with e-Textiles describe when "crafting electronics, students can construct and literally see the connections between physical actions, visual patterns, and abstract ideas; between electrical components that they design and make by hand and relevant theories from physics, electrical engineering, and computing" (Buechley et al., 2013b, p. 2). E-Textiles are specifically listed in some curriculum and syllabus documents, in a New South Wales, state-based syllabus in Australia for example: 'e-Textiles' are a required area of study (NESA, 2019). E-textiles integrate soft-circuitry—where circuits are hand-stitched into a textile rather than soldered to activate LED's, sensors, and link to programmable microcontrollers and power sources. The Arduino Lilypad (Buechley et al., 2013a) system was specifically developed by Leah Buechley, an MIT researcher, as a wearable component that could be used in educational settings. The development of the

Arduino Lilypad (2007–2012) power and circuitry components with sew-on points using stainless steel conductive sewing thread increased availability in educational settings. The design of the LilyTiny sewable microcontroller in combination with high quality creative commons online tutorials, like SparkFun (2022) made e-Textiles project's—even for those educators with little pre-existing (theoretical or practical) knowledge of electronic circuitry—much easier to implement (Lovell et al., 2022). For an historical overview of the development of e-Textiles, refer to Hughes-Riley et al. (2018) and a review of e-Textiles in education, refer to Peppler (2015). E-textiles use conductive thread to integrate electronic components into textiles, by "attaching the electronics onto the surface of a textile" (Hughes-Riley et al., 2018). However, the objective for smart, interactive textiles is that all components (including electronics and power sources, sensors etc.) become pliable, washable, wearable, and made entirely from textile materials (Kirstein, 2013). Wearable technologies are used in healthcare and sporting applications (Muhammad Sayem et al., 2020; Wu & Luo, 2019) meaning that in class students can learn about broader applications for e-Textile's technologies for user-centred design projects whilst integrating electronic textiles into simple projects themselves. For example, e-Textile sensors are being integrated into prosthetic limbs by Biomedical Mechanical Engineers, to restore touch sensation to amputees (Searle et al., 2021). Related research includes conductive polymers in development as a 'smart' biomaterial for tissue engineering: "these versatile polymers can be… electro-spun into [biocompatible] microfibers" (Balint et al., 2014). Explorations into 'smart textiles', use the conductive properties of e-Textiles to support 'smart' applications including wearable sensors in high-performance apparel applications. For an overview of wearable textile electronics refer to Heo et al. (2018) and Dias (2015).

E-textiles are being used in STEM programs in Australia, the UK, Canada (Hughes & Morrison, 2018), and the US (Kafai et al., 2014). Many models for the integration of e-Textiles have been published, along with research outcomes on e-Textiles in STEM education (Hébert & Jenson, 2020). Educators in Science, Engineering, Fashion, Computing, Mathematics, and Technologies are using e-Textiles through design-based learning (DBL) and problem-based learning (PBL) approaches (Table 11.3). Educators are also extending student learning through research-inspired speculative design thinking (combining DBL and PBL) by asking students to consider how smart textiles could assist a user in a future-based design scenarios (von Mengersen, 2018) and design fictions (Hardy, 2018). One approach to problem-based learning used in e-textiles education is 'learning by fixing design problems', related to both the physical problems which require solving, stitching soft circuits, the circuitry issues, and coding problems with 'de-bugging' Arduino programming (Lui et al., 2017). E-textiles provide a multitude of opportunities for the interdisciplinary learning approaches implied by STEM education and in diverse contexts (Hughes & Morrison, 2018). Berzowska (2013) and Peppler (2015) provide additional examples of how e-textiles are used in society, education, design, research, and pedagogy. Nugent et. al. (2019) provide an overview of how textiles in STEM education supports learning and attitudes. However, implementation of e-Textiles is not without its challenges, for a brief overview of these refer to Hughes and Morrison (2018), and many educators offset the practical limitations with research investigations into how the wearable technologies can and will support human wellbeing.

The literature review (Table 11.2) demonstrated that e-Textiles have been used widely in education contexts since approximately 2010 (Jayathirtha & Kafai, 2020), when sew-in, textile specific electronic components became available. In education, e-Textiles are utilized in various discipline areas where aims and pedagogical approaches vary. For a snapshot of examples (refer to Table 11.3). Whilst a comprehensive review of the varied disciplines, approaches, and educational aims lies beyond the scope of this discussion, examples have been selected to demonstrate a range of applications. For an insight into the breadth of e-Textiles refer to the e-textiles network website (E-Textiles_Network, 2022b) and their annual conference: E-Textiles: International Conference on the Challenges, Opportunities, Innovations, and Applications in Electronic Textiles (E-Textiles_Network, 2022a). A prominent

Table 11.3 Current e-Textile applications in education

Discipline	Focus	Pedagogical approach	Article title	References
Science	Electricity	Intervention Inquiry-based learning (IBL)	Electrifying Engagement in Middle School Science Class: Improving Student Interest Through E-textiles	Tofel-Grehl et al., (2017)
Technology	Fashion design and virtual prototyping	Transdisciplinary learning Task-oriented and explorative-oriented learning	Towards new fashion design education: learning virtual prototyping using E-textiles	Han et al., (2021)
Engineering	Folding and modifying structures	Engineering design process Integrated engineering education curricula	Bringing E-Textiles into Engineering Education	Eisenberg et al., (2013)
Maths	Mathematics and computational thinking	Applied mathematics	Digital Making in Elementary Mathematics Education	Hughes et al., (2017)
Computing	Computational thinking & iterative design thinking	Inquiry-based learning (IBL)	Teaching computational thinking with electronic textiles: modelling iterative practices and supporting personal projects in exploring computer science	Fields et al., (2019)

research voice in this community is Dr Hughes-Riley, a Senior Research Fellow in Electronic Textiles (2022b) where he acts as a research lead for the Advanced Textiles Research Group (ATRG, 2022) and has authored 30 peer reviewed journal articles, primarily in the fields of E-textiles. Hughes-Riley's (2022a) range of publications further demonstrates a diversity of applications for e-Textiles, with research focused on advanced textiles especially electronically active wearable technology. His work aims to extend the range of potential applications for conductive threads and electronic yarns for medical applications including smart knitted wearables, solar harvesting, temperature control and monitoring, knitted electrodes, washable and durable electronic yarns, vibration sensing, integration of pressure-sensors and thermographs, designed to extend e-textile product lifetimes and capabilities. Many of these publications provide useful case studies of user-centred design for e-Textiles, and his article 'a historical review of the development of electronic textiles' (Hughes-Riley et al., 2018) offers a valuable overview of the field for educators.

Whilst the use of electronic textiles in education has become widespread, some educators and researchers also offer critique's including practical concerns about the limited applications of e-textile kits in design applications, like the relatively high cost of electronic components, perceptions of how it may continue to foster gender-bias in textiles by perpetuating a: "serious cyborg" v's "frivolous fashion" (Wilkins, 2020, p. 37) dichotomy along with debates about the ethics, reliability and durability of wearable technologies. In many programs electronic textiles are used to 'tick-the-box' for a STEM requirement—however, are they really STEM as these projects combine textiles and electronics (two areas of technology), but usually don't explicitly include science, math, or engineering components in the learning design? These projects can be tokenistic, not aiming to solve user-centred design problems, and the level of programming attained is often 'trivial' (Eisenberg et al., 2013). Textile teachers are encouraged to use e-textiles as an accessible and practical example of smart textiles. E-textiles are seen to interest females in STEM and contribute to increased gender equity in STEM disciplines (Searle et al., 2019). The literature indicates textiles and STEM can provide an inspirational pathway and starting point that might be used to increase student confidence in STEM disciplines, potentially leading to further STEM studies during their education. In Australia this discussion is also political as recent government funding initiatives including: "advancing women in STEM strategy" focused on "enabling STEM potential through education" aims to bring young women into STEM (Australian Government, 2022). At this stage the logic of this aspiration is clear and the rhetoric prevalent in government documents, so it will be good to see how the research continues to evolve and if a more widespread use of e-Textiles does lead to an increase in young women's self-efficacy in STEM and influence their educational and career pathways. As Wilkins reminds us: "many STEM programs offer Wearable Technology options at an attempt to gain gender diversity, but robotics teams are seen as hard science, and wearable tech as Frivolous Fashion. Instead of textile futures being drawn out from their original creations and forms, they are retrofitted to existing technologies, creating limits that are further resistant to change" (Wilkins, 2020). So, whilst e-textiles are now widely used in educational contexts—do they cover the requirements of a STEM project, or do they remain an introduction? If so, what else could be offered?

Many educators from STEM and STEAM disciplines are using e-Textiles, implementing the learning activities through the lens of their discipline areas. In this way the use of e-Textiles mirrors a criticism of STEM education which suggested that it is often not as well integrated as it could be—where the educator who has initiated the task will focus on their area of expertise primary so the focus will usually be one of the four areas. The task will therefore be taught through a 'science' or 'math' or 'computing' lens and less often an engineering or design and technologies lens. With this, it depends upon the school and educator context—it may be possible to address some of these limitations in future with a focus on how initial teacher education programs implement STEM—ideally, supporting Design and Technologies teachers to become the STEM leaders in their schools and initiate STEM activities which include textile application. To be effective, Textiles needs to be either integrated (Leung, 2020) through an "overarching learning framework" (Burrows et al., 2006) in STEM, as with any other discipline area or opportunities for interdisciplinary 'interaction' used more effectively in design education contexts (Williams, 2011). Opportunities to improve access to STEM activities including e-Textiles in both Initial Teacher Education (ITE) and Professional Development (PD) programmes could have a positive influence in the development of individual educators' self-efficacy and confidence to perceive themselves as STEM leaders. Educators are using a range of design-inspired pedagogical approaches in STEM including Inquiry-based learning (IBL), Problem-based learning (PBL), and Design-based learning (DBL), (see Table 11.4).

Shirley and Kohler (2012) reason that STEM is an "integral part" (Shephard, 2023) of textiles and apparel curricula. Textiles Literacy is described as foundational knowledge for fashion production including the role of textiles, fashion, and apparel curricula (Shephard, 2023). Shepard's (2023) research indicates that textiles literacy can be fostered through inquiry-based learning (IBL) in textiles courses. 'Textile literacy' has been linked to STEM literacies (Tang & Williams, 2019) including knowledge and skills related to the STEM disciplines, higher order thinking related to STEM interaction (Williams, 2011) alongside textiles specific discipline knowledge and what Buckley et al. outline as student's incremental learning and understanding of 'implicit theories of intelligence in STEM education' (2018).

11.4 How Do STEM Principles Apply to Textile Design?

The STEM acronym denotes the integration of Science, Technology, Engineering, and Mathematics in an educational context. Textiles are well suited to STEM applications, as they already integrate knowledge and skills from all four disciplines (Shirley & Kohler, 2012). Sophisticated textiles design requires an interdisciplinary approach (Shephard, 2023) as per the examples listed below in (Table 11.5).

Table 11.4 Current use of Textiles in/for STEM education

Discipline	Focus area	Pedagogical approach	Example	Application notes	References
Textile Science	Biomimicry, biomimetics, bioinspiration as a design problem-solving approach	Problem-, inquiry-, and integration-based approaches	Bioinspiration as a method of problem-based STEM education: a case study with a class structured around the COVID-19 crisis	Mask design and biological systems relevant to filtration and transparency	Snell-Rood et al. (2021)
Textile Bio Engineering	Bioengineering Bioplastics to replace petroleum-based synthetics for textile production Biopolymer components, biofabrication		Bioengineering textiles across scales for a sustainable circular economy	Development of bio-based recyclable fibres at nano-, micro-, and macroscale to improve environmental impacts and technical performance	Schiros et al. (2021)
Textile Engineering	Mathematics, physics, chemistry/ material science, and electro-technics	Design-Based Learning (DBL) Linked with problem and project-based learning	E-learning Instruments for Design-based learning in textiles	Manufacturing Chain of smart textiles: fibres and yarns, materials and methods, virtual models, design prototypes, manufacture prototypes, data processing and testing	Rădulescu et al., (2021)

(continued)

Table 11.4 (continued)

Discipline	Focus area	Pedagogical approach	Example	Application notes	References
Textile Technology Chemistry, Knitting, Printing	Advanced functional materials 3D/4D printing Knitting Shape memory polymers	Design-Based Learning (DBL)	4D printing of shape memory materials for textiles: mechanism, mathematical modelling, and challenges	4D textiles embrace the interaction of elastic textiles, Smart textiles	Biswas et al., (2021)
Textiles, Apparel, and Fashion Mathematics	Garment manufacturing methods including the application of specific geometry standards	STEM principles	Clothing and Textiles: Reinforcing STEM Education through Family and Consumer Sciences Curriculum	Development of the sewing pattern requires the application of specific geometry standards	Shirley and Kohler (2012)

Table 11.5 The relationship between STEM principles and textiles design

Science	Textile science, chemistry, e.g. dye/fibre affinity, textile testing—properties and performance, physics, e.g. the LZR racer swimsuit
Technology	Textile technology, textiles & design, product design and development, e.g. biomimetic design in Velcro
Engineering	Textile engineering, electronic/smart textiles, material engineering, textiles—e.g. non-woven filters in engineered systems
Mathematics	Design principles, e.g. proportions, spatial awareness, optimization, e.g. zero-waste pattern making, mathematical modelling, e.g. the hyperbolic shapes of Euclidian geometry

If design and manufacture in Textiles already utilize all areas within STEM—why is this not commonly perceived? A STEM principles framework can demonstrate to Textile's students the breadth of the discipline and encourage them to work collaboratively rather than in isolation and to think beyond fashion applications, to consider that textiles can be redesigned from its chemical building blocks onwards. All textiles are products of a myriad of textile technologies. Textiles provide a material framework for many products from automotive to architecture, agriculture to healthcare (von Mengersen, 2017). Textile's technologies have shaped humankinds' capacity to navigate the harsh climates of their world, beginning with string twisted from plant fibres and animal gut. Humankind's use of textile string has a 52,000-year history (with a new discovery in 2020 indicating string was used by Neanderthals approximately 120,000 years ago) (Bar-Yosef Mayer et al., 2020). Researchers have claimed recently that Neanderthal fibre technology discoveries have "cognitive and behavioural implications" (Hardy et al., 2020). The range of textiles and their uses is infinitesimal as variants continue to evolve through textile chemistry and material engineering research and development in response to diverse and demanding applications. Textiles are a distinctive material category. Some Textiles are formed into fashion, draped, cut, stitched, and manipulated, but many other textiles are utilized in product design. Textile innovation can occur at any stage of the manufacture from: fibre–yarn–thread–fabric–patterns–construction–finishing. In the NSW state-based syllabus Textiles Technology includes textile science, material chemistry, textile testing, and lab-based methods that test the properties and performance of textiles and aspects like dye-fibre affinity as well as industry study; and Textiles & Design focusses on design for textile prints, textile products, and textile art. Textiles in this state-based syllabus is multifaceted discipline area that straddles science and design, and where educators are routinely introduce students to both scientific and design based textile principles. In the textiles discipline, educators advise that it is necessary to understand textile material science to be able to design effectively, as it is the science which helps us determine the specific attributes of a material and successfully apply that knowledge in designed products. Technology education also provides a framework for students and educators to view textile and fashion product design through a user-centred lens providing a sociological and ethical viewpoint as well as technical knowledge. Consequences of the make-take-use-dispose model in textiles and fashion require redress and social rehabilitation. Effective design, manufacture and textiles sustainability utilizes all areas within STEM (Elmogahzy, 2019; Shirley & Kohler, 2012), vital to the development of a sustainable circular economy, including research into bioengineered textiles (Schiros et al., 2021).

11.5 Which Textile Case Studies Could Be Used in STEM?

Textiles can be integrated by educators into STEM programs to support education in all the aspects of STEM. An example is hyperbolic crochet, where mathematician Daina Taimina (Henderson & Taimina, 2001) created an elusive visual mathematical model for illustrating the concept of hyperbolic space in geometry, demonstrating the infinite planes of hyperbolic geometry using crochet and its inherent capacity for crenulation (Taimina, 2014). The resultant models are also forms of biomimicry in that they demonstrate how organic shapes like coral grow (Taimina, 2018). Biomimicry is being used to enhance creativity in design education (McGlashan, 2018) and is an example of interdisciplinary solutions, that enable students to think beyond what they know and cross-pollinate ideas from other disciplines.

What remains fascinating from the perspective of a textile educator is how the integration of many innovative technologies like e-textiles require traditional technologies like hand-stitching or crochet, or how an ancient fibre crop can rival the properties and performance characteristics of carbon fibre. Hemp, woven on a multi-axial loom and cast into components in a sustainable prepreg rivals the strength-to-weight ratio of carbon fibre panels used for Formula One racing cars (a textile technology that Henry Ford originally experimented with). Thus, links between emerging and traditional textiles technologies can be explored alongside examples of how Textiles are used in STEM education. Less well-known aspects of textiles can be introduced, like the evolution of self-healing fabrics for instance. Innovation in Textiles Technologies are driving what has been called the 'Materials Revolution' (Quinn, 2010). The fields of Textile Engineering and Textile Chemistry are transforming how Textiles continue to adapt through technological innovations in architecture, healthcare, science, engineering, automotive, art and fashion. For example, Janet Echelman's (2011) large-scale public sculptural string installations, whilst in theory, are based on traditional net weaving techniques (used for centuries in fishing nets); in practice, have required entirely new textiles threads to be designed along with specific Computer Aided Design (CAD) modelling and (CAM) manufacturing techniques—capable of rendering complex spatial orientations of her softly moving forms—to evolve along with her vision. Achieving this vision requires Echelman to work in all areas of STEM (Echelman et al., 2018) in collaboration with aeronautical and mechanical engineers, mathematicians, weavers, and textile chemists. Echelman's sculptures which require a mix of hand and machine manufacturing methods, also provide a good example of how collaboration between art and science is necessary to achieve specific design constraints. Science and the Arts contribute to STEM education in textiles and many examples of STEM can be observed in visual arts education contexts. Braund and Reiss (2019) provide an overview of how the Arts contribute to Science and Science education cognitively, collaboratively, and pedagogically which is relevant to Technologies education. STEM approaches are also being used in Textile research to solve problems like the proliferation of microplastic textile particles in wastewater (Curlin et al., 2022). From Personal Protective Equipment (PPE) like facemasks (Snell-Rood et al., 2021) to Fashion, Textiles—among humankinds first technologies are working harder and getting smarter, and as we demand more, the value of textiles—and all the hidden work they do in all areas of the STEM domain continues to evolve.

Textiles have organic attributes, elasticity for instance, that can come from the fibre itself (like handle or absorbency) the yarn (strength and durability) or the fabric construction method (stretch from knitting) (Kadolph & Marcketti, 2017). Textiles are often constructed through repetitive stitch-by-stitch-based construction techniques (weaving and knitting); other more organic forms are achieved through crochet. With the aid of three-dimensional body scanning, 3D and 4D knitting, three-dimensional all-in-one or seamless knitted garments are being produced made-to-measure (no more ill-fitting t-shirts). Mimicry is a principle deeply embedded in the history of textiles—synthetic textiles still aim to replicate the inherent properties of natural fibres, especially wool and silk. Synthetic dyes aim to copy the startling colours found in the natural environment. Intricate hand-worked textile techniques are of mimicked by printing techniques, including direct digital printing. A story told to students of textile mimicry is of the history of the paisley shawl, intricately hand-woven versions, leading to embroidered and printed copies, and its consequent social devolution. This affinity for mimicry makes textiles a well-suited vehicle for biomimicry inspired design. Velcro is the most common example of this—where the designer observed how burs 'locked' onto his dogs' fur. Biomimicry focuses on design which observes how things work in the natural world. The LZR (pronounced laser) racer swimsuit (Sportsmatik, 2017) is another example, Australian gold medal winner Ian Thorpe was famously disqualified for wearing this suit, because it was deemed to give him an unfair advantage. The streamlined suit reduced drag significantly (tested in a NASA wind tunnel) by using ultrasonic bonding technique instead of stitching panels, and etched polyurethane knit fabric with a textured pattern inspired by sharkskin (Speedo, 2009).

During the acceleration of fast-fashion over the past 20 years, mainstream fashion seemed to forget about textile diversity—focusing its energy on a conceited use of mostly one material from one source: Polyester. Fashion became cheaper, the turn-around was faster, however, polyester is currently made from a non-renewable resource. Polyester is derived from crude oil, and consequently now, the fashion industry is in competition with the automotive industry for its share of the stocks that remain. Ironically, fashion needs textiles and a STEM disposition now to solve its endemic problems. Like most problems the polyester paradox is nuanced, because whilst polyester is derived from a non-renewable source, it can be recycled with little loss of quality (Fletcher, 2013). Until recently, the fashion industry ignored this impending issue, and now due to its lack of vision—and divorce from textiles looks to the scientific community for a bioplastic solution. Biosynthetic fibres grown or harvested (seaweed) from algae offer one promising alternative. Algae-based bio synthetics are being considered in many industries as a replacement for plastics (Schiros et al., 2021).

According to Li Edelkoort (fashion forecaster and fashion/textiles academic), there are very few people in the fashion industry who "know anything at all about textiles". She describes in her Anti-Fashion manifesto (Edelkoort, 2017) (and related presentations) how textiles knowledge in the industry has been decimated by economic rationalization in design courses—stripping them of textile science laboratories and workshops, and other related ateliers (fashion and textile workshops). Couture houses like Chanel have been buying up specialized ateliers for years—like Lesage (embroidery) in order to preserve them (including the libraries, materials, tools and legacy of skills and techniques), now Edelkoort says: "they're buying textile mills" (Chin, 2011), because they know that if they don't know won't be able to continue to supply high quality, luxury products they're known for—products like a Chanel jacket made from fibre-dyed woven Boucle' and the Barrie knitwear cashmere mill in Scotland which produces their quintessential two-tone cardigan. This acute lack of knowledge and large-scale textile mill closures sets up an unrealistic proposition of asking designers to design with material they know little about or with a very limited palette. In Australia, like many other countries, Textile education (focused on fashion design, textile design, textile art or initial teacher education) has moved away from its applied science core—to focus on design—meaning understandings of the properties and performance, and capability of textiles have been diminished. The aspects of Textiles including textile chemistry including fibre and dye chemistry, technical aspects of textile fibre, yarn, and fabric manufacture, textile analysis and testing are taught to a limited extent if at all in many Initial Teacher Education (ITE) programs. Initial Teacher Education in Textiles in Australia used to include extensive modules on textile science, now only some courses run small textile science and testing unit within a larger design module. Consequently, students are abstractly applying general knowledge rather than experience born out of practice, experimentation and testing. Even if students are fortunate enough to encounter the knowledge theoretically in their tertiary education studies, they are very unlikely to have an opportunity to apply their knowledge in practice. This severely limits their career opportunities—a background in Textile science is prerequisite for Textile's conservation, analysis, chemistry, engineering, research, and development industries, and higher education courses for instance. Disconcertingly the job opportunities that do exist are in inverse proportion to the skills, knowledge, and experience provided by most Fashion and Textile courses. Graduates from these programs (approximately 30,000 from the US annually) would mostly need to undertake extensive additional specialized studies to apply to become a textile chemist, engineer, researcher, or conservator. This situation varies slightly in each country, in Australia, for instance, we have three research and development centres for textiles, but a dearth of textile engineers, due to no specialized university course being available, we employ chemists or scientists, or textile engineers who have been educated oversees—and yet Textiles is part of the technologies secondary curriculum; whereas in the USA, textiles is not widely offered in ITE programmes or high schools as part of a technologies program—yet they do have tertiary courses like Textile Engineering, Chemistry, and Science programmes (Wilson_College, 2023). Even from these contradictory positions, we can surmise that there is a clear need for Textiles in STEM education in a ways that could build our global problem-solving capacity in textiles—but at this time it is an aspirational objective in both industry and education. Increasingly due to the demands for smart, technical textiles and sustainability, multi-disciplinary knowledge is urgently required in textiles industry and education contexts (Kirstein, 2013).

11.6 Science v's Art: Why is There Opposition Between Science and the Arts?

Textiles in STEM education has been inhibited by a perceived incompatibility between Science and Art (Braund & Reiss, 2019). Textiles are seen as being linked with Art and Design and as incompatible or even in opposition to science, math, and engineering. This false perception pervades society and education that they're deeply contradictory in nature. Fashion and textile design honours students, writing practice-led research these which informed the conceptual development of their textile or fashion collections often focus on the false dichotomy of Art and Science, offer an insight into this. These students aim to interrogate a perception of Fashion being analogous to art and creative thinking in opposition to science being analogous to logical and critical thinking. Some of them had experienced prejudicial remarks from friends and family dismissing the seriousness of their studies and undermining their academic dispositions and the hard work necessary to complete a unique textile design or fashion collection. They wanted to challenge the notion that textiles and fashion were not informed by research practices nor even thoughtful or challenging. Many of these students felt belittled by society and even their family and friends for choosing to do something perceived to be ‘light-weight’, non-critical and superfluous. They were seen to be wasting their intellectual capacities, and yet, these students were enrolled in the second most highly sought-after fashion degree program in Australia, where places are highly competitive, and the ATAR necessary for entrance hovered between 96–98 out of 100. So, crucially, they did have a choice, and based on the evidence of their academic acumen, could have enrolled into a Science, Engineering, Math, or Medical program. However, they elected not to—and were being derided for electing a more creative path. What they found during their research was that effectively science required a lot of risk-taking, and that in many ways, in terms of the work involved: testing, revising, retesting, evaluation, critical, and creative thinking, there were more similarities than differences. Both fields require research, testing, involved inherent risk-taking and innovation, alongside an ability to look beyond that which already existed, crucially, they also both involved investigations using emerging technologies. Students who did well in the honours program undertook research into cross-disciplinary areas—often into architecture, science, mathematics, psychology, literature, and history. Fashion and textiles are falsely perceived as historical, static, superfluous, and gratuitous. Textiles are often perceived as having only aesthetic value—where they’re practical worth is of equal or in most cases greater value and yet, often taken for granted, “in a post-industrialized world, fabric has become ubiquitous and inexpensive” (Gordon, 2011, p. 6). Textiles hold great value to humankind in all aspects of their lives, not only to provide daily comfort to the body and in domestic environments, but visible in all transport, work, and leisure contexts (Gordon, 2011). There are an increasing number of human endeavours that rely upon the physical properties and performance characteristics of textile technologies in less visible applications. This perception influences student’s choice of elective units in Stage 5 (Years 9–10, age 14–16) and 6 (Year 11–12, age 16–18) in Australia. Initial Teacher Education (ITE) students in this context indicated that they felt that Textiles & Design (Stage 6 elective) was perceived to be for ‘want-to-be-fashion-designers’ and believed that it had a much broader appeal and should work harder to demonstrate its practical worth and become less gender biased (Personal communication, 16th May 2022). It is also pertinent to consider that the preceding elective is called Textiles Technology (Stage 5) and very much focused on textile science, especially chemistry of dye and textile testing to learn how to identify specific properties and performance of textiles and industry studies. In NSW, Australia the subject content knowledge in these elective subjects is guided by an Australian National Curriculum (ACARA) in Design and Technologies, and a State-based curriculum and subject specific syllabus documents, New South Wales Educational Standards Authority (NESA). All aspects of Textile and Fashion design and manufacture are informed by scientific, mathematical, technological principles, and the principles because textiles are designed and manufactured in complex engineered systems (Shephard, 2023; Shirley & Kohler, 2012). Using STEM principles in Textile's education—even when describing basic concepts like the role of STEM in textile and fibre manufacturing processes (Table 11.6) could counter the narrow perceptions held by some members of society regarding its limitations as a material for design, its range of purpose and practical functional applications.

In Australia a narrow perception of Textiles also leads students, when they're electing to do Textiles subjects, to deliberately move away from science and mathematics-based electives. We should ask the question why parents and teachers are encouraging this dichotomy when there is clear evidence that electing to do Textiles and Design in NSW for instance, can enhance their results (Educonsulting, 2023). Many students who select Textiles as an elective do so based on an interest in fashion and fashion design—and some will undertake further study in fashion design. Australia graduates many fashion designers (like the USA)—at a time when job opportunities for designers in either Australian Fashion, Textiles, Clothing, or Footwear industries continue to decline. However, because they disassociate fashion and textiles from STEM, tending to 'drop' the STEM related subjects and focus on design.

Fashion design has become the focus of higher degree offerings in Textiles, where a fashion course may offer a Textile design stream. These degrees teach creative expression, fashion illustration, textiles surface design and fabric manipulation, and some conceptual narrative, pattern making, skills for manufacture, with some exposure to emerging or experimental technologies. However, many don't offer in-depth insight into sustainability, or advanced technical skills, only limited insights into industry structures, limited business knowledge and acumen, limited coverage of textile science, properties, and performance analysis of textiles, including textile testing, a narrow sense of possible career pathways, some graphic design skills, some creative and critical writing skills, limited historical research into fashion and textiles, limited curatorial or textile conservation skills, limited insight into the breadth of textile science or textile engineering, esp. as they're applied in industries other than

Table 11.6 The role of STEM disciplines in Textiles and Fibre Manufacturing processes

Fibre category	Fibre type	Manufacturing processes
Natural cellulosic fibres	Cotton, linen, flax, hemp	Fibre crops are grown using agricultural and scientific principles
Natural protein fibres	Wool, silk	Fibres are grown using the combined principles of animal husbandry and agriculture
Synthetic fibres	Polyester, acetate, nylon	Fibres are synthetically manufactured from chemicals
Regenerated fibres	Viscose, rayon	Fibres are engineered through a combination of mechanical and chemical processes from a variety of crops and vegetable materials sources

fashion. Some programs connect students with international artisans through international experience units (practical print, dye, or weave workshops) with the aim to create more sustainable networks of makers and support traditional textile industries.

As well as being maligned for wanting to 'create' in a world focused on consumption: Why would you want to make something when you buy it? These want-to-be fashion designers find themselves feeling guilty and accused of being morally culpable for perpetuating an unsustainable system. The fast-fashion phenomenon enabled accelerated mass production on a monumental and unsustainable scale (Fletcher, 2013). Consequentially, the value of textiles has been undermined in society. The value of clothing has diminished, and people are less inclined to care for or repair or modify their clothing—and even if they're willing to do so, the requisite skills are less available in society (Fletcher, 2016). Lockdown and the Covid pandemic have been described as a 'wake-up call' for the fashion industry with many now asking for a 'new textiles economy' (Ellen_Macarthur_Foundation, 2023). Meantime, NGOs have been creating fashion guides to inform consumers about brand practices from an environmental and ethical perspectives for nearly ten years now, like: the Ethical Fashion Guide, Baptist World Aid (Nimbalker et al., 2018) and Textiles 2030 (WRAP, 2023), post-Covid becoming more widely known.

A STEM-based approach to Textile's education could shift the focus from perpetuating problems in the fashion and textiles industry to solving them. If student's self-efficacy in design and 'textiles literacy' (Shephard, 2023) can be extended beyond working with known materials in a flawed system. Then, their knowledge energy and enthusiasm can be directed to responsive problem solving. New material and systemic approaches are required to redesign both the textiles and fashion industries for a sustainable future.

Mark Liu makes a case for the increasing number of STEM skills required by contemporary Fashion designer's, stating that: "the future of fashion requires science, technology, engineering, and mathematics (STEM skills) more than ever" (Liu, 2016). In his doctoral work Liu designed an entirely new way to measure the curves on the body influenced by non-Euclidian geometry and focused on developing zero-waste fashion (Liu, 2022). Zero-waste fashion is innovative, on as it goes against the waste produced by European-fashion cutting methods, and reconsiders Japanese patterns like kimono which are made from narrow widths of cloth, essentially the fabric has been designed to create the cloth required for the pattern, resulting in virtually no material waste. When fashion students move into self-directed post-graduate study in design they begin to look beyond their discipline and consider design problems. Two innovative fashion designers work consciously engage with STEM principles through interdisciplinary working relationships: Iris van Herpen is credited with "transforming fashion by working between fashion, design, technology, and science" (HIGH, 2022) and Issey Miyake was known for incorporating mathematical models and designing the entire garment from a piece of cloth beginning with the fibre and yarn (Mikaye, 2022). Herpen is known for bringing complex 3D print (HIGH, 2022) into garments both work collaboratively—as textile and fashion researchers—who work from the textile material to the form (rather than a fashion-school model of shaping existing fabrics). Miyake created limited collections and

Herpen bespoke pieces for individual clients in a couture model. Edelkoort describes their work as a bridge between high tech and slow craft, she would like to see this way of working become more common in fashion design programs—working from the textile using a combination of traditional and experimental methods. Interdisciplinary STEM collaborations are vital because leaders in textile research are often from a science background because those trained in fashion and textiles do not have the knowledge, skills, or experience to do this. Techniques like 3D knitting (seamless knitting) (Sheikh et al., 2020) construct textiles from yarn directly to product—missing the entire step of fabric manufacture (Miyake also experimented with combining knitting and weaving—usually discrete textile construction techniques into one garment—with the aid of a mathematician who specialized in non-Euclidian geometry). Edelkoort has started "to believe that textile will be the dominant mode of manufacturing in the future: because if a product design starts from the textile—the designer can influence the materiality through 3D manufacturing" (Chin, 2011). Whist Miyake and Herpen's products remain exclusive—a revolution is taking place with sports shoes—a product most people own: Nike (Nike Lab) and Adidas (Adidas Future Team) run research centres focused on creating using raw materials with 3D printing and 3D knitting (*Nike flyknit*) because they've demonstrated that design solutions emerge from re-thinking the entire process, from raw materials through to product manufacture. The most innovative fashion outcomes evolve from interdisciplinary and collaborative work. This is what the STEM model offers fashion and textiles—a collaborative model where students are encouraged to work in interdisciplinary ways.

11.7 Conclusion

STEM knowledge, skills and dispositions are urgently required in textiles and fashion. Knowledge of textiles technologies, material technologies, engineered systems, design thinking, user-centred problem-based learning can develop student's ability to authentically contribute to interdisciplinary approaches to systems redesign for a circular economy (Ellen MacArthur Foundation, 2018).

11.8 A Short Manifesto for Textile's Educators

What if we embrace this opportunity to reimagine textiles in education and industry, now?

What's preventing Textile's inclusion in design and technologies education for textile's materials revolution has something to offer all disciplines?

Why aren't Textiles and Fashion designers encouraged to become STEM literate?

What if we used STEM to provide a framework for interdisciplinary Textiles?

What if we graduate less fashion designers?

What if instead we graduate more fashion technicians, textile engineers, and chemists, textile sculptors, sustainability consultants, environmental engineers, textile mathematicians, advanced pattern makers, textile technologists, textile fibre, fabric and form researchers, textile conservators?

What if we teach them to design beyond the surface—beyond print and garment forms... let's start with redesigning the chemical building blocks of textiles, diversifying the source materials and reimagining garment versatility through by designing for longevity by creating textiles with modifying properties and performance characteristics?

What if craft and couture collaborate with computer aided design and manufacturing?

What could be manufactured locally?

Why don't we use body-scanning technology to make clothes that fit?

Educational researchers concur that "STEM education is not a well-defined experience" (Margot & Kettler, 2019), however, in design and technologies education contexts, textile educators have the advantage of design-inspired pedagogical approaches, including problem-based learning (PBL), which can scaffold learning and teaching and aligns with STEM principles. PBL asks students to identify with a design problem or need, based on user-centred principles of design—outlining a specific user and their contextual needs. The student is encouraged to consider a wide range of possible solutions which consider different discipline areas—through an interdisciplinary approach. In this instance, the value of an interdisciplinary approach is the skills, knowledge, and experience brought from a range of disciplines to assist students and educators to work collaboratively to find a solution. A wave of change is occurring in textiles and fashion industries. Consumers are seeking sustainability in textiles and fashion, so at this time of restructure we need to reconsider design, production, values, and patterns of consumption (Gardetti et al., 2013). Local and international legislation is evolving, and frameworks are being developed to support the necessary changes through transition pathways. Australia has launched a national clothing product stewardship scheme (AFC, 2023) proposing a levy as an intervention measure that will provide funding towards solving systematic industry issues. Behind the changes are many interdisciplinary voices, seeking change. These changes can be better reflected in textiles and fashion education contexts where design is extended to include PBL learning approaches and STEM principles.

In conclusion, a STEM principles learning approach based on 'interaction' (Williams, 2011) in Textiles technologies and design education focused on improving both 'textile' and STEM literacies (Shephard, 2023), can contribute to designing solutions for an industry in crisis and heal the rift between textiles, fashion, and education. Ideally, this could reduce the *practice gap* between curriculum and industry (Hughes et al., 2011) by placing students and their educators within a global community of practice seeking to contribute to more sustainable, preferred futures.

References

AFC. (2023). *National clothing product StewardShip scheme*. https://ausfashioncouncil.com/product-stewardship/

ATRG. (2022). *Advanced textiles research group*. Nottingham Trent University. https://www.ntu.ac.uk/research/groups-and-centres/groups/advanced-textiles

Australian_Government. (2022). *Advancing women in STEM strategy*. Department of Industry, Science and Resources. Retrieved from https://www.industry.gov.au/publications/advancing-women-stem-strategy

Balint, R., Cassidy, N. J., & Cartmell, S. H. (2014). Conductive polymers: Towards a smart biomaterial for tissue engineering. *Acta Biomaterialia, 10*(6), 2341–2353. https://doi.org/10.1016/j.actbio.2014.02.015

Bar-Yosef Mayer, D. E., Groman-Yaroslavski, I., Bar-Yosef, O., Hershkovitz, I., Kampen-Hasday, A., Vandermeersch, B., Zaidner, Y., & Weinstein-Evron, M. (2020). On holes and strings: Earliest displays of human adornment in the middle palaeolithic. *PLoS ONE, 15*(7), e0234924.

Berzowska, J. (2013). E-textile technologies in design, research and pedagogy. In L. Buechley, K. Peppler, M. Eisenberg, & Y. Kafai (Eds.), *Textile messages: Dispatches from the world of e-textiles and education* (pp. 171–179). Peter Lang.

Biswas, M. C., Chakraborty, S., Bhattacharjee, A., & Mohammed, Z. (2021). 4D printing of shape memory materials for textiles: Mechanism, mathematical modeling, and challenges. *Advanced Functional Materials, 31*(19), 2100257.

Braund, M., & Reiss, M. J. (2019). The 'great divide': How the arts contribute to science and science education. *Canadian Journal of Science, Mathematics and Technology Education, 19*(3), 219–236. https://doi.org/10.1007/s42330-019-00057-7

Buckley, J., O'Connor, A., Seery, N., Hyland, T., & Canty, D. (2018). Implicit theories of intelligence in STEM education: Perspectives through the lens of technology education students. *International Journal of Technology and Design Education, 29*(1), 75–106. https://doi.org/10.1007/s10798-017-9438-8

Buechley, L., de Boer, S., Goldfein, J., & Qiu, K. (2013a). *Sew electric: A collection of DIY projects that combine fabric, electronics, and programming*. HLT Press.

Buechley, L., Peppler, K. A., Eisenberg, M., & Kafai, Y. B. (2013b). *Textile messages: Dispatches from the world of e-textiles and education*. Peter Lang.

Burrows, V., Oehrtman, M., & Lawson, A. (2006). Development of an integrated learning framework for STEM learning. In *2006 Annual Conference & Exposition*.

Chin, A. (2011). *Li Edelkoort: Talking textiles at Milan design week 2011*. Design Boom. https://www.designboom.com/design/li-edelkoorts-talking-textiles/

CO. (2022). *Faces and figures: Who makes our clothes*? http://www.commonobjective.co/article/faces-and-figures-who-makes-our-clothes

Curlin, M., Pusic, T., Vojnovic, B., & Vincic, A. (2022). STEM approach in assessment of microplastic particles in textile wastewater. *Tehnički vjesnik, 29*(5), 1777–1781. https://doi.org/10.17559/TV-20220121153959

Dias, T. (2015). *Electronic textiles: Smart fabrics and wearable technology*. Elsevier Science. https://books.google.com.au/books?id=DsqcBAAAQBAJ
E-Textiles_Network. (2022a). *4th International Conference on the Challenges, Opportunities, Innovations and Applications in Electronic Textiles*. https://e-textilesconference.com/
E-Textiles_Network. (2022b). *E-textiles network website*. https://e-textiles-network.com/
Echelman, J. (2011). *Taking imagination seriously*. @tedtalks. https://www.ted.com/talks/janet_echelman_taking_imagination_seriously
Echelman, J., Sarkisian, M., Long, E., Beghini, A., & Wang, N. (2018). The dream catcher: Design of a rope net sculpture. *Proceedings of IASS Annual Symposia, 2018*(5), 1–8. https://www.ingentaconnect.com/content/iass/piass/2018/00002018/00000005/art00010
Edelkoort, L. (2017). *Anti-fashion: A manifesto for the next decade*. The Business of Fashion. https://www.youtube.com/watch?v=LV3djdXfimI
Educonsulting. (2023). *Subjects with bad reputations*. https://www.datainschools.com.au/blog/badreps
Eisenberg, M., Eisenberg, A., & Huang, Y. (2013). Bringing e-textiles into engineering education. In L. Beuechley, K. Peppler, M. Eisenberg, & Y. Kafai (Eds.), *Textile Messages: Dispatches from the world of e-textiles and education* (pp. 121–129). Peter Lang.
Ellen MacArthur Foundation. (2018). *Ellen MacArthur Foundation: Design technology education resources*. Ellen MacArthur Foundation. Retrieved 15 June from https://kumu.io/ellenmacarthurfoundation/educational-resources#circular-economy-educational-resources/design-technology?focus=2
Ellen_Macarthur_Foundation. (2023). *A new textiles economy: Redesigning fashion's future*. https://ellenmacarthurfoundation.org/a-new-textiles-economy
Elmogahzy, Y. E. (2019). *Engineering textiles: Integrating the design and manufacture of textile products*. Woodhead Publishing.
Fields, D. A., Lui, D., & Kafai, Y. B. (2019). Teaching computational thinking with electronic textiles: Modeling iterative practices and supporting personal projects in exploring computer science. In *Computational thinking education* (pp. 279–294). Springer.
Fletcher, K. (2013). *Sustainable fashion and textiles: Design journeys* (2nd ed.). Taylor and Francis.
Fletcher, K. (2016). *Craft of use: Post-growth fashion*. Routledge, Taylor & Francis Group.
Gale, C., & Kaur, J. (2002). *The textile book*. Berg.
Gardetti, M. A., & Torres, A. L. (Eds.). (2013). Sustainability in fashion and textiles: Values, design, production and consumption (1st ed.). Routledge. https://doi.org/10.4324/9781351277600. eBook Published 1 October 2017
Gordon, B. (2011). *Textiles: The whole story: Uses, meanings, significance*. Thames & Hudson.
Han, A., Wohn, K., & Ahn, J. (2021). Towards new fashion design education: Learning virtual prototyping using E-textiles. *International Journal of Technology and Design Education, 31*(2), 379–400. https://doi.org/10.1007/s10798-019-09558-w
Hardy, A. (2018). Using design fiction to teach new and emerging technologies in England [Article]. *Technology & Engineering Teacher, 78*(4), 16–20. https://ezproxy.acu.edu.au/login?url=https://search.ebscohost.com/login.aspx?direct=true&db=a9h&AN=133398159&site=ehost-live&scope=site
Hardy, B. L., Moncel, M. H., Kerfant, C., Lebon, M., Bellot-Gurlet, L., & Mélard, N. (2020). Direct evidence of Neanderthal fibre technology and its cognitive and behavioral implications. *Scientific Reports, 10*(1), 4889. https://doi.org/10.1038/s41598-020-61839-w
Hébert, C., & Jenson, J. (2020). Making in schools: Student learning through an e-textiles curriculum. *Discourse: Studies in the Cultural Politics of Education, 41*(5), 740–761. https://doi.org/10.1080/01596306.2020.1769937
Henderson, D. W., & Taimina, D. (2001). Crocheting the hyperbolic plane. *The Mathematical Intelligencer, 23*(2), 17–28. https://doi.org/10.1007/bf03026623
Heo, J. S., Eom, J., Kim, Y.-H., & Park, S. K. (2018). Recent progress of textile-based wearable electronics: A comprehensive review of materials, devices, and applications. *Small (weinheim an Der Bergstrasse, Germany), 14*(3), 1703034. https://doi.org/10.1002/smll.201703034

HIGH, H. M. O. A. A. (2022). *How Iris van Herpen Transformed Fashion: Working between fashion, design, technology and science*. Google Arts & Culture. https://artsandculture.google.com/entity/iris-van-herpen/m0h_9fwd

Hughes, C., & Bell, D. (2011). Underpinning the STEM agenda through technological textiles? An exploration of design technology teachers' attitudes. *Design and Technology Education: An International Journal, 16*(1).

Hughes, C., Bell, D., & Wooff, D. (2010). Sustaining the essence of design and technology: The case for textile technology. In *Design and Technology Association International Conference*

Hughes, C., Bell, D., & Wooff, D. (2011). *Reducing the practice gap between the design and technology curriculum and the needs of the textile design/manufacturing industry*. Goldsmith's University.

Hughes, J., Gadanidis, G., & Yiu, C. (2017). Digital making in elementary mathematics education. *Digital Experiences in Mathematics Education, 3*(2), 139–153.

Hughes, J., & Morrison, L. (2018). The use of E-textiles in Ontario Education. *Canadian Journal of Education, 41*(1).

Hughes-Riley, T. (2022a). *Research publications for Theodore Hughes-Riley*. Nottingham Trent University. https://llr.ntu.ac.uk/rpd/researchpublications.php?pubid=9d99816d-0c69-4d3c-b2ea-76c767c2d9e3&pubid=9d99816d-0c69-4d3c-b2ea-76c767c2d9e3

Hughes-Riley, T. (2022b). *Theodore Hughes-Riley*. Nottingham Trent University. https://www.ntu.ac.uk/staff-profiles/art-design/theodore-hughes-riley

Hughes-Riley, T., Dias, T., & Cork, C. (2018). A historical review of the development of electronic textiles. *Fibers, 6*(2), 34. https://www.mdpi.com/2079-6439/6/2/34

Jayathirtha, G., & Kafai, Y. B. (2020). Interactive Stitch sampler: A synthesis of a decade of research on using electronic textiles to answer the who, where, how, and what for K–12 computer science education. *ACM Transactions on Computing Education (TOCE), 20*(4), 1–29.

Kadolph, S. J., & Marcketti, S. B. (2017). *Textiles* (12 ed.). Pearson.

Kafai, Y., Fields, D., & Searle, K. (2014). Electronic textiles as disruptive designs: Supporting and challenging maker activities in schools. *Harvard Educational Review, 84*(4), 532–556. https://doi.org/10.17763/haer.84.4.46m7372370214783

Kirstein, T. N. (2013). *Multidisciplinary know-how for smart textiles developers*. Woodhead Publishing

Leung, A. (2020). Boundary crossing pedagogy in STEM education. *International Journal of STEM Education, 7*(1), 15. https://doi.org/10.1186/s40594-020-00212-9

Liu, M. (2016). Why STEM subjects and fashion design go hand in hand. https://www.abc.net.au/news/2016-08-25/why-stem-subjects-and-fashion-design-go-hand-in-hand/7784834

Liu, M. (2022). *Zero-waste fashion*.

Lovell, E., Buechley, L., & Davis, J. (2022). *The LilyTiny: A case study in expanding access to electronic textiles CHI Conference on Human Factors in Computing Systems*. https://dl.acm.org/https://doi.org/10.1145/3491101.3503579

Lui, D., Anderson, E., Kafai, Y. B., & Jayathirtha, G. (2017). Learning by fixing and designing problems: A reconstruction kit for debugging e-textiles. In *Proceedings of the 7th Annual Conference on Creativity and Fabrication in Education*, Stanford, CA, USA. https://dl.acm.org/https://doi.org/10.1145/3141798.3141805

Margot, K. C., & Kettler, T. (2019). Teachers' perception of STEM integration and education: A systematic literature review. *International Journal of STEM Education, 6*(1), 2. https://doi.org/10.1186/s40594-018-0151-2

McGlashan, A. (2018). A pedagogic approach to enhance creative Ideation in classroom practice. *International Journal of Technology and Design Education, 28*(2), 377–393. https://doi.org/10.1007/s10798-017-9404-5

Mikaye, I. (2022). *ISSEY MIYAKE INC*. https://www.isseymiyake.com/en/

Muhammad Sayem, A. S., Hon Teay, S., Shahariar, H., Luise Fink, P., & Albarbar, A. (2020). Review on smart electro-clothing systems (SeCSs). *Sensors, 20*(3), 587. https://www.mdpi.com/1424-8220/20/3/587

NESA. (2019). *Textiles technology 7–10|Syllabus*. Sydney, Australia. Retrieved from https://www.educationstandards.nsw.edu.au/wps/wcm/connect/9a0b6b2f-bb31-41e3-ac0f-a29abcf96da4/textiles-technology-years-7-10-syllabus-2019.pdf?MOD=AJPERES&CVID
Nimbalker, G., Mawson, J., Harris, C., Rynan, M., Sanders, L., Hart, C., & Shove, M. (2018). *Ethical fashion guide 2018: The truth behind the barcode*. Baptist World Aid Australia. https://baptistworldaid.org.au/resources/2018-ethical-fashion-guide/
Nugent, G., Barker, B., Lester, H., Grandgenett, N., & Valentine, D. (2019). Wearable textiles to support student STEM learning and attitudes. *Journal of Science Education and Technology, 28*(5), 470–479. https://doi.org/10.1007/s10956-019-09779-7
Paul, R. (2019). *High performance technical textiles*. Wiley. http://ebookcentral.proquest.com/lib/acu/detail.action?docID=5720857
Peppler, K. (2015). *A review of e-textiles in education and society* (pp. 268–290). https://doi.org/10.4018/978-1-4666-8310-5.ch011
Quinn, B. (2010). *Textile futures: Fashion, design and technology* (English ed.). Berg.
Rădulescu, I., Ghiţuleasa, C., Visileanu, E., Almeida, L., Malengier, B., Stjepanovič, Z., Blaga, M., & Dufkova, P. (2021). *E-learning instruments for design based learning in textiles.*
Schiros, T. N., Mosher, C. Z., Zhu, Y., Bina, T., Gomez, V., Lee, C. L., Lu, H. H., & Obermeyer, A. C. (2021). Bioengineering textiles across scales for a sustainable circular economy. *Chem, 7*(11), 2913–2926. https://doi.org/10.1016/j.chempr.2021.10.012
Searle, K., Tofel-Grehl, C., & Breitenstein, J. (2019). Equitable engagement in STEM: Using e-textiles to challenge the positioning of non-dominant girls in school science. *International Journal of Multicultural Education, 21*(1), 42–61.
Searle, T., Sencadas, V., Greaves, J., & Alici, G. (2021). Room-temperature self-healing piezoresistive sensors. *Composites Science and Technology*. https://doi.org/10.1016/j.compscitech.2021.108856
Sheikh, J. A., Waheed, M. F., Khalid, A. M., & Qureshi, I. A. (2020). *Use of 3D printing and nano materials in fashion: From revolution to evolution*. Advances in Design for Inclusion.
Shephard, A. J. (2023). Textiles literacy: Implementing inquiry-based learning in a basic textiles class. *Clothing and Textiles Research Journal, 41*(1), 71–88.
Shirley, L. M., & Kohler, J. (2012). Clothing and textiles: Reinforcing STEM education through family and consumer sciences curriculum. *Journal of Family & Consumer Sciences Education, 30*(2).
Snell-Rood, E. C., Smirnoff, D., Hunter, C., Chapman, K., Kirscht, E., & Stretch, E. (2021). Bioinspiration as a method of problem-based STEM education: A case study with a class structured around the COVID-19 crisis. *Ecology and Evolution, 11*(23), 16374–16386. https://doi.org/10.1002/ece3.8044
SparkFun, E. (2022). *LilyPad Basics: E-Sewing*. https://learn.sparkfun.com/tutorials/lilypad-basics-e-sewing/all
Speedo. (2009). *Designing Speedo LZR Racer: Speedo's innovative and controversial line of high-end swimsuits—Operations case studies—Case Study|Operations|Case Study*. https://www.icmrindia.org/casestudies/catalogue/Operations/OPER075.htm
Spendlove, D., & Barlex, D. (2011). Researching STEM? *Design and Technology Education: An International Journal, 16*(1).
Sportsmatik. (2017). *LZR Racer X Swimwear: Sensor technology, swimming, features*. @sportsmatik. https://sportsmatik.com/sports-corner/sports-technology/lzr-racer-x-swimwear
Taimina, D. (2014). *Crocheting hyperbolic planes|Daina Taimina|TEDxRiga*, TEDxRiga. https://www.youtube.com/watch?v=D-AHvZqbMT4
Taimina, D. (2018). *Crocheting adventures with hyperbolic planes: Tactile mathematics, art and craft for all to explore*. CRC Press.
Tang, K. S., & Williams, P. J. (2019). STEM literacy or literacies? Examining the empirical basis of these constructs. *Review of Education, 7*(3), 675–697.
TBRC. (2022). *Textile global market report 2022*. https://www.thebusinessresearchcompany.com/report/textile-global-market-report

Techno-Espace. (2022). Clothing Technology Laboratory, Collège LaSalle Montreal. https://www.lasallecollege.com/about-us/facilities/techno-espace

Tilton, M. (2018). *Get schooled in outdoor product design.* https://www.rei.com/blog/stewardship/get-schooled-in-outdoor-product-design

Tofel-Grehl, C., Fields, D., Searle, K., Maahs-Fladung, C., Feldon, D., Gu, G., & Sun, C. (2017). Electrifying engagement in middle school science class: Improving student interest through e-textiles. *Journal of Science Education and Technology, 26*(4), 406–417. https://doi.org/10.1007/s10956-017-9688-y

Ul-Islam, S., & Butola, B. S. (2018). *Advanced textile engineering materials.* Wiley, Incorporated. http://ebookcentral.proquest.com/lib/acu/detail.action?docID=5494683

USU. (2022). *Outdoor Product Design & Development (OPDD).* https://opdd.usu.edu/index

VMR, V. M. R. (2022). *Technical Textile Market size worth $ 285.88 Billion, Globally, by 2028 at 5.15% CAGR.* @PRNewswire. https://www.prnewswire.com/news-releases/technical-textile-market-size-worth--285-88-billion-globally-by-2028-at-5-15-cagr-verified-market-research-301454624.html

von Mengersen, B. (2017). Textiles, a dynamic exchange: Challenges and opportunities. In M. J. de Vries (Ed.), *Handbook of technology education* (pp. 1–15). Springer International Publishing. https://doi.org/10.1007/978-3-319-38889-2_26-1

von Mengersen, B. (2018). Speculative writing: Enabling design thinking. In *36th International Pupils' Attitudes Towards Technology Conference.*

Wilkins, L. M. (2020). E-textiles: Power and resistance. *Virtual Creativity, 10*(1), 35–49. https://doi.org/10.1386/vcr_00016_1

Williams, P. J. (2011). STEM education: Proceed with caution. *Design and Technology Education: An International Journal, 16*(1). https://ojs.lboro.ac.uk/DATE/article/view/1590

Wilson_College. (2023). *B.S. Textile Engineering.* NC State University. https://textiles.ncsu.edu/academics/undergraduate/textile-engineering/

Wooff, D., Hughes, C., & Bell, D. (2011). *Does the Science, Technology, Engineering and Mathematics (STEM) agenda need separate curricular input in the realms of undergraduate Initial Teacher Education (ITE)*? Goldsmith's University.

WRAP. (2023). *Textiles 2030.* https://wrap.org.uk/taking-action/textiles/initiatives/textiles-2030

Wu, M., & Luo, J. (2019). Wearable technology applications in healthcare: A literature review. *Online Journal of Nursing Informatics (OJNI), 23*(3). https://www.himss.org/resources/wearable-technology-applications-healthcare-literature-review

Chapter 12
Case Studies of STEM Projects

HildaRuth Beaumont

Abstract This chapter will describe and analyse approaches schools have taken to the implementation of STEM through a case study approach. There are two case studies, and each will be presented through a set framework in order to identify the differences and similarities in the approaches taken in two different jurisdictions and scrutinise them with regard to possible STEM futures.

Keywords STEM (Science Technology Engineering Mathematics) · Curriculum · Scenario building

12.1 Introduction

This chapter is in four Sections. Section 12.1 is a preamble that establishes possible futures for STEM education which will be used to scrutinise the STEM education case studies that appear later in the chapter. Section 12.2 details the framework through which the case studies will be presented. Section 12.3 will present the case studies. Section 12.4 will present a commentary on the studies with regard to the extent to which they align with the possible futures established in Sect. 12.1.

12.1.1 Preamble

Writing in 2013 (Banks & Barlex, 2014), John Holman, National STEM Director for the English Government between 2006 and 2010, was clear with regard to the meaning of STEM education to government and employers.

> … it means an area of school, college and university curriculum that is of great economic importance. All over the world, developed and developing countries see Science, Technology,

H. Beaumont (✉)
University College London, London, United Kingdom
e-mail: hildaruthbeaumont@btinternet.com

W. Fox-Turnbull and P. J. Williams (eds.), *Locating Technology Education in STEM Teaching and Learning*, Contemporary Issues in Technology Education,
https://doi.org/10.1007/978-981-97-1995-2_12

> Engineering and Mathematics education as fundamental to a successful industrial base. Prowess in STEM education is the new educational 'arms race'. (p. xi)

He also notes that

> … In secondary schools Science, Technology, Engineering and Mathematics are usually taught as separate subjects with science and mathematics often getting the lion's share of time and resources. (p. xi)

Writing some seven years later, Jim Al-Khalili, who was a contributing author to the Royal Society's 2014 Vision for Science and Mathematics Education Report, notes (Banks & Barlex, 2021) that … *science, technology, engineering, and mathematics are mostly taught as separate subjects* … and he laments that pupils do not appreciate the interconnectedness of what they are learning or how these subjects overlap in the real world. He goes further in supporting the case for STEM education being important for all young people not just those entering a technical career.

> In an age when all information is literally at our fingertips, it is vital that STEM teaching also focuses on the necessary skills needed to navigate through a world where problem-solving and rational thinking skills are more important than ever. In addition, STEM teaching must encompass softer skills, often acquired in group-based work, such as communication skills or an appreciation of the values and ethics of the application of new scientific knowledge in technology and an appreciation of the role of scientific evidence. (p. vii)

One way to explore possible STEM futures and the role of the T in these futures is the 2 × 2 matrix technique which develops four contrasting scenarios relevant to a particular area of interest by placing two factors that influence the future of the issue under study on two axes which cross to form four quadrants. A simple example is shown in Fig. 12.1.

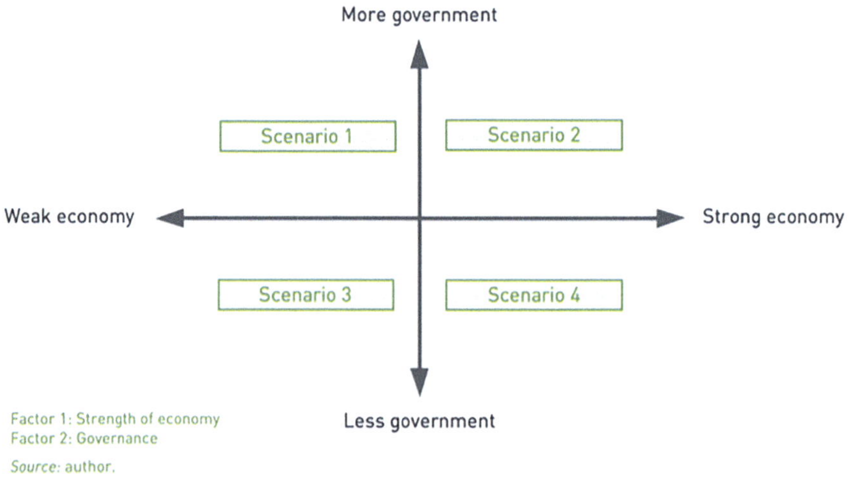

Fig. 12.1 A sample scenario matrix (Rhydderch, 2017)

The 2 × 2 approach was formalised during the 1990s by the consulting firm Global Business Network (GBN). In 1996, an account of the 2 × 2 approach was published under the heading 'Steps to Developing Scenarios' as an appendix to Peter Schwartz's *The Art of the Long View.* Although sometimes criticised, notably for reducing the variety of drivers of change to two dimensions (axes), this approach remains widely used in all sectors, as it effectively marries the analytical element with 'space' to expand the drivers and develop scenarios that incorporate both qualitative and quantitative elements (Rhydderch, 2017). The choice of the influencing factors on the axes is of course crucial.

The comments of John Holman and Jim Al-Khalili above indicate that some schools organise their curricula such that the STEM subjects operate in isolation from one another, the so-called silo approach, whilst other schools go to great lengths to ensure that there is collaboration and cooperation between the STEM subjects. Hence, the vertical axis of the 2 × 2 matrix to establish four future scenarios (Fig. 12.2) will be the extent to which STEM subjects operate in isolation as opposed to the extent to which they operate collaboratively; these are labelled S.T.E.M (taught in isolation) and STEM (taught collaboratively) at the extremes of the vertical axis.

There have long been tensions between providing general education as opposed to vocational education in England. The case for general education has been made through the introduction and successive revisions of a National Curriculum in

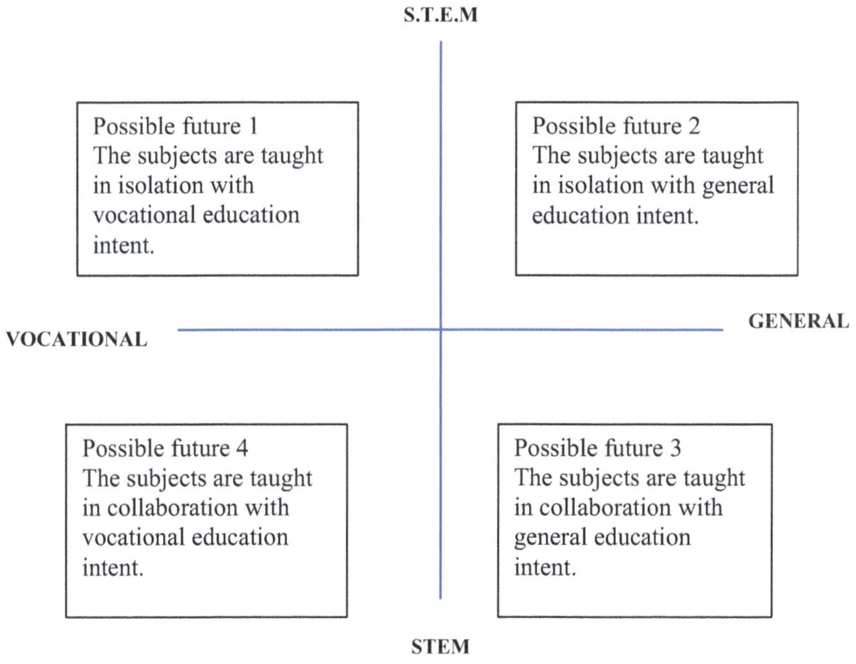

Fig. 12.2 Four possible STEM FUTURES from axes of uncertainty concerning isolation/collaboration and vocational/general education

England. This has involved defining what should be taught in terms of individual subjects—science, mathematics, history, English etc. through the ages 5–14 years and then a choice of a reduced number of subjects between the ages of 14 to 16 years. Within this choice, there are opportunities to continue studying some of the subjects previously studied and also other subjects leading to vocational qualifications (see for example https://btecworks.com/about/). However, there are those who believe that this approach does not provide sufficient vocational education and that after the age of 14 years, young people should be able to decide what sort of occupation they wish to enter and attend a school that caters for that career path. The University Technical Colleges (see https://www.utcolleges.org) aim to provide this sort of learning. Funded partly by the government but also by local industry and working with local universities they aim to develop a cohort of young people ready to enter the workforce in local businesses and industry. Given this division of opinion and the 'STEM arms race' that John Holman referred to above the horizontal axis of the 2×2 matrix to establish four future scenarios will delineate the extent to which the subjects are seen as providing vocational education or general education The extremes of the horizontal axis are labelled Vocational and General.

Note that the possible futures operating in each quadrant are not mutually exclusive, and two or more of these scenarios could operate simultaneously in a single secondary school. It is also important to realise that these futures do not necessarily represent what will happen; they are a product of the chosen uncertainties and choosing different uncertainties would result in different futures, but they do give us the opportunity to explore the case studies in the light of the dichotomies facing STEM education.

12.1.1.1 Possible Future 1

This future manifest itself through teaching the subjects more or less in 'silos' particularly valuing those subjects which are gatekeeper subjects providing qualifications that enable progression into further study. Whilst this reinforces the status of science and mathematics, it does little for design and technology or engineering as these do not function as gatekeeper subjects.

12.1.1.2 Possible Future 2

This future appears on the surface to be very similar to Possible Future 1 but it is likely that the make-up of the classes taking design and technology and engineering options will be different. This is because at 'option time', when subjects are chosen by learners aged 14+ for further study, design and technology and engineering will have been promoted on general education grounds for all learners as opposed to appealing just to those the learners who wish to pursue a particular trade or technical career pathway. Hence, it is likely that classes will include learners who have shown aptitude in academic subjects as well as those who have been successful in 'practical' subjects. As in Possible Future 1, the subjects will be taught in isolation.

12.1.1.3 Possible Future 3

As in Possible Future 2, the make-up of the classes post 14+ is likely to include learners who have shown aptitude in academic subjects as well as those who have been successful in 'practical' subjects. However, in this scenario, definite attempts will have been made by the teachers to utilise the learning in some STEM subjects to support the learning in other STEM subjects. In some situations, there will have been timetable suspensions, so-called drop-down days, or even weeks, in which learners collaborate in tackling an open task, sometimes of their own choosing and in so doing use their learning from across the STEM subjects. In such cases, the learners will probably need to be supported by a team of teachers from different STEM subjects.

12.1.1.4 Possible Future 4

In this future teachers will utilise the learning in some STEM subjects to support the learning in other STEM subjects. Given the vocational intent, they are likely to involve their learners in multi-disciplinary events that promote STEM careers.

12.2 The Framework Through Which the Case Studies Will Be Presented

This structure will consist of the following elements.

- A description of the school or college providing the context of the study.
- Description of the STEM curriculum model used for T in STEM by the school.
- Comments from teachers about the activities.
- A closer look at some of the activities that comprise the case study.

12.3 The Case Studies

12.3.1 Case Study 1

Nottingham High School is an independent day school for boys and girls in Nottingham, England, with an Infant and Junior School (ages 4–11) and Senior School (ages 11–18). There were 738 students in the 2019–2020 academic year, of whom 151 were in the sixth form, studying for advanced certificate examinations.

The pupils come from a range of professional and business backgrounds, with a majority from white British families alongside a number from other ethnic groups. Nationally standardised test data provided by the school indicate that the ability of the pupils is above average. The school has identified 92 pupils as having special educational needs and/or disabilities (SEND), which include dyslexia and dyspraxia, 44 of whom receive additional specialist help. English is an additional language for 61 pupils, whose needs are supported by their classroom teachers.

The curriculum is delivered under the title of 'D&T: Design Engineering.' Students in Years 7 and 8 have two 1 h lessons a fortnight and Year 9 students, should they opt for D&T have 3 × 1 h lessons every two weeks. In Years 10 and 11, students have 5 × 1 h lessons per fortnight. Years 12 and 13 students study for OCR[1] Design Engineering and have 5 h a week of lessons. The curriculum is designed around core concepts - Engineering Principles, Systems Principles, designing, making, designing and making, CAD/CAM and considering the consequences of all of these. The curriculum for students aged 11–14 (Key Stage 3, Years 7–9) is based on four types of learning activity; Designing, Designing and Making, Making and Considering wider consequences (Barlex and Steeg, 2017a, 2017b). The 'technology' part of this programme of study is twofold: (1) 3D CAD/CAM and (2) use of programmable microcontrollers in designed products. Schemes of learning are designed to be interleaved across lessons and with increasing complexity and challenge over year groups. The curriculum is summarised in Tables 12.1, 12.2, and 12.3.

CADCAM has become a significant feature in most school D&T curricula with the government funding several schools to pilot 3D printing (DfE 2013). Such interest and support may be seen as supporting a STEM education agenda with 3D printing becoming more recognised as having manufacturing as well as prototyping applications (The Economist 2017). Hence, the comments of this teacher on CADCAM are significant with regard to the role of D&T within STEM.

> I believe that CAD has to some extent democratised designing by taking away the barriers some students find when communicating their ideas. Students have gained experience in seeing the direct link between creating 3D CAD data and how this can be outputted to a manufacturing system such as 3D printing. CAD/CAM has worked particularly well for my students where there are constraints for students to meet, for example, a part that is designed and printed by the student has to assemble to another part. TinkerCAD has been a successful introduction to CAD/CAM with students working through selected tutorials to learn the basics features before putting these into practice in a design challenge to design and laser cut, or 3D print a fidget spinner. Students progress to OnShape. When students are proficient using OnShape 'features' they are able to fairly rapidly create designs and iterations that; most students would struggle to create so quickly using hand drawing skills; adding an element of creativity that students couldn't have explored when hand drawing.

The Head of Department Ed Charlwood is clear that the curriculum he has designed does not involve any explicit links or collaboration with other STEM subjects and whilst he appreciates the potential for these he is adamant that in his situation *'D&T, science and mathematics more than "have their hands full" with*

[1] *OCR* is a UK awarding body, providing qualifications for learners of all ages at school, college, in work or through part-time learning programmes; see https://ocr.org.uk.

Table 12.1 D&T curriculum for year 7

Year 7: Autumn term *Context: Engineering*	
Designing activities Recreate a famous bridge in 3D CAD (TinkerCAD)	**Making activities** Egg drop challenge Aluminium foil boat challenge
Designing + making activities Fidget spinner (3D printed) Truss bridges	**Considering wider consequences activities** Bicycle analysis Design Museum: see https://designmuseum.org/exhibitions/designer-maker-user/crowdsourced-wall
Year 7: Spring term *Context: Hand-held lights*	
Designing activities Design a torch for a particular person, in a specific place for defined extra task Simulate the torch circuit in TinkerCAD Circuits and PhET	**Making activities** Solder a solar PCB torch, see https://kitronik.co.uk/products/2175-kitronik-solar-torch-kit and assemble an enclosure
Designing + making activities Design and 3D print a carabiner to meet a set of technical requirements	**Considering wider consequences activities** Product analysis of a Maglite (ACCESSFM—aesthetics, consumer, cost, environment, size, safety, function and materials)
Year 7: Autumn term *Context: Skateboards*	
Designing activities A template to aid making a penny skateboard A skateboard park (TinkerCAD)	***Making activities*** Assemble a skateboard deck, trucks and wheels A phone stand from folded sheet material
Designing + making activities A penny skateboard deck	Considering wider consequences activities Product analysis of a scooter with light up LED wheels (ACCESSFM) Consequences of disruptive forms of personal transportation, e.g. electric scooters

teaching to meet National Curriculum and public examination requirements' and he sees time spent adapting curricula to enable links and/or collaboration as *'better spent on improving teaching and learning within the individual subjects'*. He is firm in his belief that *'despite operating in a silo his curriculum makes a significant contribution to the STEM education agenda.'*

An interesting feature of this curriculum is the Considering Consequences of Technology sessions which take place each term throughout each of Years 7, 8, and 9. These embrace the following broad topics:

- Electrification of transportation (including the impact of electric scooters).
- The impact of robotics on society.
- The consequences of using particular materials.

Table 12.2 D&T curriculum for year 8

Year 8: Autumn term *Context: Robotics*	
Designing activities MicroBit code in MakeCode 3D CAD robot design challenge in TinkerCAD	**Making activities** LittleBits products Robot Arm: hydraulic/pneumatic
Designing + making activities Lego Spike Prime zip wire challenge Vibrating cupbots	**Considering wider consequences activities** Robotics case study (Boston Dynamics) Robotics debate
Year 8: Spring term *Context: Mechanisms*	
Designing activities Cardboard modelling and ergonomic testing	**Making activities** A ping pong ball shooter
Designing + making activities Design and prototype a range of targets using Lego BriQ motion mechanisms and MicroBits	**Considering wider consequences activities** Properties of materials Forces, simple machines/mechanisms Product analysis: apple master (peeler and corer)
Year 8: Autumn term *Context: Bronze CREST award (pulley powered drag racer)*	
Designing activities Prototype a range of solutions to optimise the performance of a 'drag racer car' cardboard and CAD modelling	**Making activities** Lego belt and pulley train A simple phone/tablet stand
Designing + making activities To design and make a drag racing car, powered by pulleys, to go as fast as possible	**Considering wider consequences activities** How to eliminate the use of fossil fuels > move to renewable sources of energy Case study: Bloodhound LSR Lifecycle of aluminium

- New methods of manufacturing including micro-manufacturing and the new products and service they might afford.

The consequences of adopting a particular technology will not treat all members of society equally. Some will lose out while others will benefit. There will almost certainly be unintended consequences. Thinking about this encourages students to see the world through others' eyes and explore their own values. This is important if they are to be able to think constructively about the sort of world in which they want to live.

Table 12.3 D&T curriculum for year 9

Year 9: Autumn term *Context: 3D CAD and CAM*	
Designing activities Design a range of 3D chess pieces	**Making activities** Slice and 3D print a chess piece
Designing + making activities An 'egg mobile' crumple zone to protect an egg Design and prototype a 'better' toothbrush	**Considering wider consequences activities** Product analysis task of a 3D printer Disruptive technologies and micro-manufacturing Consequences of plastic products
Year 9: Spring term *Context: Lighting*	
Designing activities Design (not make) a lamp based on a contextual challenge, considering a user, the place it will be used in and the specific task it is required for Reverse engineer a 3D CAD lamp bracket design	**Making activities** Make joint samples Making an adjustable lamp from a range of parts
Designing + making activities Additional parts to hold the light source and for frame junctions	**Considering wider consequences activities** Lighting case study of existing consumer products, e.g. dLightS1
Year 9: Autumn term *Context: Speakers*	
Designing activities Design a 'fantasy boombox'	**Making activities** A speaker cube A slotted laptop stand A layered circuit enclosure
To design and make a Bluetooth speaker with an upcycled part	**Considering wider consequences activities** Product disassembly (e.g. FlipJam speaker) teardown analysis e.g. see https://www.ifixit.com/Teardown/iPad+Pro+11-Inch+Teardown/115457 repairability/circular economy

12.3.2 *Case Study 2*

Lewis Girls School is an 11–18 comprehensive school and currently has 702 students, with 136 students currently attending the sixth form. It is situated in Ystrad Mynach, Wales and attracts students from the surrounding 40 square miles. In total, 6.7% of students have a special education need; 24.3% of students receive free school meals and for 1.1% of students English is an additional language. The new Welsh Curriculum (see https://hwb.gov.wales/curriculum-for-wales/science-and-technology/descriptions-of-learning/) is divided into six areas of learning and experience (AoLEs): Expressive Arts, Humanities, Health and Well Being, Languages, Literacy and Communication, Mathematics, and Numeracy and Science and Technology. The school has organised time for the Science and Technology AoLE according to the following components:

Design Technology) (DT)—1 h per fortnight in year 7 and 2 h per fortnight in years 8 + 9.

- STEM—1 h per fortnight each year
- Textiles—1 h per fortnight each year
- Digital Design—2 h per fortnight each year
- Food—1 h per fortnight in years 7 + 9
- Science— 5 h per fortnight each year

Note the use of the term Design Technology (DT). This is common in many schools across England and Wales although the correct term to use for the subject, as defined in National Curriculum legislation in England (DfE 2014) and justified at some length by David Layton in the Interim Report (DfES 1988) is in fact Design & Technology (D&T). The way these components play out in each of year 7, 8 and 9 as described in Tables 12.4, 12.5 and 12.6 with the content organised according to an overall theme for each year: Year 7 Sustaining Planet Earth, Year 8 Protecting Planet Earth, Year 9 Planet Earth and Beyond.

The order in which the topics are taught is flexible as Matthew James the Head of Science and technology explains …

> In Food, pupils learn about growing their own food in their garden and how pH levels in soil affect what vegetables and fruit grow well in that type of soil as well as how to affect/change the pH levels in the soil to allow for better growing. This involves an introduction to what the pH scale is before the activity. At about the same time in science, pupils are learning in more detail about the pH scale. Both subjects teach these topics in their own way but then connect back through the pupil's AoLE personal journal (called iDiscover) to complete certain tasks based on what has been learnt. It doesn't matter who gets to the pH part of either the food project or the science project first. If food teaches it first, the pupils gain a basic understanding which they revisit in more detail in science and have seen how the knowledge can be applied practically in the real world. If science teaches it first the pupil has a much greater understanding of what is happening and why when they revisit it in food.

Matthew is also convinced that traditional subject boundaries are being eroded. He cites as a good example of this the Year 9 activity *Are aliens among you?*

> Pupils are given stories which provide problems that a robot on Mars has to solve. The pupils have to programme a small robot to complete these tasks on a school-made map of Mars. An example is programming the robot to go to set grid references (numeracy + programming skills) to conduct a geo-prospecting exercise looking for potential signs of life. In science, they have at around the same time been considering the possibility of 'Alien' life.

An interesting feature of this curriculum is its sheer breadth which to some extent is a consequence of the new Curriculum for Wales and the descriptions of learning for the science and technology AoLE. However, it would be possible to interpret these descriptions in terms of previous curricula i.e. physics, chemistry, biology, design and technology and computing. And these items could well operate in silos. However, Matthew James and his teaching team have resisted this temptation and conceptualised their curriculum in a way which cuts across the traditional single-subject boundaries and integrates elements from across the subjects. An example of this is in the STEM unit which deals with vertical farming. The students experiment

Table 12.4 Science and technology curriculum for year 7

Year 7 sustaining planet Earth	
Component(s)	Topic/Content
DT + Science	Personal protective equipment and health and safety
DT	Introduction to hand tools used in the workshop
Science	The pH scale
DT	Making a planter in which students grow their own plants
STEM	Vertical farming with the study of forces, hydroponics and structures
Digital design	Data analysis of STEM and science experiments with spreadsheets and data bases
DT	Plants and photosynthesis (linking with planter work)
Textiles	Fabrics from plants and animals
DT	Natural resources used in products
Food	Bees, pH and seasonal vegetables
Food	The journey of food from garden to table
Digital design	An introduction to stop frame animation
Science	Working scientifically
Digital design	Programming a virtual greenhouse to optimise growing conditions
Science + STEM	Earth's forces—an introduction to gravity
Textiles	Fairtrade with regard to the fashion industry
Textiles	Animal rights in the context of producing clothing
Science	Earth's living things—an introduction to classification
Science + Food	Acids and alkalis
Textiles	Sustainable fabric products
Digital Design	Websites—an introduction to safe and critical use

with designing and building towers to hold weight, the more weight the more hydroponics can be grown on their tower helping to provide food for the country. They compare the potential crop yield of a flat field compared to a tower in the field and the footprint of the tower, then how many towers would fit in the field and as a result how much more food could be grown in that way compared to standard agriculture.

12.4 Commentary

Initial scrutiny of Case Study 1 curriculum reveals it is very much in a silo, with no explicit links to maths or science which would seem to place it in Possible Future 1 with the aim to enable students to enter an engineering or design profession of some sort. However, further inspection reveals that although quite deliberately engineering-focussed focused it does have some breadth caused by the use of the four different

Table 12.5 Science and technology curriculum for year 8

Year 8 protecting planet Earth	
Component(s)	Topic/Content
DT	Making an eco-passive giant bamboo speaker
DT + Textiles	Social, moral and ethical design in the context of clothing and accessories
Textiles	Eco fabrics through designing and making a fabric bag
DT	Solar powered lighting
DT + Science	Natural resources
Science	Biofuels
DT	Raw materials for manufacturing
DT + Textiles	The 7 Rs for sustainability—rethink, refuse, reduce, reuse, repair, regift, recycle
STEM	Energy and energy production
STEM	Energy transfer to drive toy boats—steam versus elastic
Science	Communication systems on Earth
Science	Disease
Science	Cloning
Science	Bio-plastic production
STEM	Insulation—its importance in combating global warming
Science	Genetic engineering
DT + Digital design	16 Bit game design and programming
Textiles	Re-used clothing turning an old item of clothing into a new product

teaching activities especially considering wider consequences. Hence, it is preferable to see this curriculum as being in Possible Future 2 with the result that at option time D&T appeals to a wider range of pupils including those who have shown aptitude in academic subjects as well as those who have been successful in 'practical' subjects.

Scrutiny of Case Study 2 immediately places the curriculum on the right-hand side of the 2 × 2 scenario grid. Many of the topics/content focus on issues concerning sustainability and societies' response to the problem of global warming and these are relevant to all members of society and students at school whatever occupation they might pursue. So, although this curriculum will undoubtedly prepare and predispose some young people to enter a STEM career this is not its main function. Hence, it is preferable to see this curriculum as being in Possible Future 3 and as indicated in the commentary on Scenario 3 at the beginning of this paper such an approach will require a team of teachers from different disciplines working together. And this is exactly what Matthew James has set up.

Although there is a stark difference between the curriculum of the two case study schools both represent robust representations of the T in STEM. In the first case study in which the T component of STEM operates in a silo, the curriculum provides learning experiences that are highly focused on content that is easily identified as

Table 12.6 Science and technology curriculum for year 9

Year 9 planet Earth and beyond	
Component(s)	Topic/Content
STEM	Rocket design—CREST award[2]
DT + STEM	Intellectual property
Food	Nutrition
Science	Digestion
Science	Combustion
DT + Science	Getting back to Earth—entry, decent, landing
Science	Insulation (revisiting year 8)
DT	Commemorative product
DT	3D CAD and 3D printing habitat design
Science	Genetic Modification
Science	Photosynthesis—revisiting year 7
STEM	Programming drones and rovers
DT	Jewellery products based on found items
DT	Smart materials
Food	Sources of proteins—insects
Digital design	Programming Lego EV3 tasks to solve problems on Mars
STEM	Are aliens among you? (programming task using Micro:bits and a model Mars base
Textiles	Alien inspired soft toys
Science	Infrasound and ultrasound

technology (design & technology). An audit of the curriculum against the presence of Big Ideas (Barlex and Steeg, 2017a, 2017b) indicates that all are present in abundance: materials, manufacturing, functionality, design, and critique. It must be acknowledged that the focus of the curriculum is unashamedly 'engineering' and hence omits food and textiles which some may see as a limitation. In the second case study every attempt has been made to link the learning across the STEM subjects and the resulting curriculum has significantly more breadth than that in Case Study 1 and this perhaps gives the curriculum greater appeal with regard to a general education suitable for all learners. However a criticism of this breadth is that it leads to a lack of depth within specific technology (design and technology) subject knowledge.

[2] The CREST Awards scheme is the British Science Association's flagship programme for young people, providing science enrichment activities to inspire and engage 5-to-19-year olds. See https://www.crestawards.org

References

Banks, F., & Barlex, D. (2014). *Teaching STEM in the secondary school* (1st ed.). Routledge.
Banks, F., & Barlex, D. (2021). *Teaching STEM in the secondary school* (2nd ed.). Routledge.
Barlex, D., Steeg, T. (2017). *Big Ideas for design & technology* Available at this url: https://dandtfordandt.files.wordpress.com/2017/05/big-ideas1.pdf. Accessed October 6, 2022.
Barlex, D., & Steeg, T. (2017). *Re-Building D&T v2.* Available at this url: https://dandtfordandt.files.wordpress.com/2017/05/re-building-dt-v21.pdf Accessed October 6, 2022.
Department for Education and Science and Welsh Office. (1988). *National curriculum design and technology working group interim report.* HMSO.
Department for Education. (2013). *3D printers in schools: uses in the curriculum Enriching the teaching of STEM and design subjects* Available at this url: https://assets.publishing.service.gov.uk/government/uploads/system/uploads/attachment_data/file/251439/3D_printers_in_schools.pdf. Accessed October 14, 2022.
Department for Education. (2014). *National curriculum in England: secondary curriculum* Available at this url: https://www.gov.uk/government/publications/national-curriculum-in-england-secondary-curriculum. Accessed October 14, 2022.
Rhydderch, A. (2017). *Scenario building: The 2x2 matrix technique.* Available at this url: https://www.futuribles.com/en/group/prospective-and-strategic-foresight-toolbox/document/scenario-building-the-2x2-matrix-technique/. Accessed October 6, 2022.
The Economist. (2017, June 29). *3D printers will change manufacturing* Available at this url: https://www.economist.com/leaders/2017/06/29/3d-printers-will-change-manufacturing?utm_medium=cpc.adword.pd&utm_source=google&ppccampaignID=18156330227&ppcadID=&utm_campaign=a.22brand_pmax&utm_content=conversion.direct-response.anonymous&gclid=CjwKCAjwkaSaBhA4EiwALBgQaChJM-HKKjlR5U1cOhXIZu1hqriLkqKG4crCnNWBbwzho6m6Aa_PmhoCijcQAvD_BwE&gclsrc=aw.ds. Accessed October 14, 2022.

HildaRuth Beaumont has worked in education for some 50 years and for most of that time was known as David Barlex. David became an acknowledged leader in design & technology education, curriculum design and curriculum materials development. Recently David identified disruptive technologies relevant to the school curriculum and ways in which young people may be engaged with these. His most recent publication with Frank Banks *Teaching STEM in the secondary school* was very well received with a second edition being published in 2020. Since 2021 David has found that he is much, much happier and at ease with himself if dressed and presenting as female. Hence, she is now living full time as HildaRuth and continuing with both academic writing and consultancy under the name of HildaRuth Beaumont.

Chapter 13
Integrating Technology in STEM with Integrity: Avoiding the Mucky Brown Paint

Wendy Fox-Turnbull

Abstract This chapter explores and discusses a range of issues associated with the integration of science, technology, engineering, and maths in the classroom. It will argue for an approach that enables authentic and appropriate representation of each discipline and the ways in which approaches can be students centred. The main implication of this approach for technology is that it puts it centre stage as student work towards the development of a technological artefact to meet an identified need or opportunity. Disciplines must be considered and specifically taught to develop skills and knowledge required by the students to reach their intended goal—a successful outcome to their technological practice. However, if purposeful teaching fails to occur students find themselves busy but not necessarily learning to their potential in the STEM disciplines.

Keywords Integration · Multidisciplinary · Interdisciplinary · Transdisciplinary · Curriculum integrity

13.1 Introduction

In Chap. 2, Lee introduces us to the rationale and theory that underpins integration and in Chap. 7 Hacker outlines the integrated nature of STEM education through several possible models. The purpose of this chapter is to highlight several specific approaches used to integrate STEM subjects in the compulsory schooling sector (Grades K-12, Years 1–13) and to identify and discuss potential issues that can occur with integration in this sector. In the writing of this chapter, I draw on my personal experiences as a trained teacher with 25 years of experience in teaching and researching technology education in the initial teacher education sector.

W. Fox-Turnbull (✉)
The University of Waikato, Hamilton, New Zealand
e-mail: Wendy.fox@waikato.ac.nz

W. Fox-Turnbull and P. J. Williams (eds.), *Locating Technology Education in STEM Teaching and Learning*, Contemporary Issues in Technology Education,
https://doi.org/10.1007/978-981-97-1995-2_13

Integration occurs when a number of different subjects or academic disciplines are taught through a project or learning task. The tasks should be designed to facilitate the development and application of knowledge through authentic contexts. STEM is learning to solve issues and problems and designing and creating artefacts (products and systems) by drawing on and integrating knowledge from science, technology engineering, and mathematics disciplines.

How and why integration occurs and what form integration takes varies. Variations occur for many reasons and may be based on people's perceptions and understanding of technology, technology education, their competence in other contributing disciplines, and/or the nature of integration. When teachers do not have sound content knowledge of the disciplines involved students' learning can be compromised. Much of the literature identified discusses the nature or approaches to integration in research (Menken & Keestra, 2016; Pohl et al., 2017; Stember, 1991; Wohl, 1955).

The rationale for subject integration is strong. It can facilitate the development of meaningful programmes of work for students by increasing the relevance of learning and situating theoretical learning in authentic and relevant contexts. This approach is known to increase motivation and the value of learning (Bellanca & Brandt, 2010; Fox-Turnbull, 2003; Hill, 2017; Roberts et al., 2018) and to situate discipline learning in relation to other disciplines and the lives of learners (Pitt, 2017). Integration can also decrease unnecessary duplication across disciplines. For example, let us take the scientific concept of 'fair testing'. In a non-integrated environment, students are taught to undertake a fair test in science to complete an experiment task set for them. In technology, students are taught how to test a range of fabrics for durability and need to be taught how to do this so that only one variable is changed in each round of testing. In an integrated environment with a carefully planned programme, these lessons could have been combined into one with identified learning outcomes for both science and technology thus saving precious time for further learning. Unnecessary duplication can be decreased as students work within authentic projects to develop critical conceptual knowledge related to the 'big understandings' also known as core concepts, principles and theories that serve as a focus for learning, leading to enduring understandings central to disciplines and transferable to new situations (Gilbert, 2005; Murdoch & Hornsby, 2003).

In the school curriculum teachers and principals are often frustrated by a crowded curriculum. Schools are under constant pressure to keep pace with change and are frequently required to add new content to what they teach (OECD, 2020). This expansion puts pressure on education systems and teachers alike. Such pressure increases the temptation to 'cover lots with little depth'.

A well-managed curriculum focuses on 'big understandings' or conceptual ideas implemented using an integrated approach to teaching and learning and can offer a more authentic and motivational solution to the issue of overcrowding (Bellanca & Brandt, 2010; OECD, 2020). For example, recently I have been working on a project related to teaching junior children, aged 5–6 years, many of whom identify as Māori, the sequencing aspect of computational thinking as a part of digital technologies situated within technology education in The New Zealand Curriculum (Ministry of Education, 2017). The 'big understanding' for this unit is whakapapa—a Māori term meaning to recite in proper order. Māori use the concept of whakapapa to recite their genealogies and legends. Reciting *whakapapa* is central to all Māori institutions and an important skill that reflects the importance of genealogies in Māori society in terms of leadership, kinship, and status.

The concept of sequencing is critical to computer programming and mathematics. Rather than teachers teaching sequencing in mathematics and then again in technology they will be taught through the big understanding of whakapapa the importance of order and sequence, with the goal of developing an online 'tour of their school' for newcomers. Thus, integration also has the potential to develop students' ability to transfer knowledge and skills across learning contexts.

What should integration look like in STEM when it upholds the integrity of technology education as described above? Imagine approaches to integration as a continuum, from simple to complex. There are three approaches on this continuum: multidisciplinary, interdisciplinary, and transdisciplinary integration, with multidisciplinary integration as the simplest approach and transdisciplinary being the most complex. When deploying a multidisciplinary approach to integration, teaching and learning occur in a single theme, which is explored within a range of discrete disciplines. Discipline knowledge remains discrete with little opportunity given for students to transfer knowledge to new situations. The interdisciplinary approach is positioned in the middle of the continuum. It begins with a teacher-identified context, and students or groups of students identify an issue or problem and then design and develop possible solutions. At the other end of the continuum, a transdisciplinary approach engages students in authentic, often self-identified issues or problems, which students seek to solve by drawing on a range of discipline knowledge—as and when needed. Teachers guide their students to explore issues in depth and investigate, design and develop solutions. Using this approach, teachers can facilitate learning across a range of disciplines, often deliberately teaching curriculum knowledge on a 'just-in-time basis', to meet the emerging or future needs of their students.

Figure 13.1 illustrates each level with a different colour representing learning within disciplines. You will see in the right-hand 'transdisciplinary' column that the curricula A & B represented by blue and yellow are merged into green representing new learning and understanding through merged understandings from multiple learning areas as well as knowledge and skills of a non-academic nature such as collaboration or empathy.

Each comes with its advantages and challenges. The following sections describe and illustrate each of the three approaches of integration and discuss some advantages and potential issues associated with each.

13.1.1 Multidisciplinary Integration

The multidisciplinary approach, also known as a thematic approach involves learning in discrete learning areas on the same theme or topic. In the traditional classrooms of

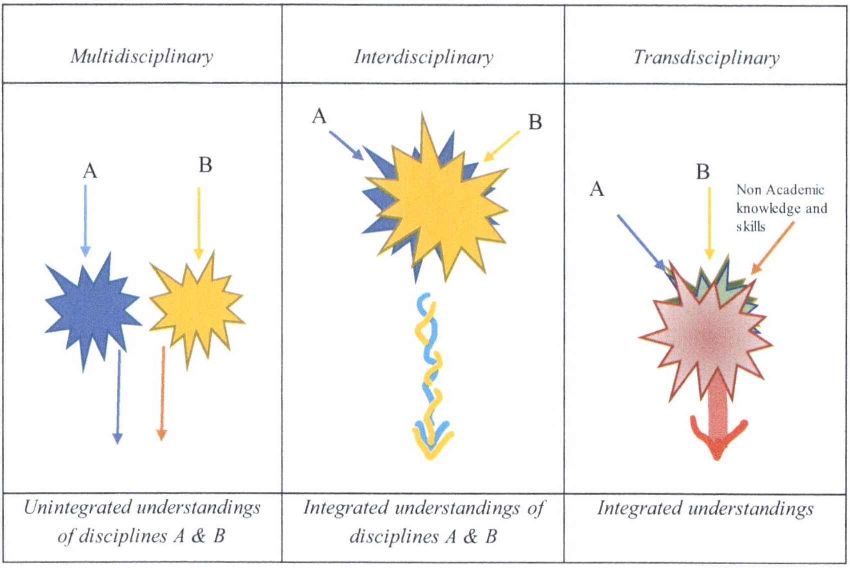

Fig. 13.1 Approaches to integration

the nineteenth and twentieth centuries student learning was largely based on instruction where students were required to learn discrete facts in distinct subjects for ease of understanding or to pass a test. It was subsequently assumed that students could and would reassemble the school knowledge to apply to complex problems outside school when required. However, the school curricula failed to assist students in reassembling their learned facts into useful coherent bodies of knowledge thus the concept of curriculum integration and connecting school subjects gained momentum in the early 1990s (Wicklein & Schell, 1995).

The multidisciplinary approach is the lowest level of integration and typically involves teachers gathering resources, websites and books that they use to plan many activities that are fun and engaging for students. This approach is called a thematic approach to teaching. When deploying a multidisciplinary approach to integration, teaching and learning occur in a single theme, which can be explored within a range of discrete disciplines. In primary school, units of work may include a range of activities and learning in different discrete curriculum areas, centred around a theme such as 'toys' (Fox-Turnbull et al., 2021). Learning in this scenario could involve students reading about toys, writing about their favourite toy, drawing or painting pictures of toys, counting toys, surveying peers for their favourite toy, playing with toys and maybe even making toys as illustrated in Fig. 13.2. In a multidisciplinary approach to integration apart from the context or 'topic', there are no further explicit connections across disciplines. In a secondary scenario, Wicklein and Schell (1995) discuss a multidisciplinary approach that used a technology survey tool to investigate a biological phenomenon with the assistance of the mathematics department for statistics interpretation. In this scenario, technology is reduced to the use of tools.

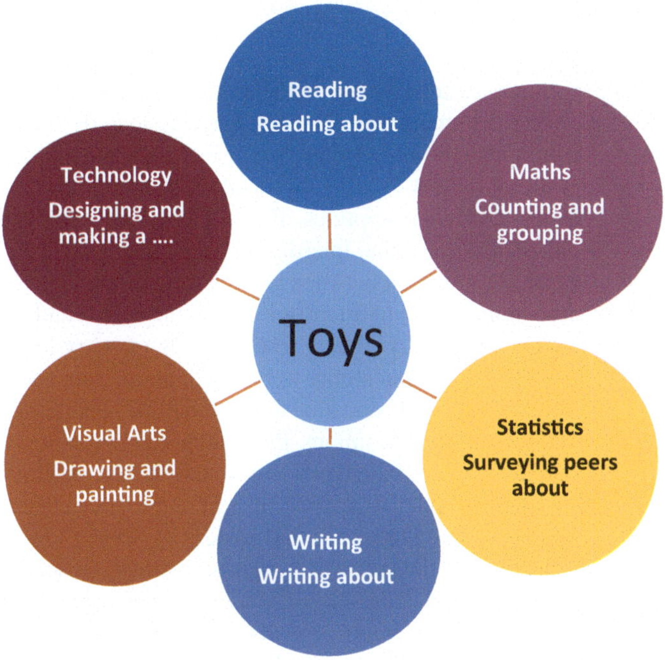

Fig. 13.2 Multidisciplinary approach to a unit on toys

There are several issues with this approach for students' learning and for technology education. The first is the lack of depth and understanding developed in each discipline. Units are not based on major concepts and provide little opportunity to transfer skills and understandings between different academic disciplines or into the students' world beyond school. Another issue is that students can become bored with the topic. An issue for technology education is that teachers, particularly in primary schools may believe they are integrating technology into their classroom programme or unit if their students undertake an activity on a digital device when in fact it is no more the integration of technology than using a pen or pencil as a writing tool. Another challenge with the multidisciplinary approach in technology is that students may be involved in designing and making an artefact but have no knowledge or understanding that they are engaged in technology practice, nor develop their understanding of the discrete components or aspects of the technology curriculum, nor how each contributes to technology practice. This becomes problematic for teachers who are required to progress students in technology. One further issue is that students may not have the opportunity to develop their understanding of the nature of technology, which in the New Zealand curriculum is a major aspect of learning in technology. Students may fail to understand the ethical, moral and environmental issues associated with their activity and how their technological outcome impacts and influences people and society because they are not encouraged to investigate potential social, ethical and environmental implications of the made outcome.

13.1.2 Interdisciplinary Integration

An interdisciplinary approach to teaching and learning is positioned in the middle of the continuum. When deploying this approach teachers identify the knowledge and skills needed to complete the intended project or to design and make a potential solution. It begins with a problem or issue that facilitates the linking of skills and understandings between different disciplines. The idea of interdisciplinary integration is not new. Wohl (1955) stated that interdisciplinary integration is a reciprocal interchange of ideas and information between two specialties. Wohl gives us some insight into early ideas of the nature of interdisciplinary integration and although he writes within the context of interdisciplinary research at the tertiary level there are several critical ideas with equal applicability to the integration of technology, engineering, maths and science in STEM education today. Interdisciplinary integration in the early days had its critics and its champions. Early supporters suggested that it was best when it arose spontaneously, rather than when contrived. The creation of formal organisation and legislation for interdisciplinary integration between disciplines did not produce effective conversations between disciplines.

Stember (1991) identified a three-pronged argument for interdisciplinary integration: intellectual—the bringing together of academic knowledge from relevant disciplines; practical—a range of practical skills required to develop a potential solution, and pedagogical- a teaching approach that is a mix of specific teaching episodes, inquiry, and investigation and needs-based or 'just-in-time teaching'. All three are equally applicable today. Motivation for interdisciplinary integration must be intrinsic. It is not unusual for the critics of interdisciplinary integration to suggest that its supporters aim to abolish discipline divisions, however, this is not the case. Interdisciplinary integration cannot occur without the necessary contributing specialisations. Synthesis is not automatic nor spontaneous, but it must involve joint action by specialists. Interdisciplinary integration needs a commitment to work together collaboratively (Menken et al., 2016a, 2016b).

In the school context interdisciplinary projects offer students more than one perspective on the same subject. It may begin with a 'big understanding' that threads and guides students' learning. After exploration and investigation, students or groups of students identify more specific issues or problems within the given context and then design and develop possible solutions. STEM education should reflect the working world within which many parameters constantly change such as cost, materials, time availability and cultural values. Creating an interdisciplinary approach is an almost natural way to solve complex STEM-related problems (Klaassen, 2018). In this approach, there is an emphasis on students' needs and making meaningful connections between the curriculum and their lives. The inter-relationship of important concepts is explicit, as are the cross-curricular skills such as the key competencies (Fox-Turnbull et al., 2021).

In the classroom, units of work typically start with an issue that will contribute to a broader programme of work to develop an enduring understanding transferrable to other contexts. Students then research and identify a specific issue and need within the given context, explore, design, and develop a solution, often presented to an audience at the end of the unit as a 'grand finale'. When planning an interdisciplinary approach to curriculum integration, teachers consider and identify specific skills and knowledge from a range of disciplines. The knowledge and skills identified are deemed as necessary for the students to successfully complete and present their project 'grand finale'. Students understand what and why they are learning, as it all relates to their current study.

Technology education lends itself to this approach as it allows students to make enterprising use of technological academic and practical knowledge and skills, together with those of other disciplines. Graphics and other forms of visual representation offer important tools for the exploration and communication of design ideas. Skills and knowledge from other areas such as Science, Maths, English, Arts and Social Studies also be authentically engaged. Again, we use the context of toys to illustrate this approach (Fig. 13.3). The teacher takes the students on a visit to a local playcentre that is struggling for resources. The students observe and talk to children and staff at the centre, thus identifying their shortage of toys. Back at school students decided to design and make toys for the playcentre. For the students to be successful the teacher identified a range of skills and knowledge needed. These include understanding the role and function of toys, drawing designs, identifying, and investigating the nature and properties of suitable materials, measuring, cutting and joining materials, understanding what makes toys attractive to the children and constructing toys from drawings to meet identified attributes. A big understanding for a unit such as this could be the 'educational value of play' or 'helping others'. It is important that these are identified by teachers well in advance and are selected with the needs of the students at the forefront of decision-making. Once identified it should drive learning and action and be evident in the classroom conversation.

This approach reflects the belief that the construction of knowledge and active involvement is essential for effective learning (Blythe, 1998; Kuhlthau et al., 2007; Murdoch, 2004). Within this approach, students' learning is guided and systematic with learning proceeding through a number of teaching/learning phases similar to that of guided inquiry. It is very different from 'open' discovery learning as teachers have a major and continuing responsibility to structure activities sequenced to maximise the skill development and thinking processes of their learners (Fox-Turnbull et al., 2021). This approach involves students developing deep learning through the process of self-motivated inquiry developing their 'big understandings' about the world and how it functions. It is centred on both process and content, with students taking considerable ownership and responsibility for their learning and 'grand finale' projects.

One issue is that the initial problems to be investigated and solved can be ill-defined and vague (Wohl, 1955). Another issue with this approach, especially in the classroom is that sometimes learning and activities are not structured nor targeted. Students may be left to their own devices with very little specific teaching in all or some of the necessary disciplines (Fox-Turnbull et al., 2021). This issue is exacerbated when teachers do not have comprehensive curriculum and pedagogical content knowledge, meaning some opportunities for learning specific skills, and knowledge may be missed. Discrete curriculum knowledge can be in danger of disappearing.

Fig. 13.3 Interdisciplinary approach to an inquiry on toys

In such situations, students may be very busy and having fun making their 'grand finale', but without focus on curriculum content or process knowledge, the quality of students' learning and the intended outcomes may be diminished. 'Mucky Brown Paint Syndrome' is a term used to describe this phenomenon (Fox-Turnbull, 2012). Imagine each discipline represented by a different coloured paint. When taught in a planned and structured manner using an interdisciplinary approach the teacher will facilitate learning in a range of disciplines to enable students to learn and deploy specific knowledge and skills to assist them with their project. Each discipline, while remaining discrete, contributes to the students' ability to develop a quality outcome or 'grand finale'. Imagine vibrant swirls of colour as portrayed in Fig. 13.4), with each maintaining its integrity while enhancing and supporting its neighbour. However, if learning is not discrete and clearly articulated to the students, colours (disciplines) blend, and lose identity, with mucky brown paint emerging as illustrated in Fig. 13.5. This means that although students are engaged in a range of activities, they are not necessarily aware of the disciplinarity of the knowledge and skills they are learning which may become an issue as they progress through their studies and are required to be aware of and draw on knowledge and skills from specific disciplines.

Interdisciplinary integration is the most relevant approach to the integration of STEM subjects in mainstream education. STEM integration using an interdisciplinary approach supports learning as the brain seeks patterns, meaning and connectedness. To avoid 'Mucky Brown Paint Syndrome' within STEM the integrity of each

Fig. 13.4 Merging of discipline skills and knowledge (Paint) with integrity

Fig. 13.5 Merging of discipline skills and knowledge (Paint) without integrity

discipline must be maintained within projects and teachers must have strong content and pedagogical content knowledge in all four disciplines or work in teams of discipline experts to ensure that their programme planning and teaching makes content knowledge and connections obvious to their students.

13.2 Transdisciplinary Integration

At the other end of the curriculum integration continuum is the transdisciplinary approach within which students come across a range of curriculum areas appropriate to real-life contexts. It aims to facilitate learning processes between science and society. Transdisciplinarity is a means to bridge between a Western way of thinking and more holistic forms of indigenous knowing that do not separate scientific and spiritual thinking. To be truly transdisciplinary there must be an intention for improvement in the situation of the societal problem identified at the outset of the project. Artefacts must be designed and produced and there must be mutual and transformational learning for both learners and practitioners about the societal problem (Jahn et al., 2012), similar to that illustrated in Fig. 13.5.

A transdisciplinary approach engages students in authentic, often self-identified issues or problems. Students seek to solve problems by drawing on a range of relevant knowledge and skills—as and when needed. In this approach, students direct their own learning and with that gain additional social skills and a deeper understanding of their world and the way it works (Zimmerman, 1990). Using this approach, teachers can facilitate learning across a range of disciplines, often deliberately teaching curriculum knowledge on a 'just-in-time basis', to meet the emerging or current needs of their students either individually or in small groups. Typically, when a study begins, the community becomes the classroom and learning involves real-life problem-solving. Students' learning is self-directed and they may work as individuals or in small interest groups. This approach is an example of full and more complex curriculum integration and underpins the philosophy of teaching evident in a range of schools alternative to the mainstream, such as Christchurch's Ao Tawhiti Unlimited Discovery School (Fox-Turnbull et al., 2021). Again, we illustrate this approach through the context of toys (Fig. 13.6). This starts with a student arriving at school upset about the fact so many toys are made of unfixable plastic, which are thrown away when broken. The student knows that plastic can be harmful to the environment and asks the teacher if the class or a small group can work together to develop a solution to the issue. A thorough investigation of the issue begins by identifying and researching a range of relevant questions before determining a possible solution—the development of a range of environmentally friendly toys.

This is an exciting approach to learning but does come with challenges. Students do not work in specifically identified disciplines, but instead develop the knowledge and skills required to solve their issues. This becomes problematic for teachers when they are required to report on students' learning in discrete disciplines or prepare students for formalised discipline-based assessment. Tracking students' learning in all disciplines is also an issue for teachers as students within one class may be working on a range of issues at any one time and learning in different disciplinary contexts. There is also a danger that the students themselves are unaware of discrete discipline knowledge when at some stage they must select or identify specific 'subjects' for secondary assessment or further study for example. However, given the school-based and life experiences students assimilate by the time they select pathways for further

Fig. 13.6 Transdisciplinary approach with issue about plastic toys

study, it is likely that many will have little difficulty in determining programmes of potential interest and be rich in disciplinary knowledge.

13.3 Conclusion—Avoiding the Mucky Brown Paint: Implications for the Classroom

In summary the three approaches to integration discussed in this chapter have distinctive characteristics and offer students considerably different learning experiences. Table 13.1 offers a summary of the key ideas of each approach.

So, what does this mean for technology education? Interdisciplinary and transdisciplinary approaches are integral parts of the technological innovation cycles and can bridge the gap between research, industry and education (Ehlen, 2015; Klaassen,

Table 13.1 Summary of the characteristics of three approaches to integration

Multidisciplinary	Interdisciplinary	Transdisciplinary
Topic identified by teacher and same across disciplines	Context is identified by the teacher in most cases until senior secondary. Learning can occur across topics and varies as long as skills and knowledge work towards developing the final intended outcome or 'grand finale'	Context is driven by students to make a meaningful difference in the world
Perspectives used to illustrate a theme	Several disciplines contribute to the solution of a problem Differences in disciplinary knowledge are privileged	Involves collaboration with non-academic stakeholders
Revisiting major concepts, not a core instructional goal	Several disciplines contribute to solving a problem	Technology is seen as the central agent in STEM integration
Students are taught in discrete disciplines and are not taught knowledge transfer	Students are aware of the disciplines they are working across and are encouraged to transfer knowledge and skills across disciplines	Students are not taught in disciplines. They develop the knowledge and skills required to solve their issue
Disciplinary interrelationships may not be apparent	Focus on disciplinary interrelationships is critical to success	Learning that knowledge and skills occur in discrete disciplines is incidental
Limited attention to learning transfer	Curriculum and instruction are intended to increase learners' coherence and transfer across and between disciplines	Unifying ideas revisited in multiple contexts to facilitate a more holistic understanding
Assessment discrete and progress easily monitored and reported	Assessment is discrete but monitoring can be difficult as students work across projects within one context	Formative and Self-assessment with focussed feedback assist students to drive for success. Monitoring and reporting in discrete disciplines is very difficult
Students can become bored	Students are likely engaged and motivated if the initial context piques students' interests	Students are very engaged and motivated as they drive their own learning

2018). We talk about the need for technology education to be holistic rather than fragmented. By holistic we mean the need for students to develop an understanding of the links within technological practice and to see how key ideas reoccur in different contexts rather than having a fragmented view of each part in the process. To develop this understanding, students need to be aware of the whole technology process and how the components link together and are best undertaken in situations authentic to practice. Thus, students are aware of what a technologist or team of technologists is involved in during technological development. This often means that technology is very comfortable as the central agent in STEM integration because students are aware of the authenticity of the disciplinary knowledge they draw on. For example, they learn how to speak in public because they need to present their ideas to the school's Board of Trustees or governing body or calculate the area of their designed garden to calculate the materials required.

In today's educational environment where senior secondary assessment processes and practices determine future pathways for students in discrete discipline areas, the interdisciplinary level of integration is most suited. It allows students, with the assistance of their teachers, to identify and draw on specific and relevant disciplinary knowledge for authentic purposes. In New Zealand, the Ministry of Education has signalled very clearly that technological activities, including digital technologies, must occur in authentic situations (Ministry of Education, 2017). This means that T in STEM is not only the use of digital devices or other technological tools. Students must be embedded in authentic technological contexts to develop and practice skills and develop knowledge that is contextualised and integrated with a range of other discrete disciplines, for reasons that are real and make sense to them. There are implications here for the direction of professional development for primary and STEM teachers. Teachers need to be informed about technological practice and the technological areas or contexts and make links with authentic practice facilitating sound decision-making for their students.

References

Atkinson, S. (1999, Spring). Key factors influencing pupil motivation in design and technology. *Journal of Technology Education, 10*(2).

Bellanca, J., & Brandt, R. (Eds.). (2010). *21st century skills: Rethinking how students learn*. Solution Tree Press.

Blythe, T. (1998). *The teaching for understanding guide*. Jossy-Bass.

Ehlen, C. (2015). Co-creation of innovation: Investment with and in social capital. In *Studies on collaboration between Education–Industry–Government*.

Fox-Turnbull, W. (2003). *The place of authentic technological practice and assessment in technology education* [Thesis for Master of Teaching and Learning, Christchurch College of Education]. Christchurch.

Fox-Turnbull, W. (2012). Learning in technology. In P. J. Williams (Ed.), *Technology education for teachers*. Sense Publishers.

Fox-Turnbull, W., Reinsfield, E., & Forret, A. M. (2021). *Technology education in New Zealand: A guide for teachers*. Taylor & Francis Group. http://ebookcentral.proquest.com/lib/waikato/detail.action?docID=6605959

Gilbert, J. (2005). *Catching the knowledge wave? The knowledge society and the future of education*. NZCER Press.

Hill, A. M. (2017). Authentic learning and technology education. In M. de Vries (Ed.), *Handbook of technology education* (pp. 473–488). Springer.

Jahn, T., Bergmann, M., & Keil, F. (2012). Transdisciplinarity: Between mainstreaming and marginalization. *Ecological Economics, 79*, 1–10.

Klaassen, R. G. (2018). Interdisciplinary education: A case study. *European Journal of Engineering Education, 43*(6), 842–859. https://doi.org/10.1080/03043797.2018.1442417

Kuhlthau, C., Maniotes, K., & Caspari, A. (2007). *Guided inquiry: Learning in the 21St century*. Libraries Unlimited Inc.

Menken, S., & Keestra, M. (Eds.). (2016). *An introduction to interdisciplinary research*. Amsterdam University Press.

Menken, S., Keestra, M., Rutting, L., Post, G., de Roo, M., Blad, S., & de Greef, L. (2016). An introduction to interdisciplinary research: Theory and practice.

Ministry of Education. (2017). *Technology in the New Zealand Curriculum*. Ministry of Education. Retrieved 20 February from http://nzcurriculum.tki.org.nz/The-New-Zealand-Curriculum/Tec hnology

Murdoch, K. (2004). *Classroom connections—Strategies for integrated learning*. Eleanor Curtain Publishing.

Murdoch, K., & Hornsby, D. (2003). *Planning curriculum connections whole school planning for integrated curriculum*. Eleanor Curtain Publishing.

OECD. (2020). *Curriculum overload: A way forward*. O. Publishing.

Pitt, J. C. (2017). Teaching science and technology. In M. de Vries (Ed.), *Handbook of technology education* (pp. 33–38). Springer.

Pohl, C., Truffer, B., & Hirsch Hadorn, G. (2017). Addressing wicked problems through transdisciplinary research. In J. T. K. R. C. S. P. R. Frodeman (Ed.), *The oxford handbook of interdisciplinarity* (2 ed., pp. 319–331). Oxford University Press.

Roberts, T., Jackson, C., Mohr-Schroeder, M. J., Bush, S. B., Maiorca, C., Cavalcanti, M., Craig Schroeder, D., Delaney, A., Putnam, L., & Cremeans, C. (2018, 2018/09/21). Students' perceptions of STEM learning after participating in a summer informal learning experience. *International Journal of STEM Education, 5*(1), 35. https://doi.org/10.1186/s40594-018-0133-4

Stember, M. (1991). Presidential address: Advancing the social sciences through the interdisciplinary enterprise. *The Social Science Journal, 28*(1), 1–14.

Wicklein, R., & Schell, J. W. (1995). Case studies of multidisciplinary approaches to integrating mathematics, science and technology education. *Journal of Technology Education, 6*(2), 59–76.

Wohl, R. R. (1955). Some observations on the social organization of interdisciplinary social science research. *Social Forces, 33*(4), 374–383.

Zimmerman, B. J. (1990). Self-regulated learning and academic achievement: An overview. *Educational Psychologist, 25*(1), 3–17. https://doi.org/10.1207/s15326985ep2501_2

Wendy Fox-Turnbull is an Associate Professor at the University of Waikato and Associate Dean-Academic in the Division of Education. Wendy is the editor of Australasian Journal of Technology Education and on Editorial Board of International Journal of Work Integrated Learning and co- editor European Journal of STEM Education. In 2019 she was guest editor for Design and Technology: An International Journal Primary Edition. In 2018 she was the invited section editor for the International Handbook in Teacher Education, Section Technology Teacher Education. Wendy was chair of the Technology Education New Zealand (TENZ Council) from 2006 to 2018 and has convened three TENZ conferences: TENZ 2005, TENZ2015 and TENZ/ICTE2017 and one International Technology Research conference (PATT) in 2013. Research special interests include authentic learning in technology education, the place of women in technology related STEM careers, the role and nature of effective conversations in learning and teaching and learning approaches for the 21st Century. Wendy is a registered and certified teacher.

Chapter 14
STEM Projects: Outside of the Curriculum

Kuen-Yi Lin and Yu-Jen Sie

Abstract Taiwan is one of the highly developed economies, with many important multinational corporations that influencing technological developments, and has attached great importance to technology education. STEM education has been a pivotal learning area in the latest technology education curriculum in Taiwan. In recent years, STEM education has not only been increasingly emphasized in formal education, but also in informal education. Opportunities for informal STEM learning experiences are rapidly expanding in Taiwan. This chapter will center on discussing the relevant findings, to examine the characteristics and values of the STEM projects in informal education. Meanwhile, this chapter will also consolidate and summarize past research and critical online resources, to discuss and analyze several types and examples of STEM projects in informal education. Moreover, the advantages and issues of STEM projects in informal education will be studied in the chapter as well, for the purpose of evaluating the status of STEM projects in informal education. Specific conclusions and future suggestions on STEM projects in informal education will be drawn upon the discussion.

Keywords STEM projects · Informal education · STEM education

14.1 Introduction

Science, technology, engineering, and mathematics (STEM) education is an interdisciplinary teaching approach that provides students with an understanding of these disciplines, how they relate to each other, and how they can be applied in real-world settings. Technology plays a key role in STEM education because it is a cross-disciplinary integrator that supports the design process by allowing students to

K.-Y. Lin (✉) · Y.-J. Sie
National Taiwan Normal University, Taipei, Taiwan
e-mail: linkuenyi@ntnu.edu.tw

Y.-J. Sie
e-mail: umi.sie168@gmail.com

W. Fox-Turnbull and P. J. Williams (eds.), *Locating Technology Education in STEM Teaching and Learning*, Contemporary Issues in Technology Education,
https://doi.org/10.1007/978-981-97-1995-2_14

create, test, and create solutions to real-world problems (International Technology & Engineering Educators Association, 2020).

STEM education continues to attract global attention, such that many researchers are increasingly involved in K–12 and university STEM education studies and have produced valuable findings (Dika & D'Amico, 2016; Ku et al., 2022; Xu et al., 2020; Yang et al., 2023). Some of these researchers have focused on formal STEM education, such as the format offered in the K–12 curriculum (Lin et al., 2021); other researchers have concentrated on informal education, such as camps, clubs, and competitions (Ku et al., 2022; Schwab et al., 2018). Hooper-Greenhill (2007) noted that informal education contributes to an individual's understanding of diversity, including their attitudes, emotions, values, and beliefs. Thus, informal education helps to prepare students for their future. According to these studies, informal STEM education meets the diverse needs of students; however, specific practical characteristics differ among countries.

Taiwan, officially the Republic of China, has a high level of economic freedom. Taiwan's gross domestic product ranks 5th in Asia and 18th in the world in terms of purchasing power parity. The International Monetary Fund and the World Bank have classified Taiwan as a developed, high-income economy. Taiwan is home to many multinational corporations, such as Taiwan Semiconductor Manufacturing Corporation, one of the most advanced microchip technology manufacturers worldwide. In addition to pursuing technological developments, Taiwan's multinational corporations collaborate with educational institutes at all learning stages to provide diversified professional development opportunities in artificial intelligence and robotics in the form of STEM courses, camps, competitions, and seminars in informal educational settings. Many Taiwanese scholars have devoted themselves to STEM education research over the past decade and have published influential findings in numerous international journals (Irwanto et al., 2022), such as the *Journal of Research in Science Teaching*, the *International Journal of Science Education,* the *International Journal of Technology and Design Education*, the *Journal of Engineering Education,* the *International Journal of Science and Mathematics Education*, and the *International Journal of STEM Education* (Cheng et al., 2022; Halawa et al., 2020; Ku et al., 2022; Lin & Tsai, 2009; Tsai & Chang, 2005; Yang & Li, 2018). An analysis of the recent status of STEM projects and specific suggestions for the future development of STEM in informal educational settings is presented in the following sections.

14.2 Objectives of STEM Projects in Informal Education

A search of STEM education-related research articles in Scopus revealed that 70% focused on university education, whereas approximately 30% focused on K–12 education. K–12 STEM education research has shown that STEM projects in informal education are mainly designed to meet the diverse needs of school clubs (Chittum et al., 2017; Christensen et al., 2015; Gomoll et al., 2016), school academics (i.e., universities, primary schools, and secondary schools), corporate camps (Nugent et al.,

2010; Roberts et al., 2018), teaching aids from publishers (Cheng et al., 2023), museums (Anderson et al., 2017; McCreedy & Dierking, 2015), STEM competitions (Ku et al., 2022; Stewardson et al., 2019; Üçgül & Altıok, 2017), and web-based learning platforms (Gamage et al., 2022; Gilliam et al., 2017). According to previous research, STEM projects in informal education have the following main objectives.

Objective 1 To offer STEM projects in informal education with features that satisfy the needs of different countries and cultural backgrounds.

The development and direction of STEM projects in informal education differ among countries, cultural backgrounds, and participant needs. In the United States, for example, STEM education is treated as policy and as an investment in human resources for the nation's future competitiveness. Participating organizations include educational institutions, research institutions, and non-profit organizations at the state and federal levels. For instance, twenty-first Century Community Learning Centers work with local museums and libraries to promote STEM courses. Additionally, adult and continuing education STEM programs are offered by community colleges to meet the demands of students and the workforce (Sergeyev et al., 2019).

Objective 2 To create diversified STEM projects that provide all students with sufficient experiential learning opportunities to promote interest in STEM fields (Denson et al., 2015).

There are numerous considerations regarding the content and pedagogy of STEM courses in informal settings when designing the courses; these considerations are based on variations in students' identities, backgrounds, prerequisite knowledge, skills, attitudes, values, and other personal characteristics. There is a need to design a learning environment and activities that address the abovementioned differences and provide opportunities for informal experiential learning.

Objective 3 To strive for education equity in informal STEM education programs, considering the educational needs of disadvantaged or under-represented groups (e.g., gender and race) (Atkins et al., 2020; Singer et al., 2020; Long III et al., 2016).

The shortage of STEM talents is a global concern. Thus, there is a need to provide opportunities for STEM education, particularly among women, individuals with disabilities, and members of under-represented and disadvantaged groups. The provision of these opportunities could increase participation in STEM fields, while addressing deficiencies in STEM teaching positions and STEM talent (Gomoll et al., 2016).

Objective 4 To develop advanced STEM projects (e.g., robotics) that can deepen STEM learning among exceptional students (Üçgül & Altıok, 2017; Witherspoon et al., 2016).

Gifted and talented education is valued in all countries, and differentiated learning offers the potential for the development of advanced STEM programs (Lundgren et al., 2011). These programs are beneficial because they inspire highly motivated students to choose STEM fields and professions (Stoeger et al., 2016).

14.3 STEM Projects in Informal Education

Informal STEM education can be defined as lifelong learning in science, technology, engineering, and mathematics that takes place across a multitude of designed settings and experiences outside of the formal classroom (Allen & Peterman, 2019, p. 18). Informal STEM education may also provide deep opportunities for self-directed learning. Both formal and informal STEM education programs are influenced by national curriculum standards, age-based curricula, and high-stakes assessment practices (Allen & Peterman, 2019). Informal STEM education takes place across many settings, including the following: school clubs, school/corporate camps, teaching aids from publishers, museums, STEM contests, and web-based learning platforms. More details are provided in the following sections.

14.3.1 School Clubs

Clubs help students make connections among STEM disciplines while promoting student interest and engagement (NAE & NRC, 2014). Students can achieve their learning goals by connecting their learning experiences to different learning settings. There is evidence that after-school robotics clubs play an important role in stimulating student interest in STEM fields. Informal learning settings provide students with an early learning experience that encourages the development of personal interests. Peers are particularly influential: peer reviews serve as positive motivators for strengthening an individual's STEM identity and interest in STEM careers (Gomoll et al., 2016). Moreover, research has revealed that the placement of after-school STEM programs in school clubs affects student willingness to participate in STEM fields in future (Chittum et al., 2017). Thus, students' early STEM learning experiences in informal education build self-confidence, particularly in STEM disciplines (Chittum et al., 2017). Informal education is particularly valuable in the early stages of STEM learning, considering that the informal setting reduces barriers to learning and provides hands-on experiences that promote interest and self-efficacy in STEM, as well as the continued pursuit of STEM career goals.

In Taiwan's new curriculum guidelines, "Topics in Robotics" is included as an elective course in the technology learning area, which allows robotics learning and teaching to be implemented in high-school programs. Robotics is a topic that integrates the knowledge and skills of individual STEM areas (i.e., science, technology, engineering, and mathematics) in forms such as mechanical engineering, electrical engineering, and computer science & information engineering. Therefore, Taiwan's new curriculum guidelines enable high schools to integrate knowledge from different domains through a flexible program, while inspiring students to cultivate future technology talents through project-based discussion and practical problem-solving. For example, the abovementioned FRC is regarded as a high-level national high-school robotics competition in Taiwan. Each year, participating teams from locations throughout Taiwan, many including self-organized school clubs, meet to compete in the FRC (Fig. 14.1). Students and instructors from school FRC robotics clubs must form their own teams and raise funding to compete. Students are then responsible for designing, developing, and building the robots for the competition; they must also test their work within designated competition themes.

In addition to promoting FRC in school clubs, some high-school teachers use funds awarded by local governments to support their own classroom FRC teams. Other teachers who have not received such funds may integrate contest content into their curriculum using existing classroom supplies. When high-school students gradually develop their interests in STEM through this type of diversified learning experience, either formal or informal, they become more likely to pursue related disciplines (Chittum et al., 2017; Christensen et al., 2015).

Technology in STEM school clubs can be used to enhance learning and engagement, as well as help facilitate communication and collaboration between students and instructors. Components of technology in STEM school clubs include (1) robotics and coding, (2) interactive simulations and animations, (3) communication and collaboration platforms, (4) data analysis and visualization tools, and (5) project management and tracking software. Many STEM school clubs focus on teaching students coding and robots, which can help them develop problem-solving and critical thinking skills. Interactive simulations and animations can help students understand complex STEM concepts and visualize difficult processes or systems. Video conferencing, chat tools, and other online collaboration platforms can be used to facilitate communication and teamwork among students and mentors. Spreadsheets, statistical software, and data visualization tools can be used to help students analyze

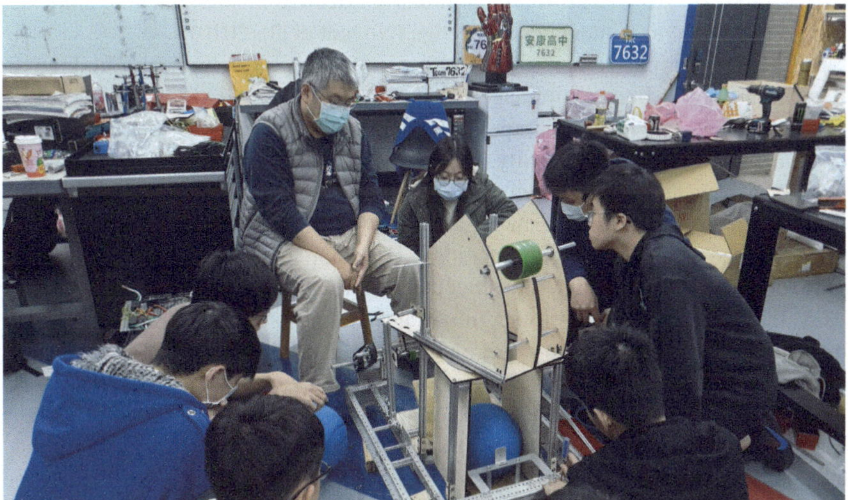

Fig. 14.1 FIRST robotics competition club at An Kang senior high school. *Source* https://www.youtube.com/watch?v=rj3FQVjET4Y

and interpret data, which can be key to understanding complex STEM concepts. Finally, STEM school clubs often involve large projects that are complex and time-consuming. Project management and tracking software can be used to facilitate the sharing of and collaboration on documents and resources and help manage and track project progress.

14.3.2 Schools and Corporate Camps

STEM projects outside of the curriculum can complement courses that are not available via formal education. Camp-based activities significantly impact students' attitudes toward STEM. These types of activities help to nurture students' STEM interests and tend to reduce gender and racial disparities, while providing a more diverse workforce in STEM fields. Roberts et al. (2018) found that the knowledge and skills learners gained from specially designed games and challenges at STEM camps strongly affected their learning in formal settings. Many higher education institutions organize summer or winter STEM camps to provide a few days of experiential courses. In these camps, instructors lead small groups of students in project-based explorations of STEM topics. The STEM series courses also provide learning content at various levels for their students, thereby accommodating a wide range of abilities. Short-term courses focus on increasing students' STEM knowledge and interests (Barker & Ansorge, 2014; Sullivan, 2008; Williams et al., 2014); these courses often use integrative tasks (outside of a formal educational setting) to prepare students to manage open-ended situations while nurturing their collaborative skills. Long-term training courses in STEM are particularly effective in strengthening students' abilities to acquire and integrate knowledge (Nugent et al., 2010).

STEM camp topics include Mechanism and Structure, Logical Thinking in Programming, and Interdisciplinary Problem-Solving. "STEAM Kinetic Mechanism Camp–The Moving Landscapes" (Fig. 14.2) is a unique STEM camp hosted by National Taiwan Normal University (introductory video for basic class: https://www.youtube.com/watch?v=rImY5rggmA4; film of works: https://www.youtube.com/watch?v=7_-qIiGdSsE). Using an engineering design process, students learn about power machinery and various applications. The camp aims to help students build a foundation of scientific inquiry and reflective integration, with a rich interdisciplinary STEM learning experience. Another important example is the summer STEM camps hosted by the university. This program offers various types of classes: introductory, advanced, English, and creative. Students can customize advanced STEM projects to meet their personal learning needs. The camps provide courses focused on various topics (e.g., bipedal robots, electronic building blocks, and bionic beasts) to enhance students' programming skills and their abilities to use interdisciplinary knowledge to solve real-life problems.

Taiwanese corporations also demonstrate their technological expertise via STEM camps, providing learning support in areas where the students have limited knowledge. Corporations and private organizations often collaborate with schools to

Fig. 14.2 STEAM kinetic mechanism camp–the moving landscapes. *Source* https://www.youtube.com/watch?v=7_-qIiGdSsE

arrange STEM camps. An example of this collaboration is the "Future Star, Smart Technology" camp that was co-organized by MediaTek Foundation, the National Taiwan Science Education Center, and other partner organizations. Based on the smart technology learning needs of high-school students, Arduino micro-control circuits with C/C++ were used as the hardware and software integration platforms in the camp. The camp also used small projects/games as a course unit to guide students in identifying the problem to be addressed, finding solutions, and exploring possible technological applications. This camp also offered courses focused on electromechanical control, the Internet of Things, and artificial intelligence-driven Internet of Things, with project-oriented approaches to enhance students' problem-solving skills.

Technology plays a similar role in STEM school clubs and camps, but there may be some differences in how it is used and integrated into the learning experience. Compared to STEM camps, school clubs may have a more consistent mentorship structure, which can help students develop their skills and knowledge over time. STEM camps may have more of a one-time or short-term mentorship experience. STEM school clubs target students in the school, while STEM camps target a wider audience that may not be limited to students in a school. Finally, STEM camps can be more flexible than school clubs; they often provide a wide range of activities and options for students to choose from and also run during summer or holidays, which provides students with an opportunity to explore STEM outside of their regular school schedule.

14.3.3 Teaching Aids from Publishers

Teaching aids are commonly used to achieve meaningful learning experiences. Students possess a high degree of learning autonomy when using teaching aids, and this approach is good for self-exploration. Teaching aids facilitate practical learning by providing innovative technological support and a flexible learning environment. Students can establish their own sequential learning experience using the scaffold provided by the teaching aids (Cejka et al., 2006). Additionally, there is evidence that using teaching aids as tutors or learning companions increases students' cognitive abilities and learning engagement (Cheng et al., 2023).

Many publishers have introduced teaching aids related to STEM education, such as wall-climbing robots, quadcopters, and program cars. For example, "geckobot" is a teaching aid that offers a broad range of exploration activities; it has 176 snap-on and assembled accessories. Students learn about the principles of air pressure and suction during the assembly of a robotic gecko with suction cup feet. Geckobot has various fine-tuned valve designs and uses a motor-driven air intake system. It can be assembled into six types of objects, including an elliptical gauge, phone holder, vacuum trailer, vacuum gun, canker-worm, and an industrial robotic arm with a suction cup. Another example is the "quadrotor aircraft," a teaching aid that introduces the use of four-blade propeller acceleration in drones. In addition to introducing the principles of multi-axis aircraft, the quadrotor aircraft provides guidance regarding aircraft construction, assembly, and operation. This teaching aid also helps students to learn about aircraft configuration, adjustment, and aircraft-related hardware and software. A third example is the "program car," a programmable toy car that operates independently of any mobile device. Using program commands, the car can communicate with a computer and plan routes for task completion. The car contains 10 plastic sensor pads that can be used to program basic commands such as forward, reverse, left turn, right turn, stop, headlight, warning light, horn, loop, and endpoint. The simple "hydraulic ratchet car" (Fig. 14.3) has a custom-designed body, in which a rack is driven forward through a syringe that pushes a plastic plate on the ratchet wheel, thus moving the car forward. In this design, students must transfer force through the syringe to enable the conversion of straight-line motion into rotary motion. Ultimately, they learn to steer the car using the syringe. The "wired remote control car" (Fig. 14.4) is a teaching aid that allows students to control the direction of the motor and the car body by running the electrical current forward and backwards using various hardware parts (e.g., batteries, motors, wires, and forward and reverse switches). By interacting with the wired car, students can enhance their knowledge of hardware parts, learn to use a triple electricity meter, and learn to assemble a wired remote control car.

Technology produces benefits in STEM teaching aids through gamification and personalization. It can be used to make learning more engaging and interactive, which can help students stay focused and motivated. By tracking student progress, it can also be used to provide personalized learning experiences and tailored feedback and recommendations to further education.

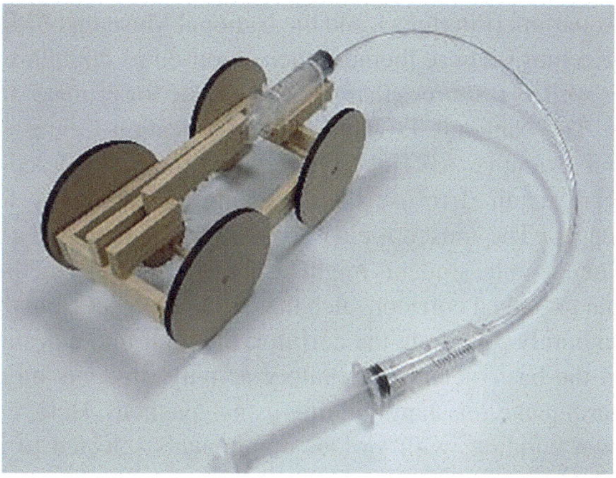

Figure 14.3 Hydraulic Ratchet car. *Source* https://sites.google.com/view/everybodygood/woodcrafts/ratchet

Fig. 14.4 Wired remote control car. *Source* https://sites.google.com/view/everybodygood/woodcrafts/EasyCar

14.3.4 Museums

Museums play an important role in STEM education by offering local community resources and learning opportunities, as well as various STEM projects based on student learning status (Adams et al., 2014). Museums also deliver authentic interactive experiences through their exhibits (Anderson et al., 2017) and help to establish STEM literacy through learning, teaching, and educational activities, with a focus on basic skills and foundation learning. Science museums under the Ministry of Education in Taiwan include the National Taiwan Science Education Center (Taipei), the National Museum of Natural Science (Taichung), the National Science and Technology Museum (Kaohsiung), the National Museum of Marine Biology and

Aquarium (Pingtung), and the National Museum of Marine Science and Technology (Keelung). These themed science museums contain rich and extensive collections, as well as resources that are appropriate for primary and secondary school students.

The National Taiwan Science Education Center organizes special exhibitions. For example, a "Tinkering" exhibition was offered. In the exhibition, tinkerers were required to use their thinking and creativity to solve various STEM challenges. The museum also provides many distinctive STEM learning experiences. For example, students might be asked to design a method for using any type of material to build a corridor, such that a round object from the vertex on a pegboard could smoothly roll along the corridor, then fall into a designated entrance. Additionally, in the basic science laboratory activity, students might be challenged to build an earthquake-resistant building using spaghetti. Here, each group would design their own building, with various components selected to impart optimal strength. The structures are then tested in an earthquake simulator with gradually increasing earthquake magnitude; this provides the students with practical experience regarding the principles of structural resistance to seismic activity by observing the destruction of an earthquake-resistant house.

The National Science and Technology Museum offers various collections and research findings to develop experiential STEM activities, such as the "Weights and Measures" exhibition. In this activity, students learn about the scientific principles involved in the development of weights and measures and the International System of Units, as well as the operation, principles, and limitations of weights and measures. When measuring different objects, students determine the appropriate metrics to measure length, weight, and capacity; they also compare and convert the measured values of different metrics. The examples provided are all based on local resources, with a focus on STEM knowledge or competencies.

Overall, technology plays a significant role in many STEM museums, as it can be used to enhance the experience and make it more engaging and interactive for visitors. It can be used to create interactive exhibits that allow visitors to engage with STEM concepts in a hands-on way, such as virtual reality simulations, touchscreen displays, and interactive games. Videos, audio, and animation can be used to create engaging and informative multimedia displays that can help visitors understand complex STEM concepts. Some STEM exhibits may use technology to enhance the visitor experience by overlaying virtual content on real-world environments through their smartphones or tablets. Sensors and IoT devices can be used to gather data about the exhibit and visitors, which can help organizers and researchers understand how visitors interact with the exhibit and make improvements for future exhibitions. Finally, mobile apps may enhance the visitor experience by helping visitors navigate through the exhibits and learn more as they explore.

14.3.5 STEM Contests

There is evidence that competing in open contests enhances passion and motivation for STEM among the participants (Hazari et al., 2022). In STEM competitions, students gain STEM disciplinary knowledge while improving their collaboration, communication, presentation, budgeting, and fundraising skills. Participation in STEM competitions helps students to improve self-efficacy, intensify STEM learning, and increase their educational opportunities (Stewardson et al., 2019).

The contests are commonly designed to challenge students at all levels. For instance, the "National Junior High School Living Technology Contest" (Fig. 14.5) is a major national contest organized by the K–12 Education Administration and the Ministry of Education. It aims to promote students' STEM interests and improve their problem-solving, creativity, and other higher-order thinking skills. It also encourages teachers to develop innovative learning activities and refine their instructional strategies with respect to learning effectiveness. This annual contest has been conducted for 8 years and is an important medium for promoting technology education to the general public. Currently, regional governments organize the preliminary round in their regions, which is divided into task-challenged and creative-design groups. Each year, the contest topics are revealed to students during the contest; the students must then design and produce their works without prior preparation. The topics are based on the present Junior High-School Living Technology Curriculum and the Junior High-School Living Technology Textbooks approved by the Ministry of Education.

"Technology Education Creative Practice Contest" is another example of a STEM contest in Taiwan. It is divided into primary and junior high-school groups. The contest topics are related to real-life situations, and competitors can submit their works in any form (e.g., electronic device, data processing, application, or analysis). Competitors can also analyze and solve problems using the Internet of Things, artificial intelligence, virtual reality, and big data. The topics presented are cornerstones of Taiwan's Curriculum Guidelines for 12-year Basic Education. The assessments consist of seven aspects: (1) presentation of computational thinking, including disassembly, algorithms, and data processing; (2) programming, including modularity, efficiency, and operational stability; (3) creativity and practicality of problem-solving; (4) development of creative thinking, problem-solving, and computational thinking; (5) materials, tools, and machines used in the production process; (6) proposal; and (7) division of work, and quality and assurance performance.

Additionally, the "National High School Living Technology Contest" aims to inspire student interest in creative design and engineering design, as well as provide an opportunity for students to further explore STEM subject matter and implement problem-solving skills in their lives. The contest also motivates high-school students to value technological research, creative thinking, and the observation, analysis, and production skills required in creative design. Students work in teams to solve problems identified in their lives.

This contest includes two rounds. There are three categories in the first round; four to five competitors form a team in the task-challenged group, whereas three

Fig. 14.5 National junior high-school living technology contest

to four competitors form a team in the creative-design group. The creative design group is focused on proposing innovative design solutions to designated problems. The competitors must submit proposals, and they advance to the next round if their proposal is approved. In the second round, approved competitors must complete the assigned task. Each year, participating students encounter various tasks and design themes, including objectives, design concepts, power and control applications, design drawings, performance calculations, and construction planning. These STEM-related contests examine students' abilities to successfully complete STEM projects in various real-life scenarios and highlight their learning outcomes in the competitions. Additionally, students gain experience interacting with their peers and meeting complex challenges using an interdisciplinary approach.

During the contest, students are given the opportunity to use technology tools and resources to prototype and test solutions to the challenges presented in the contest. This opportunity can help students develop their technical skills and understand the principles of technology design. Students learn how to use technology to gather data, analyze it, and create solutions. This process can include learning how to use software, programming languages, hardware, sensors, and data visualization and analysis tools.

14.3.6 Web-Based Learning Platforms

The effectiveness of the Internet as a learning aid is evident. With advancements in computer hardware and software, the Internet has become a very popular learning tool; it can be used to meet individual learning needs. This was particularly evident in 2020 during the widespread outbreak of coronavirus disease 2019. STEM learning resources can be assessed through various channels, such as shared platforms, social media, mobile applications, and audio and video streaming. The Internet is an effective environment for promoting STEM interest and learning because of its high degree of flexibility and personalization, as well as its easy access for students who are under-represented (Gamage et al., 2022; Gilliam et al., 2017).

Sharing platforms are useful for informal STEM education. The first example provided here is the "STEM Collaborative Problem-Solving Skills Assessment System" (Fig. 14.6) which provides eight scenarios from daily life. Students must use computers to collaborate and solve tasks (e.g., using a microwave oven, building a house, and designing a desk). Another example is the "Junyi Academy," a web-based learning platform that provides many introductory STEM learning resources. It records each student's learning history and analyzes the data to provide prompt feedback for personal learning. Additionally, "Machine Learning for Kids" is a web-based STEM learning platform that uses technology as the main theme. It offers an easy-to-use environment that helps students learn to build machine-learning models; the students must solve real-life problems in various contexts through problem-based projects and modular exercises focused on machine learning. The final example is "The Robot Factory," an online robotic creation kits platform. Students become familiar with the machine elements during the design and production stages. The robots are then driven in simulated situations in which performance feedback is provided to guide improvements in robot design. Students gain subject knowledge, solve authentic problems, and develop creative thinking during the robot creation process.

Technology plays a significant role in many STEM web-based learning platforms, including. Interactive simulations and animations can help students understand complex STEM concepts and visualize processes or systems that are difficult to see. Videos, interactive tutorials, and online quizzes can be used to supplement and enhance classroom instruction and can also be used as self-paced learning resources for students. Technology can be used to provide personalized learning experiences by tracking student progress and providing tailored feedback and recommendations to further education.

Fig. 14.6. STEM collaborative problem-solving skills assessment system. *Source* Lin et al. (2015)

14.4 Advantages of and Challenges with STEM Projects in Informal Education

This section focuses on the possible advantages and disadvantages of promoting STEM projects in informal education. The advantages include (1) the ability to propose more diverse, distinctive, thematic STEM projects without the constraints of formal education (Denson et al., 2015), (2) the ability to plan a series of STEM projects that enable students to learn in a more systematic manner (Tam et al., 2020), (3) the ability to adopt diversified learning strategies (e.g., game-based or inquiry-based formats) to promote student interest in STEM projects, and (4) access to various electronic components (e.g., Micro.bit and Arduino) to enrich the results of STEM projects (Tam et al., 2020).

14.4.1 Advantage 1: Diverse STEM Projects

In the study by Chittum et al. (2017), students felt empowered by participating in after-school STEM courses where they were able to assume professional roles and design devices that could effectively overcome engineering challenges (Gilliam et al., 2017). Engineering design-oriented STEM courses, often used in STEM camps, are project-based activities that guide students to use knowledge from various disciplines to solve real-life problems. Therefore, informal educational settings provide an open, creative, and supportive learning environment, in which students can create unique STEM projects and develop interests in unfamiliar fields without the constraints of formal educational settings.

14.4.2 Advantage 2: Experience-Based STEM Projects

An organized series of experiential activities is a feature of informal STEM projects. A structured curriculum helps students to manage their own learning process and facilitates their development of necessary interdisciplinary competencies, interests, and ways of thinking. The study by Dan et al. (2017) presents an example of a combination of formal and informal educational settings, in which students use active examination to solve problems through analysis, design, practice, idea sharing, and evaluation of approach effectiveness.

14.4.3 Advantage 3: Multi-Strategy STEM Projects

In informal settings, instructors can integrate diversified instructional strategies (e.g., a game-based approach, an inquiry-based approach, and small-group challenges) to guide students' scientific investigations, help students connect to real life, and maintain student engagement and interest in learning (Borda et al., 2020; Kelley, 2010; Lou et al., 2013).

14.4.4 Advantage 4: STEM Projects with Diverse Teaching Materials

In informal settings, learning outcomes are typically presented using various electronic components and devices, such as augmented reality and virtual reality (Goff et al., 2018). Roberts et al. (2018) created a STEM summer program during which the participants were interviewed, with the intention of understanding their views about STEM learning after participation in informal learning. The students felt that they had enhanced their STEM knowledge by engaging in hands-on learning and completing multiple challenges in an authentic STEM environment. Notably, this STEM project provided content, environments, and materials that are not accessible to most secondary students, as well as the opportunity to interact with STEM professionals. The students also felt that informal STEM learning supported their formal education by offering many opportunities for immersion (Jiang et al., 2021; Roberts et al., 2018).

Despite the above-mentioned benefits of informal STEM education, several issues require further consideration. (1) Electronic components can be expensive and may not be affordable for some groups of students. (2) The structures of some camps and teaching aids are not consistent with the concept and spirit of STEM education, despite claims that they constitute STEM projects. (3) Guidance from professional facilitators may be absent from STEM projects conducted through camps, teaching aids, museum exhibitions, and similar interfaces. Accordingly, students may not be able to engage in interdisciplinary learning during those projects. (4) Although some STEM projects in informal education consider the needs of different cultures, genders, and races, most STEM projects do not successfully address diversity.

14.4.5 Issue 1: Insufficient Funding

Although there is a consensus among experts regarding the need for the development of STEM, the funding required for STEM platforms is insufficient. Effective educational outreach inevitably includes expenses involved in the development of teaching plans and establishment of learning communities, as well as costs associated with functional tools and equipment (Blanchard et al., 2017). Although some organizations receive support from foundations or income from enrollment fees, the high costs of teaching aids, planning efforts, and professional connections continue to limit program implementation (Rhoads et al., 2004).

14.4.6 Issue 2: Misleading STEM Projects

Although the concept of STEM has existed for many years, the notion of STEM as an integrated discipline has gained traction much more slowly. Games, books, educational toys, and STEM camps claiming to combine nature observations and science experiments, while requiring students to solve real-life problems by hands-on exploration, are not consistent with STEM aspirations. Thus, their effectiveness has not been determined. Quality STEM materials or activities should include STEM integration and application, as well as the use of higher-order thinking. It is difficult to solve students' learning problems if none of the abovementioned criteria are included in the learning and teaching aspects.

14.4.7 Issue 3: Lack of Professional Instructors (Facilitators)

STEM education is an interdisciplinary effort. Feedback from teachers and peers is essential (Gilliam et al., 2017). Facilitators play important roles in STEM education and provide substantial motivation when students are challenged beyond their interests and skills (Gomoll et al., 2016). Additionally, students tend to report greater enjoyment of STEM learning experiences that involve professional facilitators. The facilitators provide guidance to the students, rather than didactic instruction, allowing the students to navigate the design process and make decisions. Facilitators should also stimulate critical thinking among the students, and then guide the students in active exploration, rather than allowing students to simply emulate a specific process. In this manner, students retain their enthusiasm and can become facilitators when new students join the project, which supports the overall sustainability of STEM learning and its teaching ecosystem (Dolenc et al., 2016).

14.4.8 Issue 4: Inequity of Gender and Race in STEM Education

Many studies have highlighted the need to address concerns related to participation, representation, and inclusion in STEM fields (Long III & Mejia, 2016). Persistent inequalities in students' formal educational opportunities limit their development and career choices. Furthermore, STEM projects outside of the curriculum are even less accessible than formal STEM education to certain types of students. The demand for diversity in STEM professions has become a prominent issue. Therefore, a reform process should be implemented to promote diversity among students in STEM professions. The perceptions and preferences of disadvantaged groups (with respect to factors such as class, race, gender, and disabilities) in addressing the unequal distribution of social and cultural resources should be considered during the reform process. The collaborative challenges in STEM courses become increasingly complex when students in those courses have greater age-related and cultural diversities (Duodu et al., 2017). A diverse educational environment encourages students to participate actively and express their ideas in class. By interacting with various ideas and opinions, students can vigorously participate in academic aspects while forming their worldviews and values; they can gain an understanding of the importance of reducing racial and gender discrimination, as well as the need to eliminate inequality (Bamberger, 2014). Diverse STEM talent is derived from a wide range of perspectives during STEM innovation, and educational equity is essential (Lee et al., 2020).

This chapter primarily examined Taiwan's experience in the context of STEM projects in informal education. In addition to considering the advantages and disadvantages mentioned above, it is important that experts and scholars from more countries collaborate to address issues that greatly affect STEM education; this will help to manage future challenges requiring STEM support.

14.5 Conclusions and Implications

This chapter focused on analyses of STEM projects in informal education. In addition to examining the current situation based on relevant studies and online resources, this chapter shared Taiwan's experience as a reference for the global promotion of STEM projects in informal education. The main conclusions in this chapter can be summarized as follows: First, STEM projects in informal education are flexible and have specific characteristics that require continuous promotion to meet the needs of diverse student populations and attain the goals of STEM education. Second, informal education includes school clubs, schools (universities, primary schools, and secondary schools)/corporate camps, teaching aids from publishers, museums, STEM competitions, and web-based STEM projects. Third, some of the essential criteria for the success of informal STEM education include diversified and systematic planning of themed activities, making good use of diverse learning strategies, and introducing various electronic components to enrich the results of STEM projects. Finally, informal STEM education still faces several challenges that need further consideration, including the high cost of electronic components, camps, or teaching aids that do not adhere to STEM educational best practices, lack of qualified instructors, and educational inequality.

Based on these conclusions, future promotion of STEM projects in informal education should consider the following aspects: First, informal education practices should seek to provide more STEM projects that meet the demands of a diverse student population, with respect to cultural background, gender, race, and disabilities. Second, the efficacies of many STEM projects in informal education have not been verified, but organizers may claim that their projects are beneficial to interdisciplinary STEM learning; accordingly, studies regarding the development of appropriate evaluation instruments are needed to assist in reviewing student learning performance in STEM settings.

In this chapter, several informal learning paths in STEM education have been highlighted; however, some limitations should be noted. First, in addition to the abovementioned objectives based on the literature and the authors' assessments, other central objectives of STEM projects in informal educational settings require further examination. These objectives may not be included in past research or online resources, but they can be determined through in-depth interviews with students and experts (e.g., teachers, corporations, and professionals in STEM fields) across disciplines and countries. Second, the authors have summarized six main forms of STEM projects, from the perspectives of teaching practice and previous research. However, the emphasis on these forms of STEM projects is based on their current statuses in Taiwan. Differences in STEM projects persist among countries; these differences are influenced by the educational system and the resources dedicated to STEM education. Future research should examine learning assessment incompatibility, as well as methods to appropriately design and evaluate informal learning. Third, there is a need to evaluate STEM education in an interdisciplinary manner, rather than focusing on one or two aspects of STEM education, with respect to the needs and interests of the students involved. Future investigations regarding the assessment of students' transformative competencies in STEM education are needed to identify appropriate evaluation methods and tools.

References

Adams, J. D., Gupta, P., & Cotumaccio, A. (2014). Long-term participants: A museum program enhances girls' STEM interest, motivation, and persistence. *Afterschool Matters, 20*, 13–20.

Allen, S. & Peterman, K. (2019). Evaluating informal STEM education: Issues and challenges in context. In A. C. Fu, A. Kannan, & R. J. Shavelson (Eds.), *Evaluation in Informal Science, Technology, Engineering, and Mathematics Education. New Directions for Evaluation, 161*, 17–33.

Anderson, K. L., Kaden, U., Druckenmiller, P. S., Fowell, S., Spangler, M. A., Huettmann, F., & Ickert-Bond, S. M. (2017). Arctic science education using public museum collections from the University of Alaska Museum: An evolving and expanding landscape. *Arctic Science., 3*(3), 635–653. https://doi.org/10.1139/as-2017-0010

Atkins, K., Dougan, B. M., Dromgold-Sermen, M. S., Potter, H., Sathy, V., & Panter, A. T. (2020). "Looking at Myself in the Future": how mentoring shapes scientific identity for STEM students from underrepresented groups. *International Journal of STEM Education, 7*, 42. https://doi.org/10.1186/s40594-020-00242-3

Bamberger, Y. M. (2014). Encouraging girls into science and technology with feminine role model: Does this work? *Journal of Science Education and Technology, 23*, 549–561. https://doi.org/10.1007/s10956-014-9487-7

Barker, B. S., & Ansorge, J. (2014). Robotics as means to increase achievement scores in an informal learning environment. *Journal of Research on Technology in Education, 39*(3), 229–243. https://doi.org/10.1080/15391523.2007.10782481

Blanchard, M. R., Hoyle, K. S., & Gutierrez, K. S. (2017). How to start a STEM club. *Science Scope, 41*(3), 88–94. https://doi.org/10.2505/4/ss17_041_03_88

Borda, E., Schumacher, E., Hanley, D., et al. (2020). Initial implementation of active learning strategies in large, lecture STEM courses: Lessons learned from a multi-institutional, interdisciplinary STEM faculty development program. *International Journal of STEM Education, 7*, 4. https://doi.org/10.1186/s40594-020-0203-2

Cejka, E., Rogers, C., & Portsmore, M. (2006). Kindergarten robotics: Using robotics to motivate math, science, and engineering literacy in elementary school. *International Journal of Engineering Education, 22*(4), 711.

Cheng, C. Y., Kao, C. P., Hsu, T. W., & Lin, K. Y. (2023). A study of the feasibility of a cross-college curriculum based on the experience of student cooperation. *International Journal of Technology and Design Education*, 33, 23–37. https://doi.org/10.1007/s10798-022-09752-3

Chittum, J. R., Jones, B. D., Akalin, S., & Schram, Á. B. (2017). The effects of an afterschool STEM program on students' motivation and engagement. *International Journal of STEM Education, 4*(1), 11.

Christensen, R., Knezek, G., & Tyler-Wood, T. (2015). Alignment of hands-on STEM engagement activities with positive STEM dispositions in secondary school students. *Journal of Science Education and Technology, 24*(6), 898–909.

Dan, Y., Zhang, X. L., & Wang, L. Y. (2017). The teaching case design of STEM based on the environment of combining museum and school-water resources project. *International Journal of Engineering Education, 33*(3), 1117–1123.

Denson, C. D., Hailey, C., Stallworth, C. A., & Householder, D. L. (2015). Benefits of informal learning environments: A focused examination of STEM-based program environments. *Journal of STEM Education: Innovations and Research, 16*(1), 11–15.

Dika, S. L., & D'Amico, M. M. (2016). Early experiences and integration in the persistence of first-generation college students in STEM and non-STEM majors. *Journal of Research in Science Teaching, 53*(3), 368–383.

Dolenc, N. R., Mitchell, C. E., & Tai, R. H. (2016). Hands off: Mentoring a student-led robotics team. *International Journal of Science Education, Part B, 6*(2), 188–212. https://doi.org/10.1080/21548455.2015.1039467

Duodu, E., Noble, J., Yusuf, Y., Garay, C., & Bean, C. (2017). Understanding the delivery of a Canadian-based after-school STEM program: A case study. *International Journal of STEM Education, 4*(1), 1–11. https://doi.org/10.1186/s40594-017-0083-2

Gamage, S. H. P. W., Ayres, J. R., & Behrend, M. B. (2022). A systematic review on trends in using Moodle for teaching and learning. *International Journal of STEM Education, 9*, 9. https://doi.org/10.1186/s40594-021-00323-x

Gilliam, M., Jagoda, P., Fabiyi, C., Lyman, P., Wilson, C., Hill, B., & Bouris, A. (2017). Alternate reality games as an informal learning tool for generating STEM engagement among underrepresented youth: A qualitative evaluation of the source. *Journal of Science Education and Technology, 26*, 295–308. https://doi.org/10.1007/s10956-016-9679-4

Goff, E. E., Mulvey, K. L., Irvin, M. J., & Hartstone-Rose, A. (2018). Applications of augmented reality in informal science learning sites: A review. *Journal of Science Education and Technology, 27*, 433–447.

Gomoll, A., Hmelo-Silver, C. E., Šabanović, S., & Francisco, M. (2016). Dragons, ladybugs, & softballs: Girls' STEM engagement with human-centered robotics. *Journal of Science Education and Technology, 25*, 899–914. https://doi.org/10.1007/s10956-016-9647-z

Halawa, S., Hsu, Y.-S., Zhang, W.-X., Kuo, Y.-R., & Wu, J.-Y. (2020). Features and trends of teaching strategies for scientific practices from a review of 2008–2017 articles. *International Journal of Science Education, 42*(7), 1183–1206. https://doi.org/10.1080/09500693.2020.1752415

Hazari, Z., Dou, R., Sonnert G., & Sadler, P. M. (2022). Examining the relationship between informal science experiences and physics identity: Unrealized possibilities. *Physical Review Physics Education Research, 18*. https://doi.org/10.1103/PhysRevPhysEducRes.18.010107

Hooper-Greenhill, E. (2007). *Museums and Education: Purpose, Pedagogy, Performance (Museum Meanings)*. Routledge.

International Technology and Engineering Educators Association. (2020). *Standards for technological and engineering literacy: The role of technology and engineering in STEM education.* https://www.iteea.org/STEL.aspx

Irwanto, I., Saputro, A. D., Widiyanti, R., & M. F., & Lukman, I. R. (2022). Research trends in STEM education from 2011 to 2020: A systematic review of publications in selected journals. *International Journal of Interactive Mobile Technologies, 16*(05), 19–32. https://doi.org/10.3991/ijim.v16i05.27003

Jiang, Y., Popov, V., Li, Y., Myers, P. L., Dalrymple, O., & Spencer, J. A. (2021). "It's Like I'm Really There": Using VR experiences for STEM career development. *Journal of Science Education and Technology, 30*, 877–888. https://doi.org/10.1007/s10956-021-09926-z

Kelley, T. R. (2010). Staking the claim for the 'T' in STEM. *The Journal of Technology Studies, 36*(1), 2–11.

Ku, C.-J., Hsu, Y.-S., Chang, M.-C., & Lin, K.-Y. (2022). A model for examining middle school students' STEM integration behavior in a national technology competition. *International Journal of STEM Education, 9*, 3. https://doi.org/10.1186/s40594-021-00321-z

Lin, C.-C., & Tsai, C.-C. (2009). The relationship between students' conceptions of learning engineering and their preferences for classroom and laboratory learning environments. *Journal of Engineering Education, 98*, 193–204.

Lin, K.-Y., Lu, S.-C., Hsiao, H.-H., Kao, C.-P., & Williams, P. J. (2021). Developing student imagination and career interest through a STEM projects using 3D printing with repetitive modeling. *Interactive Learning Environments*. https://doi.org/10.1080/10494820.2021.1913607

Long, L. L., III., & Mejia, J. A. (2016). Conversations about diversity: Institutional barriers for underrepresented engineering students. *Journal for Engineering Education, 105*(2), 211–218.

Lundgren, D., Laughen, R., Lindeman, C., Shapiro, M., & Thomas, J. (2011). *Schools Like Ours: Realizing Our STEM Future*. Berrien.

Lee, M. J., Collins, J. D., Harwood, S. A., Mendenhall, R., & Huntt, M. B. (2020). "If you aren't White, Asian or Indian, you aren't an engineer": racial microaggressions in STEM education. *International Journal of STEM Education, 7*(48). https://doi.org/10.1186/s40594-020-00241-4

Lou, S. J., Chung, C. C., Dzan, W. Y., Tseng, K. H., & Shih, R. C. (2013). Effect of using TRIZ Creative Learning to build a pneumatic propeller ship while applying STEM knowledge. *International Journal of Engineering Education, 29*(2), 365–379.

Lin, K. Y., Yu, K. C., Hsiao, H. S., Chu, Y. H., Chang, Y. S., & Chien, Y. H. (2015). Design of an assessment system for collaborative problem solving in STEM education. *Journal of Computers in Education, 2*(3), 301–322.

McCreedy, D., & Dierking, L. D. (2015). Cascading influences: Long-term impacts of informal STEM programs for girls. *Informal Learning Review, 130*, 5–13.

National Academy of Engineering, & National Research Council. (2014). *STEM Integration in K-12 Education: Status, Prospects, and an Agenda for Research.* The National Academies Press.

Nugent, G., Barker, B., Grandgenett, N., & Adamchuk, V. I. (2010). Impact of robotics and geospatial technology interventions on youth STEM learning and attitudes. *Journal of Research on Technology in Education, 42*(4), 391–408. https://doi.org/10.1080/15391523.2010.10782557

Rhoads, T. R., Walden, S. E., & Winter, B. A. (2004). Sooner elementary engineering and science: A model for after-school science clubs based on university and K-5 partnership. *Journal of STEM Education: Innovations & Research, 5*(3/4), 47–52. https://eric.ed.gov/?id=EJ1063417

Roberts, T., Jackson, C., Mohr-Schroeder, M. J. , Bush, SB., Maiorca, C., Cavalcanti, M., Schroeder, DC., Delaney, A., Putnam, L. & Cremeans, C. (2018). Students' perceptions of STEM learning after participating in a summer informal learning experience. *International Journal of Stem Education, 5*(35). https://doi.org/10.1186/s40594-018-0133-4

Schwab, D. B., Cole, L. W., Desai, K. M., Hemann, J., Hummels, K. R., & Maltese, A. V. (2018). A summer STEM outreach program run by graduate students: Successes, challenges, and recommendations for implementation. *Journal of Research in STEM Education, 4*(2), 117–129.

Sergeyev, A., Kinney, M., Kuhl, S., Alaraje, N., Highum, M., & Mehandiratta, P. (2019). University, community college and industry partnership: Revamping robotics education to meet 21st century workforce needs-NSF Sponsored Project final Report. *Journal of Engineering Technology, 36*(2), 8–20.

Singer, A., Montgomery, G., & Schmoll, S. (2020). How to foster the formation of STEM identity: Studying diversity in an authentic learning environment. *International Journal of STEM Education, 7.*

Stewardson, G. A., Robinson, T. P., Furse, J. S., & Pate, M. L. (2019). Investigating the relationship between VEX robotics and student self-efficacy: An initial look. *International Journal of Technology and Design Education, 29*, 877–896. https://doi.org/10.1007/s10798-018-9461-4

Stoeger, H., Schirner, S., Laemmle, L., Obergriesser, S., Heilemann, M., & Ziegler, A. (2016). A contextual perspective on talented female participants and their development in extracurricular STEM programs. *Annals of the New York Academy of Sciences, 1377*(1), 53–66. https://doi.org/10.1111/nyas.13116

Sullivan, F. R. (2008). Robotics and science literacy: Thinking skills, science process skills and systems understanding. *Journal of Research in Science Teaching, 45*, 373–394. https://doi.org/10.1002/tea.20238

Tam, H., Chan, A. Y., & Lai, O. L. (2020). Gender stereotyping and STEM education: Girls' empowerment through effective ICT training in Hong Kong. *Children and Youth Services Review, 119.*

Tsai, C.-C., & Chang, C.-Y. (2005). Lasting effects of instruction guided by the conflict map: Experimental study of learning about the causes of the seasons. *Journal of Research in Science Teaching, 42*, 1089–1111.

Üçgül, M., & Altıok, S. (2017). You are an astroneer: The effects of robotics camps on secondary school students' perceptions and attitudes towards STEM. *International Journal of Technology and Design Education, 8*(2), 397–406.

Williams, D. C., Ma, Y., Prejean, L., Ford, M. J., & Lai, G. (2014). Acquisition of physics content knowledge and scientific inquiry skills in a robotics summer camp. *Journal of Research on Technology in Education, 40*(2), 201–216. https://doi.org/10.1080/15391523.2007.10782505

Witherspoon, E. B., Schunn, C. D., Higashi, R. M., & Baehr, E. C.(2016). Gender, interest, and prior experience shape opportunities to learn programming in robotics competitions. *International Journal of STEM Education, 3*(18).

Xu, M., Williams, P. J., Gu, J., & Zhang, H. (2020). Hotspots and trends of technology education in the International Journal of Technology and Design Education: 2000–2018. *International Journal of Technology and Design Education, 30*, 207–224. https://doi.org/10.1007/s10798-019-09508-6

Yang, K. L., & Li, J. L. (2018). A framework for assessing reading comprehension of geometric construction texts. *International Journal of Science and Mathematics Education, 16*, 109-124. https://doi.org/10.1007/s10763-016-9770-6

Yang, K.-L., Wu, H.-K., Yeh, Y.-F., Lin, K.-Y., Wu, J.-Y., & Hsu, Y.-S. (2023). Implementers, designers, and disseminators of integrated STEM activities: self-efficacy and commitment. *Research in Science and Technological Education*, 41(4), 1433–1451. https://doi.org/10.1080/02635143.2021.2008343

Kuen-Yi Lin is a Distinguished Professor and Associate Dean in the Department of Technology Application and Human Resource Development and Institute for Research Excellence in Learning Sciences at National Taiwan Normal University, Taiwan. He teaches courses in Introduction to Technology Education, Integrated STEM Research, and Instructional Strategies of Technology Education et al. to technology teachers and students. His primary research interests involve Technology Education and STEM Education.

Yu-Jen Sie is a Ph.D. candidate in the Department of Technology Application and Human Resource Development and Institute for Research Excellence in Learning Sciences at National Taiwan Normal University, Taiwan. Her research interests include STEM education, testing & assessment, and e-learning. Her PhD research focuses on the development of a STEM competencies assessment for secondary school students, the instrument for measuring interdisciplinary STEM problem-solving abilities.

Chapter 15
Experts' Views on the Role of the 'T' and 'E' in Integrated STEM Education and Implications for Out-of-Field Teaching

Jonas Hallström, Per Norström, and Konrad J. Schönborn

Abstract The STEM acronym permeates educational research and practice. While the potential pedagogical merits of STEM as an opportunity to integrate knowledge from the contributing disciplines and achieve a holistic understanding are well-documented, little is known about how out-of-field teachers contend with contributing to such a vision in practice. With an intended audience of STEM teacher practitioners in mind, this chapter focuses on international expert views of technology (T) and engineering (E) in out-of-field teaching of integrated STEM. The presented views were solicited from experienced international researchers, education practitioners, and professionals in science, technology, engineering, and mathematics. Experts' views emerged as five overarching themes that primarily identified: the importance of maintaining subject integrity, the implicit nature of technology and engineering in teaching activities, the centrality of engineering design processes, the necessity of collaboration and cooperation, and the need for specialised teacher competence. The emergent views have practical implications regarding engineering design and design-based teaching for informing curriculum design, teacher education programmes, as well as STEM textbooks and resource composition. The chapter closes by illuminating the question as to whether integrated STEM remains a sought epistemological position or only a method to teach STEM subjects, a dilemma whereupon our future work with STEM experts shall continue to explore.

Keywords Experts' views · Integrated STEM teaching · Technology education · Engineering education

---

J. Hallström (✉)
Department of Behavioural Sciences and Learning (IBL), Linköping University, Norrköping, Sweden
e-mail: jonas.hallstrom@liu.se

P. Norström
Department of Learning, KTH Royal Institute of Technology, Stockholm, Sweden

K. J. Schönborn
Department of Science and Technology (ITN), Linköping University, Norrköping, Sweden

W. Fox-Turnbull and P. J. Williams (eds.), *Locating Technology Education in STEM Teaching and Learning*, Contemporary Issues in Technology Education,
https://doi.org/10.1007/978-981-97-1995-2_15

15.1 Introduction

According to contemporary views, integrated STEM—science, technology, engineering, and mathematics—education aims to achieve learning that goes beyond the individual disciplines (e.g. Hallström, & Schönborn, 2019; Kelley, & Knowles, 2016), a STEM literacy (Tang et al., 2019). STEM literacy denotes the ability to identify, apply, and integrate knowledge from the STEM disciplines (Zollman, 2012). Integrated STEM education could, for example, provide learning that is holistic and authentic because it is based on the integration of knowledge that occurs in real-world practices in these disciplines. However, many integrated STEM initiatives are taught by a single teacher from one of these disciplines. This approach is termed 'out-of-field teaching', because the teacher is expected to teach several STEM disciplines while they may only be a specialist in one of them (Du Plessis, 2020). This chapter focuses on international expert views of technology (T) and engineering (E) in out-of-field teaching of integrated STEM. Based on these views, we suggest measures to be taken to further STEM as a fruitful concept in education. The intended audience of the chapter is pre-service and in-service STEM teachers and teacher educators.

15.2 Experts' Views on the Role of 'T' and 'E' in Integrated STEM Education

The STEM acronym has been in widespread use since the early 2000s. Even though there is consensus that it stands for science, technology, engineering, and mathematics, it has been described in different ways in textbooks, teachers' handbooks, standards, and curricula. A majority of the descriptions could be seen as points on a continuum. At one end, STEM is just a collective name for a group of separate disciplines or school subjects. At the other end, STEM is regarded as a unified body of knowledge that combines the learning from the traditional disciplines and refuses to acknowledge the borders between them. Most definitions are somewhere in between, often without explicitly stating their epistemological and sociological status (for example, see, Banks & Barlex, 2014; ITEEA, 2020; Kelley & Knowles, 2016; Ramalay, 2009).

As is the case with 'STEM', both technology and engineering are subjects that have been interpreted, described, and defined in a multitude of ways. In this chapter, we use the term technology to describe a wide range of objects (artefacts), systems, processes, and knowledge intended to solve problems or fulfil intentions mainly through the design and use of physical objects/systems (also see Mitcham, 1994; de Vries, 2016). Knowledge about technology in a school context concerns objects, systems, and processes, as well as their relation to individuals, society, and the environment (Nordlöf et al., 2021). Engineering is a subset of technology; a process and an activity that has to do with the design of technological objects and systems. A typical textbook description of engineering commences with a problem or an intention and ends with a product (sometimes also maintenance and/or recycling are included). Engineering education tends to be hands-on, with students participating in real-world engineering projects or parts thereof. Design and design processes consist of some kind of more or less ordered list of activities (sketching, simulating, evaluating, creating prototypes, etc.) that taken together lead from a problem or idea to a solution. In many interpretations of the term in the literature, the very essence of engineering is the design process (ITEEA, 2020, p. 56; Norström, 2016).

A prerequisite for successful STEM teaching is well-prepared STEM teachers. A shortage of teachers in traditionally organised subjects is a major issue in many countries (Hutchison, 2012; Saat et al., 2021; Skolverket, 2022). Of these, only a small minority have sufficient training in all STEM areas. This is a major challenge for the educational system at large. The transferability of teaching skills from one area to another is limited, which has been noticeable for example when science teachers teach mathematics (Shah et al., 2020; Vale et al., 2020) and when science teachers teach technology (Skolinspektionen, 2014). The teachers express uncertainty both concerning subject content and how to teach it. Suggestions for how to improve the situation include team teaching (El Nagdi et al. 2018), collaborations between teachers and researchers (Saat et al., 2021), and re-training of science and mathematics graduates who are supported by a team of experienced teachers during the initial teaching period (Hutchison, 2012).

We solicited a group of experts, hereafter referred to as the *STEM expert panel*, to collect views on integrated STEM as an educational enterprise at large. The panel consisted of twenty-eight experienced international researchers and practitioners in science, technology, engineering, mathematics as well as related educational disciplines. The panel members represented three STEM areas (some participants had expertise in more than one area), namely (i) researchers in the fields of science, technology, engineering, and mathematics education, (ii) STEM education actors such as teacher educators, curriculum specialists and policymakers, and (iii) STEM professionals such as engineers and designers working primarily in industry. The panel responded to an anonymous survey with three open-ended questions, of which two specifically addressed the role of the 'T' and 'E' in integrated STEM education (cf. Khatri et al., 2017; Osborne et al., 2003). An example of an included open-ended question was to ask their opinion of possible opportunities and limitations of integrating science, technology, engineering, and mathematics (STEM disciplines) in education. Since we sought the experts' unsolicited views, we deliberately did not offer definitions for neither STEM nor science, technology, engineering and mathematics, in the survey.

The survey responses were analysed through a qualitative content analysis focussed on inductive category development (Mayring, 2000). The approach included repeated reading of the data set to find and establish patterns. The repeated reading resulted in a step-by-step formulation of inductive categories, which after the last round of the analysis were reduced to five overarching themes. Each of the five themes emergent from the analysis is presented in the following sections. There we present the views of the STEM expert panel concerning the role of the 'T' and the 'E' and out-of-field teaching in integrated STEM teaching.

15.3 Subject Integrity Must Be Maintained as Part of Integrative STEM

A salient theme in the STEM expert responses that relate strongly to out-of-field teaching is the issue of subject integrity in integrative STEM education projects. One expert wrote that ‘it would be a limitation if the four separate STEM subjects lose their integrity during the integration’. Integrity may be interpreted in various ways, but according to the experts, it could take on multiple meanings in a STEM education context. As one expert suggested, if the ‘T’ or ‘E’ lost their integrity it could be that their core identities or activities were being weakened by those of other disciplines: ‘There are many opportunities for integration [...] but the parts in STEM should be kept as core parts and not be too diluted.’ Another expert expounded upon what exactly the ‘core parts’ of the STEM subjects are in the response: ‘Authentic educations in science and technology must reflect the natures of the parent disciplines as far as is practicable.’ The nature of science and technology has to do with, for instance, what science and technology are as well as what are the foundational methods for acquiring new knowledge in these disciplines (e.g. Pleasants et al., 2019; Tang et al., 2019).

In integrative STEM approaches, it is also crucial to maintain subject integrity in order to avoid what one expert calls the ‘"Mucky Brown Paint Syndrome". Failure to keep curriculum integrity. Students are busy engaged in projects but learn very little discipline knowledge or do not recognise disciplines engaged in.’ This quote demonstrates what can happen if the individual disciplines are ‘diluted’, namely that students do not recognise technology or engineering components in a STEM project. This might prevent them from learning the disciplinary knowledge that was intended. However, the STEM expert panel claims that if discipline and curriculum integrity are maintained, then integrated STEM education could lead to successful outcomes (e.g. English et al., 2015). For example, as one expert suggested, ‘this [earlier described example of a student] project certainly gave rise to genuine and significant learning outcomes that couldn’t have been obtained studying the components involved in splendid isolation only.’ This expert quote points to the importance of integrating the STEM disciplines in concrete student projects and curricula to enable learning and collaborative opportunities that are not possible through a single discipline alone, while at the same time maintaining the qualities of each contributing discipline. Overall, the contribution of each discipline towards potential ‘dilution’, could be analysed in terms of clarifying the contribution of each discrete discipline, as well as determining the out-of-area knowledge that a teacher has for being able to richly, rather than simply, integrate the disciplines.

15.4 Technology and Engineering Are Implicit in Integrated STEM Teaching Activities

The experts' responses suggest that technology and engineering education are sometimes implicit in integrated STEM initiatives, at the same time as the intended learning outcomes primarily relate to traditional science and/or mathematics content. One expert lucidly mentioned such an example:

> I'll refer to a student project in which four students worked on restoring distorted photographs. [...] Firstly, a photo is in and of itself a (visual) two-dimensional model of a three-dimensional world based on perspective constructions, which can be analysed in terms of classical projective geometry. Secondly, photos can be modelled by a variety of different mathematical tools so as to provide a digital representation of them (the JPEG algorithm is the best-known example of this), representations that can be transmitted electronically. Such representations make it possible to restore distorted images by making use of inverse Fourier and Radon transforms. The student group referred to wanted to learn to make such restorations both in theoretical and in practical terms. [...] Completing the project generated authentic experiences of coming to grips with different sorts of models. They also experienced the complex interplay between different STEM fields when going into depth with characterising, representing and manipulating images by way of mathematical representations.

Cases such as that above may imply that out-of-field teaching is less detrimental if the teachers are mathematics specialists, but potentially more difficult when they are science, technology, or engineering teachers. Conversely, some of the experts reflected a technology and/or engineering emphasis that included technological project goals and technology education contexts as pivotal for STEM teaching. In other obtained views, experts branded their STEM educational endeavours as being about science and/or mathematics education whilst they were still relevant in an engineering and technology context. Possible explanations include that these experts viewed technology and engineering either as 'applications' of science and mathematics, or as part of science and mathematics, or that their views were influenced by the rich experiences from their root discipline.

15.5 Engineering Design Processes Are Central to STEM

Teaching STEM should provide opportunities to combine knowledge from different subjects into a greater whole, and thereby develop STEM literacy. Recurring themes when the experts suggested how this could be achieved are design work, confronting authentic problems, and modelling. When working with sufficiently large and/or complex problems, design work tends to demand input from many different disciplines, which in an educational situation is believed to increase the likelihood of a holistic understanding (de Vries, 2018; Hallström et al., 2019; Snape et al., 2013).

Below are two expert examples of how the design process can be used as a framework to enhance a holistic STEM view of an engineering problem. One expert described a project which starts with a description of the everyday life of a farmer:

> One of my students identified a problem for farmers who have to go through many gates as they transverse their property. Typically, this involves stopping the vehicle, getting out and opening the gate, getting back in the vehicle, driving through, and then getting out again to close the gate. This is significant for farmers who may go through up to 20 gates/day. A solution was developed of a remote-controlled solar-powered gate opening mechanism. The student built a 1:20 scale model as proof of concept.

Even though the project at its core is an engineering project, it also demands knowledge about technology (sensors, control systems, actuators), physics (energy), and mathematics (e.g., calculations of how much time will really be saved). Another example is provided by an expert whose students worked designing clothing for cold and damp climates:

> … designing protective textiles – such as warm and dry clothing for trawlermen. Different materials would be modelled for their insulation and water-resisting properties. Then, what fabrics or fabric combinations have the desired properties and how should they be fairly investigated? Also, practicalities vs. costs: cotton wool might be a great insulator, but is it a sensible choice for the context of a robust water-proof jacket?

The choice of materials demands knowledge from several disciplines such as physics (heat transfer), chemistry (material properties, environmental aspects), and engineering (manufacturability).

15.6 Inputting the 'T' and the 'E' in STEM Demands Collaboration and Cooperation

The STEM expert panel exposed the opinion that teachers should be competent in teaching STEM disciplines as a collaborative enterprise. A typical expert viewpoint demonstrating this position is as follows:

> Coordination/cooperation means, for example, that the approach to current flow and potential differences taught in science can be approached in the same way in technology. Integrating either means one teacher able to support projects that cross STEM borders in the same way that the one teacher can teach 'General Science' – or the timetable must be structured for STEM subject teachers' team teaching.

As demonstrated in the excerpt above, collaborative and coordinated teaching should also include multiple perspectives that support the utility of one subject for supporting another. In this way, teachers will provide insights to one another drawn from each of their specialisations. Developing such competence also mirrors the view that solving relevant challenges in the world requires a multiple-perspective approach. In pursuing such collaborative teaching approaches, experts saliently expressed the danger of one discipline dominating over another, with one or two disciplines having a 'louder voice' or more priority over another. Clear examples in this respect pertained to science and mathematics dominating technology and engineering in attempted teaching integration. Taken together from a competence perspective, experts also noted that it is difficult to define a standardised way of teaching STEM, and therefore articulate what competencies represent integrated STEM teaching.

In addition, collaborative approaches that foster the 'T' and the 'E' in STEM integration require real-world validation and feedback. This view is represented by the following expert excerpt, upon considering a technology project:

> Acceptability to the society in which the technology will be used. This will be concerned with the prevailing values of the society, and what is thought to be important and worthwhile.

The quote supports the assertion that teacher competence needs to include valid means for assessing and evaluating outcomes of STEM integration. Also, as communicated by other experts, doing so is intimately related to authentic activities, wherein provided facilitator feedback informs the continuation of real-world design and implementation processes. In doing so, it is important for subsequent curriculum design to reflect such professional experience, which in turn would mirror how a professional would seek expert knowledge from another discipline to learn, share and collaborate.

15.7 Teachers Require Specialised Competence for Integrating the 'T' and the 'E' in Out-of-Field STEM Teaching

Experts revealed various views on necessary and favourable teacher competence related to integrating technology and engineering into STEM teaching. Firstly, there is a strong expert view that teaching STEM requires specialised teacher expertise and knowledge. This view is exemplified in the following expert quote:

> [If the integrated STEM approach] is done with teachers that know each area well AND the way in which the disciplines relate to each other (philosophically/ontologically and epistemologically) then it is possible that this could work, and we could develop a broad more critical STEM literacy.

The above quote supports the notion that teachers need to be familiar with each component area, as well as have knowledge about how STEM components relate to one another. As part of such expression, experts also noted that an integrated STEM is a challenge for teacher education since such programmes are often separated from traditional subject disciplines. Consequently, limitations are faced when teachers are expected to integrate the STEM components without competence from other subjects. It follows that developing such competence might be offered by STEM teachers also collaborating without strict syllabus constraints. Herein, individual and teams of teachers choose more open-ended approaches, where experience initially built during teacher education programmes is expanded upon by working collaboratively in response to 'contexts' and 'concepts' (Rossouw et al., 2011) through design-based teaching rather than only specific curriculum driven criteria.

15.8 Conclusions and Educational Implications

The STEM experts revealed five main views on the role of the 'T' and 'E' in integrated STEM teaching. Firstly, experts opined that subject integrity must be maintained in integrative STEM teaching for core content, activities, and methods of the individual disciplines not to be 'diluted'. Secondly, technology and engineering tend to be implicit in the integrated STEM teaching activities mentioned by the experts, while science and/or mathematics are made explicit. Thirdly, the engineering design process is at the same time pivotal to integrated STEM education projects, exemplified by the experts. Fourthly, to input the 'T' and the 'E' in STEM in a fruitful way there needs to be coordinated cooperation between areas of knowledge. Fifthly, teachers, therefore, require specialised competence for integrating technology and engineering in out-of-field STEM teaching, both in terms of disciplinary knowledge and skills in teaching STEM as a collaborative enterprise.

Consideration of the identified expert themes raises several implications for approaching the role of the 'T' and 'E' in integrated STEM teaching. While experts strongly asserted the requirement to maintain subject integrity, they also expressed concern that their disciplines would be overshadowed in such integration. At the same time, the examples that the experts provided were mostly of interdisciplinary engineering design projects where three or four of the STEM disciplines were represented. Therefore, the concerns of possible 'dilution' due to integration should be carefully considered, especially in out-of-field teaching situations. It is also important to illuminate the examples of already integrated, yet implicit, projects. Strategies for making such examples explicit could take the shape of integrated STEM projects that are specifically based on the engineering design process (e.g. Lin et al., 2021). Realising this strategy requires providing teachers with the necessary skills and opportunities to include their original knowledge in an integrated STEM context which encourages collaboration and promotes their creativity as a specialised competence.

Implications of the views of the international STEM expert panel for technology and engineering education are as follows:

- Technology and engineering curricula need to include an integrated STEM perspective that are at the same time mindful of the epistemologies and epistemic practices of each of the contributing subjects. A pertinent recent example is the new American *Standards for Technological and Engineering Literacy* (ITEEA, 2020), in which technology and engineering frame broader STEM perspectives.
- When aiming for an integrated STEM education, teacher education programmes should focus on the 'E' because it serves as a valid methodology for scaffolding the other disciplines in integrated STEM, particularly through the design process

(Hallström et al., 2019). It could therefore be beneficial, especially in out-of-field teaching, to develop design thinking and design-based methodologies in STEM teacher education, such as design-based teaching and learning approaches (e.g. Henze et al., 2021).

- Despite numerous research studies on integrated STEM education and various standard mandates, there is a lack of STEM textbook resources that address all of these disciplines together. There are alleged STEM textbooks with the word 'STEM' or 'STEM education' in the title, but they are frequently organised according to the traditional subjects. Those that try to integrate are primarily American and relate to the new standards mentioned above, but they tend to be written for younger audiences (e.g. Cameron, & Craig, 2016; Vasques et al., 2013).

  Proper STEM textbooks should include integrated projects. The engineering design process could act as an integrator for all STEM subjects, which could also inform textbook design in the sense that they are based on the design process. As regards technology and engineering in the context of STEM, textbook design could have a similar structure but focus more clearly on technological problem-solving. Younger learners are easier to target for STEM integration. Preschool and primary education traditionally have a higher degree of cross-curricular and interdisciplinary teaching (Hallström et al., 2012).
- To ameliorate the negative effects of single teachers engaging in 'out-of-field teaching', teacher education and school programs should be developed to facilitate STEM teacher collaboration, because integrated STEM and design-based teaching require interdisciplinary collaboration. It could be done through a three-part strategy: (i) Interdisciplinary collaboration on open-ended complex STEM problems and projects that promote fluidity, (ii) Devising a methodology for working based on the engineering design process, and (iii) Post-reflection on contribution and intersections between each discrete discipline, its methods and types of knowledge, as well as the development of a transferable and non-discipline specific set of skills (cf. Herrington et al., 2010).

The expert revelations seem to suggest that although there is no widespread, standardised way to teach integrated STEM, design-based teaching could be a valid foundation for the future development of such a standardised teaching method. For technology and engineering teachers, this is already a common way of working, although not necessarily in relation to science and mathematics. In the future, however, universities increasingly need to demonstrate linkages with industry so in that sense a full integration of the STEM disciplines in engineering design projects is desirable.

The expert views touch on the essence of the problem of integrated STEM. The expert panel claims that if discipline and curriculum integrity are maintained, then integrated STEM education could lead to successful outcomes. The question is whether this is even feasible in teaching practice. One can view integrated STEM education as a continuum with subject integrity at one end, and authenticity at the other. However, in turn, this leads to further questions: Is completely integrated STEM therefore an epistemological entity of its own, or is it merely a method of teaching the different STEM subjects? Or is it both? If not, what should it be?

A conclusion from our study is that integrated STEM education remains mainly an integrated way of teaching, whereas the contributing epistemologies of the STEM subjects are still separate (cf. Falloon et al., 2020). Tang and Williams (2019) call the separate conceptions 'S.T.E.M. literacies' and refer particularly to epistemic concerns and practices as difficult to reconcile within an integrated STEM construct. Previous research points to the advantages of introducing the engineering design process to promote students' developing STEM literacy, although so far only separate 'S.T.E.M. literacies' have been proven (e.g. Tati et al., 2017). Our future work shall continue to probe the dilemma of an integrated STEM epistemology and STEM literacy, while we continue to obtain experts' opinions to shed further light on the role of 'T' and 'E' in pursuing an integrated STEM education.

References

Banks, F., & Barlex, D. (2014). *Teaching STEM in secondary school*. Routledge.

Cameron, S., & Craig, C. (2016). *STEM labs for middle grades, grades 5–8*. Mark Twain Media.

de Vries, M. J. (2016). *Teaching about technology*. Springer International Publishing.

de Vries, M. J. (2018). The T and E in STEM: From promise to practice. In M. J. de Vries, S. Fletcher, S. Kruse, P. Labudee, M. Lang, I. Mammes, C. Max, D. Münk, B. Nicholl, J. Strobel, & M. Winterbottom (Eds.), *Research in Technology Education* (pp. 11–19). Waxmann.

Du Plessis, A. E. (2020). The lived experience of out-of-field STEM teachers: A quandary for strategising quality teaching in STEM? *Research in Science Education, 50*(4), 1465–1499.

English, L. D., & King, D. T. (2015). STEM learning through engineering design: Fourth-grade students' investigations in aerospace. *International Journal of STEM Education, 2*(14), 1–18. https://doi.org/10.1186/s40594-015-0027-7

Falloon, G., Hatzigianni, M., Bower, M., Forbes, A., & Stevenson, M. (2020). Understanding K-12 STEM education: A framework for developing STEM literacy. *Journal of Science Education and Technology, 29*(3), 369–385.

Hallström, J., & Ankiewicz, P. (2019). Laying down the "T" and "E" in STEM education: Design as the basis of an integrated STEM philosophy. In S. Pulé & M. J. de Vries (Eds.), *Proceedings PATT 37: Developing a knowledge economy through technology and engineering education, 3–6 June 2019 University of Malta, Msida Campus* (pp. 187–194). University of Malta.

Hallström, J., & Schönborn, K. J. (2019). Models and modelling for authentic STEM education: reinforcing the argument. *International Journal of STEM Education, 6*(22). https://doi.org/10.1186/s40594-019-0178-z

Hallström, J., Martinsson, B. G., & Sjöberg, M. (2012). *Att hävda och vårda ett revir: Argument, strategier och arbetsmetoder för ämnesföreningarna i biologi, historia och svenska 1960–2010*. Uppsala: Föreningen för svensk undervisningshistoria.

Henze, I., & de Vries, M. J. (2021). Design-based concept learning: An introduction. In I. Henze & M. J. de Vries (Eds.), *Design-based concept learning in science and technology education* (pp. 3–13). Brill Sense.

Herrington, J., Reeves, T., & Oliver, R. (2010). *A guide to authentic e-learning*. Routledge.

Hutchison, L. F. (2012). Addressing the STEM teacher shortage in American schools: Ways to recruit and retain effective STEM teachers. *Action in Teacher Education, 34*(5–6), 541–550. https://doi.org/10.1080/01626620.2012.729483

[ITEEA] International Technology and Engineering Educators Association. (2021). *Standards for technological and engineering literacy: The role of technology and engineering in STEM education*. https://www.iteea.org/STEL.aspx

Kelley, T. R., & Knowles, J. G. (2016). A conceptual framework for integrated STEM education. *International Journal of STEM Education, 3*(11).

Khatri, R., Henderson, C., Cole, R., Froyd, J. E., Friedrichsen, D., & Stanford, C. (2017). Characteristics of well-propagated teaching innovations in undergraduate STEM. *International Journal of STEM Education, 4*, 2. https://doi.org/10.1186/s40594-017-0056-5

Lin, K.-Y., Wu, Y.-T., Hsu, Y.-T., & Williams, P. J. (2021). Effects of infusing the engineering design process into STEM project-based learning to develop preservice technology teachers' engineering design thinking. *International Journal of STEM Education, 8*, 1. https://doi.org/10.1186/s40594-020-00258-9

Mayring, P. (2000). Qualitative content analysis. *Forum: Qualitative Social Research, 1*(2), Art. 20. https://doi.org/10.17169/fqs-1.2.1089

Mitcham, C. (1994). *Thinking through technology: The path between engineering and philosophy*. University of Chicago Press.

El Nagdi, M., Leammukda, F. & Roehrig, G. (2018). Developing identities of STEM teachers at emerging STEM schools. *International Journal of STEM Education, 5* (36). https://doi.org/10.1186/s40594-018-0136-1

Nordlöf, C., Norström, P., Hallström, J., & Höst, G. (2021). Towards a three-part heuristic framework for technology education. *International Journal of Technology and Design Education*. https://doi.org/10.1007/s10798-021-09664-8

Norström, P. (2016). The Nature of Pre-University Engineering Education. In M. J. de Vries, L. Gumaelius, & I.-B. Skogh (Eds.), *Pre-University Engineering Education* (pp. 27–46). Sense Publishers.

Osborne, J., Collins, S., Ratcliffe, M., Millar, R., & Duschl, R. (2003). What "ideas-about-science" should be taught in school science? A Delphi study of the expert community. *Journal of Research in Science Teaching, 40*(7), 692–720.

Pleasants, J., Clough, M. P., Olson, J. K., & Miller, G. (2019). Fundamental issues regarding the nature of technology. *Science & Education, 28*, 561–597.

Ramalay, J. A. (2009). The national perspective: fostering the enhancement of STEM undergraduate education. *New Directions for Teaching and Learning, 117*, 69–81. https://doi.org/10.1002/tl.345

Rossouw, A., Hacker, M., & de Vries, M. J. (2011). Concepts and contexts in engineering and technology education: An international and interdisciplinary Delphi study. *International Journal of Technology and Design Education, 21*, 409–424. https://doi.org/10.1007/s10798-010-9129-1

Saat, R. M., Fadzil, H. M., Adli, D. S. H., & Awang, K. (2021). STEM teachers' professional development through scientist-teacher-student partnership (STSP). *Jurnal Pendidikan Indonesia, 10*(3), 357–367.

Shah, C., Richardson, P., & Watt, H. (2020). *Teaching 'out of field' in STEM subjects in Australia: Evidence from PISA 2015*. GLO Discussion Paper, No. 511. Global Labor Organization.

Skolinspektionen [The Swedish National School Inspectorate]. (2014). *Teknik. Gör det osynliga synligt*. Skolinspektionen.

Skolverket [The Swedish National Agency for Education] (2022). *Anställda med lärarlegitimation och behörighet i ämne* [Employees with a teaching licence and competence in subject]. https://siris.skolverket.se/siris/sitevision_doc.getFile?p_id=551313

Snape, P., & Fox-Turnbull, W. (2013). Perspectives of authenticity: Implementation in technology education. *International Journal of Technology and Design Education, 23*(1), 51–68.

Tang, K.-S., & Williams, P. J. (2019). STEM literacy or literacies? Examining the empirical basis of these constructs. *Review of Education, 7*(3), 675–697. https://doi.org/10.1002/rev3.3162

Tati, T., Firman, H., & Riandi, R. (2017). The effect of STEM learning through the project of designing boat model toward student STEM literacy. *Journal of Physics: Conference Series, 895*(1), 1–8. https://doi.org/10.1088/1742-6596/895/1/012157

Vale, C., Campbell, C., Speldewinde, C., & White, P. (2020). Teaching across subject boundaries in STEM: Continuities in beliefs about learning and teaching. *International Journal of Science and Mathematics Education, 18*, 463–483. https://doi.org/10.1007/s10763-019-09983-2

Vasquez, J. A., Sneider, C. I., & Comer, M. W. (2013). *STEM lesson essentials, grades 3–8: Integrating science, technology, engineering, and mathematics* (pp. 58–76). Heinemann.
Zollman, A. (2012). Learning for STEM literacy: STEM literacy for learning. *School Science and Mathematics, 112*(1), 12–19.

Jonas Hallström is Ph.D. and Professor of Technology Education at the Department of Behavioural Sciences and Learning, Linköping University, Sweden, where he is also currently Deputy Head of Department. He is also Chair of the Swedish National Graduate School in Science and Technology Education Research (FontD). He presents regularly at international conferences and consults on technology education. He is one of the editorial board members of the Brill/Sense book series International Technology Education Studies. His research primarily concerns the historical emergence of technology as knowledge content in the school, the epistemology and subject philosophy of technology, various subject content (e.g. technological systems, artificial intelligence) as well as the attitudes to and knowledge of technology and technology education of students, student teachers and teachers. The research also relates to technology teaching in relation to, e.g., design, gender (girls and technology), authentic learning, models and modelling, and STEM (science, technology, engineering, mathematics) education.

Per Norström is Associate Professor (Swedish: universitetslektor) of technology education at KTH Royal Institute of Technology, Stockholm, Sweden. His research interests mainly concern pre-university engineering education and the epistemology of technology.

Konrad J. Schönborn is a Professor of visual learning and communication at Linköping University, Sweden. He is also scientific leader of the Swedish National Graduate School in Science and Technology Education (FontD). His research lies at the intersection of visualization and STEM education. Current work includes exploring the role of visualization for learning science in public spaces, and adaptive visualization environments for classroom learning. He is an Associate Editor of the Journal of Biological Education, and serves on the editorial boards of International Journal of Science Education and Biochemistry and Molecular Biology Education

Chapter 16
STEM in Senior Secondary

Joachim Svärd

Abstract A trend in tertiary engineering education is interdisciplinary education targeting real-world problems. This chapter will advise why senior secondary schools should pick up on this, giving their students a head start, and at the same time having teachers more likely to stay in teaching. Implementing STEM Education projects at a school is not the most accessible road to travel. Six significant areas that affect schools and teachers are: The individual subjects have *Standards* that need to be met, there are not enough teachers with the right *Qualifications*, Teachers don't have *Knowledge* about other subject matters, *design* of a STEM education project at an appropriate level is challenging, teachers are not used to *Cooperation*, and there is often little *Time* for planning. These challenges are manageable, and it is often up to the technology teacher to initiate such cooperation. This chapter will discuss these issues.

Keywords Technology education · High school · STEM education · Twenty-first-century skills · Interdisciplinary

16.1 Introduction

Introducing STEM Education at a senior secondary school is challenging. It is often connected with new ways of thinking, planning, and assessing education. But it is worthwhile if you are interested in having keen, engaged, and interested students with a better understanding of STEM subjects. At the same time, students are likely to gain many sought-after twenty-first-century skills (e.g., critical thinking, creativity, collaboration, and cooperation) making them ready for future studies and careers as they take on real-world dilemmas. As a bonus, teachers are likely to: remain in teaching, perceive improved teaching satisfaction, improve their teamwork skills,

J. Svärd (✉)
Linköping University Sweden, Linköping, Sweden
e-mail: joachim.svardh@gmail.com

W. Fox-Turnbull and P. J. Williams (eds.), *Locating Technology Education in STEM Teaching and Learning*, Contemporary Issues in Technology Education,
https://doi.org/10.1007/978-981-97-1995-2_16

and acquire a deeper understanding of STEM subjects' content and its pedagogical content knowledge (PCK).

16.2 A View on STEM

STEM Education (Science, Technology, Engineering, and Mathematics Education) is getting increasing attention. The hopes of policymakers, the industry, and the education sector are that STEM Education is the answer to a growing demand for an educated workforce and students with the much sought-after twenty-first-century skills. Senior secondary schools should therefore be a natural arena for STEM Education. But, it is not. This chapter will try to show some reasons senior secondary schools should take the extra effort to introduce STEM Education, why it has not already been done, and how it could be done.

The acronym STEM was first introduced in 2001 by the National Science Foundation in the US (Hallinen, 2021). Whenever STEM is mentioned, it is often just short for the subject areas: science, technology, engineering, and mathematics. We will use the expression *STEM Education* instead, highlighting the integration of individual subjects in school.

STEM Education has various definitions. The US Department of Education stated in 2016 that it must be about all four subject areas being present in a project (Shernoff et al., 2017). Others have more liberal definitions, such as being about two or more of the subjects (Kelley & Knowles, 2016; Moore et al., 2014; Sanders, 2009), sometimes even with just one STEM subject in conjunction with one or more non-STEM subject (e.g., California Department of Education, 2014, as cited in English, 2016; Sanders, 2009). In this chapter, we recognize any project containing problem-solving with a possible technical solution, as a valid STEM project. This standpoint is based on the connection between solving human needs and engineering (National Assessment Governing Board [NAGB], 2018; Shaw, 2001). The outcome of this process, a modification of the natural or designed world fulfilling the need, is technology (NAGB, 2018). In real life, non-STEM subjects, such as language or art, are natural and necessary components of any project. As such, they will mutually strengthen each other in a STEM project (e.g., Henriksen & Mishra, 2020; Lipowitz & Berkeihiser, 2019; Pearson et al., 2010).

16.3 Why STEM Education?

Employment in STEM occupations is expected to grow 50% faster than in non-STEM occupations during 2020–2030, according to the US Bureau of Labour Statistics (2021). STEM jobs are present in most areas of life, such as health, technological innovation, manufacturing, information technology, and political processes, and are associated with solutions to many of the global challenges ahead, e.g., climate change, sustainability, and resource management (Kelley & Knowles, 2016; Shernoff et al., 2017; Thomas & Watters, 2015). The 2005 report "Rising Above the Gathering Storm" (National Academy of Science, National Academy of Engineering, and Institute of Medicine of the National Academies [NASEM]) highlighted the development in US STEM Education. The demand for improvement in science and technology is advocated by industry, institutions, organizations, and nations. Besides core knowledge in the STEM subjects, the abilities to cooperate, communicate, think critically, and be creative, all being part of the "twenty-first-century skills" (Partnership for twenty-first-century learning, n.d.), are highlighted (Honey et al., 2014; Svärd et al., 2017, 2022). As recommended in this chapter, a STEM Education project will inevitably be interdisciplinary and deploy a range of relevant twenty-first-century skills. A STEM Education project will not only give the students the much sought-after experiences wanted by the industry but also readiness for the new university education at the junction between traditional Science, Technology, Engineering and Mathematics education offered at universities, and MOOCs (Massive Open Online Courses). "Instead of teaching content and skills and hoping students will see the connections to real-life application, an integrated approach seeks to locate connections between STEM subjects and provide a relevant context for learning the content." (Kelley & Knowles, 2016, p. 3).

16.4 STEM and the Future of University Studies

A growing number of tertiary education pathways can be described as bi-, multi-, or inter-disciplinary, contrasting with what has been offered previously. Traditionally, the STEM subjects are taught in isolation from the reality it will be used in, at various university programmes such as Architecture-, Industrial Design- or a Master of Science program. Students who studied chemistry could graduate with a Master of Science in Chemistry. It was as simple as that. For example, I entered university in the mid-1980s. There were just seven programs to choose from. At this time, a handful of students took the opportunity to study part of the physics- and chemistry programs. A perfect combination if you want to work with advanced chemical reaction calculations. Highly specialized and just for a few. When I looked up the university's webpage today, Chemical Engineering with Physics is just one out of about 40 different Master of Science programs at the university. Today it does not stand out, and I notice that four programs are about the environment or sustainability in some way. The programs are getting niched and increasingly multidisciplinary.

In times of individualization, a luxury reserved for a small elite not so long ago (e.g., Genov, 2014), the possibility of choosing and combining education is another driving factor for the design and implementation of STEM Education programs. A problem from an international perspective has been the compatibility between education in different countries. In Europe, for example, Bachelor-Master systems could differ between four and six years and not be compatible with other systems having several other levels (Crosier & Parveva, 2013). The Bologna process initiated in 1999 by 29 European countries ensures students a compatible educational system so that a course at any university can be a part of their academic degree. The Bologna process is governed by the European Higher Education Area [EHEA] and, in 2022, includes 49 countries (EHEA, n.d.-a). The basic framework of higher education in the Bologna process is three circles corresponding to the degrees of Bachelor (three years), Master (two years), and Doctor, similar to the North American educational system, but with a year shorter time for a Bachelor's degree (four years in the US). The Bologna process has catalyzed the harmonization of higher education, inspiring similar regional initiatives in various parts of the world, such as Southeast Asia, the Pacific, Latin America and the Caribbean, and West- and East Africa (Crosier & Parveva, 2013; Trines, 2018).

Considering this development, the Massachusetts Institute of Technology (MIT) studied the global state of engineering education (Graham, 2018). Graham anticipates rapid changes in higher education with new emerging players such as Iron Range Engineering, Singapore University of Technology and Design, and UC Chile leading the way. The report calls for a new curriculum that "emphasizes student choice, multidisciplinary learning and societal impact, coupled with a breadth of student experiences outside the classroom, outside traditional engineering disciplines and across the world" (Graham, 2018, p. iii).

Graham's conclusions are not a coincidence. Forbes, McKinsey & Company and LinkedIn (e.g., Marr, 2022) are among the experienced industry trend analysts that have hailed interdisciplinary areas over the past years; Life Science, Functional food, Game Development, e-health, artificial intelligence, innovation, and companies that address the global challenges mentioned earlier, to name a few. The common denominator here is the need for integration and application of knowledge and skills from various subject areas, including those in STEM, plus the competence and confidence to work collaboratively, think critically, solve problems, and communicate.

We find this trend all over the world. A leading German university presents itself as "an interdisciplinary integration of engineering, natural sciences, humanities, management, economics and social sciences" and a French institution "contributes to understanding and solving major global issues. Its research strategy focuses on interdisciplinary projects". An Australian university states that it "delivers influential, interdisciplinary research on the transition to a clean energy system" and in South Korea, "an understanding of the interactions between science, technology and society". US and Canadian universities and colleges could be viewed as offering a multidisciplinary education as a major area of study, e.g., engineering, could be combined with a minor in, for instance, environmental technology.

The compatibility of courses and educational programs is forcing universities and colleges to sharpen their research- and education profiles, not only to secure funds and professors but also students (Musselin, 2018), by offering more specialized and subject-integrated education, often related to local markets and real-world needs and opportunities. There is also a growing number of local universities, colleges, or branch campuses formed in recent years, often based on national educational policies. Times of changing demography and constant development of jobs also call for widening access to higher education. Continuous access to learning is in line with life-long learning, which is highlighted as a critical component of future success (Crawford, 2016; EHEA, n.d.-b; European Commission, n.d.; NASEM, 2005).

16.5 Introducing STEM Education at Senior Secondary Schools

To help students succeed in their tertiary studies in these transcendent times, preparing them for subject integration would make sense, giving them the necessary experiences in cooperative interdisciplinary learning. Technology often acts as the glue (or integrator) that facilitates and necessitates learning in other subject areas (Hallström & Ankiewicz, 2019). Because of this, it will often land on the technology teacher to initiate such projects.

However, there are more reasons to engage in interdisciplinary teachings, such as an integration of different STEM disciplines or STEM and other subjects. Studies have shown improved understanding in, for instance, various lab activities besides a feeling of reality and motivation among the students (Huri & Karpudewan, 2019; Svärd et al., 2022). There are also indications of improved teaching satisfaction, attitude toward interdisciplinary teaching and teamwork, and a decreased resistance to change among secondary teachers involved in interdisciplinary education (Al Salami et al., 2015). It is also more likely that the teachers will remain in teaching as professional development and opportunities to collaborate with colleagues are factors that have been deemed beneficial (Fulton & Britton, 2011; See also: EdWeek Research Center, 2022; MetLife, 2012).

A STEM Education project sounds great. So why does not every secondary school engage their students in STEM Education projects? As a practicing teacher, I know most colleagues have been reluctant to participate in such projects over the years. After studying papers about STEM Education, it turns out I am not alone. Numerous studies show several obstacles (e.g., Al Salami et al., 2015; du Plessis, 2020; du Plessis et al., 2014; Kelley & Knowles, 2016; Shernoff et al., 2017) to this. These are outlined below:

1. **Standards**. Man is a creature of habit, and teachers are no different. Department requirements, content standards, and examinations must be met (Shernoff et al., 2017). These requirements can appear more straightforward if subjects are taught in isolation. Teaching, as I have always done, is comfortable (e.g., Kelley & Knowles, 2016; Matorevhu, 2020).
2. **Qualification**. There is a shortage of STEM teachers (Sutcher et al., 2016). It is a global problem that many students meet teachers at senior secondary school who do not have a major, or are certified, in the discipline area they teach (Darling-Hammond, 1999; du Plessis et al., 2014). In the US, more than half of the students in public schools are taught STEM subjects by an uncertified teacher. Teacher excellence in subject content, pedagogical skill, and motivation abilities are solidly linked to student achievement (NASEM, 2005).

3. **Knowledge**. Having a certified teacher in the subject is the best predictor of student achievement (Nguyen & Redding, 2018; Rockoff, 2004). But even if a teacher is certified, one might find oneself teaching subject matters that they are not knowledgeable in to make the STEM Education project work (du Plessis, 2020). Not everyone is comfortable with this.
4. **Design**. STEM Education projects are interdisciplinary, demanding integration of the subjects. Usually, a secondary teacher is certified/ qualified in one or two disciplines. Teachers explain that they do not understand the other disciplines' content and standards well enough. Designing an educational project can be challenging. Ensuring a student-appropriate level in each of the involved subjects is even worse (du Plessis, 2020; Honey et al., 2014).
5. **Cooperation**. Being a secondary teacher has, throughout history, been a lonely job; Just you, your subject, and your students (Bess, 2000; Mirel & Goldin, 2012). Besides the isolation when teaching in the classroom, as a Science-, Engineering- or Technology teacher, it is not uncommon to be the only one teaching the subject at the school (e.g., Krakehl et al., 2020; Padwa et al., 2019). This means that teachers may be unfamiliar with collaborative and cooperative planning and delivery of project-based learning.
6. **Time**. Timetables are often tightly structured, compressed, or inflexible; thus, it is difficult to find time when all stakeholders in a project can meet or, even worse, implement a STEM Education project (Shernoff et al., 2017). How does a teacher find time to meet colleagues, understand the other disciplines, and design and implement such a project?

I am confident that most teachers can relate to these points. The obstacles are obvious. But hopefully, I have made you curious about what a STEM Education project could mean to a senior secondary school, its teachers and other staff, and students. There is no simple way to implement a STEM Education project, but there are experiences that one can draw on.

16.6 Implementing a STEM Education Project

Successful implementation of STEM Education projects at a senior secondary school often comes with a list of needs to be met. Based on studies on the implementation of STEM Education projects presented above, these are some points worth considering.

- Before implementing STEM projects, teachers must ensure they have enough time for planning and preparation. Not only does one need to get a grip on the other subjects' needs, but also the physical possibilities, schedules, the right kind of teaching space, and equipment. Teachers' lack of resources is the most frequently discussed challenge (Shernoff et al., 2017).

- Merging needs and standards from different subjects will inevitably cause friction. Working collaboratively with peers and colleagues to create a joint STEM Education project is about giving and taking. Don't go for too ambitious ideas initially; Keep it simple.
- A significant problem when initiating a STEM Education project is the students' reluctance to change (Shernoff et al., 2017). A STEM Education project is a step into the unknown for students too. Teachers mention having to deal with students' expectations and attitudes, interests, and gaps in students' understanding and abilities. Luckily studies show that students exhibit a higher degree of performance and motivation within STEM subjects when engaged in STEM Education projects (Tillman et al., 2014).
- Real-world relevance is key to a successful STEM Education project. John Dewey (1897) saw the mismatch between what was taught in school and students' experiences and knowledge out of school. Involve out-of-school stakeholders relevant to the students. These could, for instance, be a local company or a municipal institution. It could also be about solving a problem relevant to the students (e.g., climate change, keeping your environment clean, or improving life for the disabled).
- The nature of a STEM Education project is problem-based learning (PBL). PBL is a student-centered method, and the teacher needs to stand back, letting the students in command of the task, mainly offering scaffolding support. This is often hard at first for the teacher, used to being the center of attention and in command of how things are taught and learned. Will the students acquire the knowledge intended? Critique has been put forward (e.g., Kirschner et al., 2006) regarding, for instance, the effect on novice learners (e.g., students unable to draw conclusions from the task) and teachers that must go through instructions multiple times to ensure that the goals of the project are met. These are valid arguments. One way to overcome such shortcomings is to include short presentations that provide the students with additional information when it is called for, sometimes referred to as just-in-time learning (Brandenburg & Ellinger, 2003).
- Teachers may find a STEM Education project hard to grade at first. How can I see evidence of learning and understanding in a project built on a collective effort? The project's outcome is naturally one thing to evaluate, but one will find the process itself to be at least as interesting. Ask questions, and make the students explain what they are doing and why. Barlex (2008) suggests assessing three times during a project, but I view it as a continuous process of building your perception of each student's achievement. It also enables the teacher to present the students with adequate support when needed (Booth et al., 2017). A written self-evaluation of the project is often revealing, giving the teacher the students' view to compare with your own.
- Technology-, engineering- and mathematics teachers are generally more favorable to STEM Education projects than science teachers. Not everyone will eagerly join a project. Luckily STEM Education projects can enhance teacher satisfaction (Al Salami et al., 2015), which can increase interest among colleagues. Start with a small-scale project and spread the word.

- Implementing a STEM Education project is demanding and takes time. One project, even if it lasts a year, is not a game changer (Bolster, 1983; Goodlad, 1983; Tal et al., 2001). Like most things in life, practice makes perfect. Don´t forget to evaluate the outcome, not only if the students picked up the intended knowledge and skills, but also how the students perceived the project. Polish your setup and try it again.
- A commonly missed component in a STEM Education project is the composition of the student groups. Not only the number of participants in a group is interesting, as too large a group increases the risk of social loafing, but also what kind of students a group consists of. A mix of high and middle, middle and low, or just middle achievers is recommended (Blumenfeld et al., 1996).
- Do not forget the most straightforward STEM project; the one you make up yourself. It saves you all the trouble described above. You know your subjects, no one to cooperate with, and you have the time needed. An outside stakeholder, e.g., a local company that the students need to relate to, will increase the real-world relevance.

From my own experiences, this is good advice. As a technology teacher talking with colleagues at schools, one always finds someone who is keen or at least willing to give a STEM Education project a try. Once you are up and running, other colleagues usually show interest. One reason is that the students talk about it, most often favorably.

16.7 Suggestions and Recommendations

STEM education is in rising demand. Skills such as creativity, communication, collaboration, and working collectively in problem-solving are needed in the global and growing STEM job sectors. As the STEM jobs will constantly change, tomorrow's workforce must not only be able to master these skills but also relearn or learn new as reality changes—a Life Long Learning. Technical Universities and Colleges around the globe are adopting this, and interdisciplinary courses and programs, as well as education targeting real-world problems, are getting increasing attention.

From a senior secondary school point of view, it would make sense to prepare the students for such an educational environment far from traditional higher education. I have highlighted six areas that will affect any school entering this path; *Standards* in the individual subjects need to be met, there is a universal shortage of teachers with the proper *Qualifications*, Teachers lack *Knowledge* in other subjects, it is hard to *Design* interdisciplinary tasks at an appropriate level, *Cooperation* in teaching is not common, and *Time* is scarce.

These challenges aside, a STEM Education project in the senior secondary is very doable. When implementing a project, it is a good idea to think it through thoroughly. A crucial factor is that all teachers and other involved parties, such as student assistants, school caretakers, administrative personnel, and cooperating outside stakeholders such as companies or institutions, etc., feel engaged in the project. A bottom-up project is an excellent way to achieve this, where interested teachers design their STEM Education projects. Banks and Barlex (2021) recommend "Looking sideways", that is, looking at other STEM subjects' curricula and your colleagues' knowledge and experiences when constructing such a task. What goals do our subjects have in common? Are there elements in the involved subjects that will enhance the real-world relevance and heighten the students' engagement?

In contrast to other STEM subject teachers, a technology teacher is mainly concerned with the "How"—the natural way for humans to solve a problem. When our first ancestors needed to hunt an animal using a stone, they were probably most interested in *How* to throw the stone, not *Why* the throwing trajectories had not been calculated. In my experience, the technology teacher is often the natural initiator for a STEM Education project.

STEM Education projects can bring great rewards to all involved parties. The student can expect it to:

- Make them twenty-first-century-ready and accustomed to much of the demands they will face in STEM jobs and—education. Students experienced in managing their learning and being used to teamwork and communication will have a head start (Han et al., 2015; Honey et al., 2014; Lee et al., 2019).
- Increase engagement and interest in STEM subjects. STEM Education's connection to real-world dilemmas covers the gap in knowledge and experiences pointed to by Dewey a century ago (Jia et al., 2021; Kwon, 2017).
- Decrease the achievement gap as STEM projects seem to improve results for students with low academic abilities the most (Erdogan & Stuessy, 2016; Han et al., 2015; Yaki et al., 2019). Educational performance is often seen as the best predictor for completion of upper secondary school and is especially important in areas with low socioeconomic status (Lyche, 2010).
- Deepen the learning in STEM subjects (Matorevhu, 2020; Prince & Felder, 2006). Unfortunately, this does not seem to affect the outcome of the high-stakes exams that sometimes are the difference between continued academic studies or not. STEM-educated students perform equally well as students with a traditional education background, except for ones with low academic abilities mentioned earlier. One reason for this outcome could be that predominant multiple-choice exams are not sensitive to the higher-order skills the STEM students develop. Studies in the US also point at STEM-oriented high schools often being established in low socioeconomic societies. An outcome of the exams in line with areas better off could therefore be viewed as a success (e.g., Craig & Marshall, 2019; Han et al., 2015).

The success of any school and its students depends on their teachers. Keen teachers who remain teaching over time are beneficial. There is also a growing concern about a coming shortage of STEM teachers (e.g., Dolenc et al., 2021; van Rooij et al., 2020). STEM Education could improve:

- Teacher attitude toward interdisciplinary projects. Teachers at senior secondary schools showed the most positive change in attitude among K-12 teachers (Al Salami et al., 2015).
- Teaching satisfaction. As attitudes toward STEM Education projects improved, so did teaching satisfaction. Teaching satisfaction is a vital motivational factor (Al Salami et al., 2015).
- Understanding of STEM content and pedagogical content knowledge (PCK). When "looking sideways" one realizes the STEM subjects' mutual relationship and heightens the subject's relation to the real world (Shernoff et al., 2017).
- Teamwork skills among teachers. School leaders concerned with teachers' retention should support this (Park et al., 2005).
- Keeping teachers in teaching. Teachers leaving the profession is a problem. The possibilities to develop professionally and cooperate with colleagues are positive (Fulton & Britton, 2011).

16.8 A Final Reflection

When writing this chapter, I visited Chalmers University of Technology in Göteborg, Sweden, mentioned in the MIT report (Graham, 2018) as a beacon of interdisciplinary programs. I was shown their latest venture in interdisciplinarity, the Tracks-environment, and -courses (Enelund & Henricson Briggs, 2020). Tracks-courses are all interdisciplinary and about cutting-edge technology or hot societal topics (or both), designed in such a way that no matter what program a student attends, the Tracks-courses are always an option to study. The Tracks milieu is more like a makerspace than a traditional university. Professor Enelund, Director of Tracks, emphasized computational thinking and some experience working on interdisciplinary projects, preferably hands-on when asked what Senior secondary schools should equip their students with.

For anyone with some stakes in Senior secondary school, regardless of whether you are a policy maker, principal, teacher, or student; Integrated STEM Education brings on many changes for everyone involved, as mentioned, and not everything is easy to solve or get used to, but it is worth it.

References

Al Salami, M., Makela, C., & de Miranda, M. (2015). Assessing changes in teachers attitudes towards interdisciplinary STEM teaching. *International Journal of Technology and Design Education*. https://doi.org/10.1007/s10798-015-9341-0

Banks, F., & Barlex, D. (2021). *Teaching STEM in the secondary school* (2nd ed.), Routledge. ISBN: 9780367330460

Barlex, D. (2008). Assessing capability in design & technology: The case for a minimally invasive approach. *Design and Technology Education: An International Journal, 12*(2), 49–56.

Bess, J. (2000). *Teaching alone/ teaching together: Transforming the structure of teams for teaching*, Jossey-Bass, ISBN: 978-0787947989

Blumenfeld, P., Marx, R., Soloway, E., & Krajcik, J. (1996). Learning with peers: From small group cooperation to collaborative communities. *Educational Researcher, 25*(8), 37–40. https://doi.org/10.3102/0013189X025008037

Bolster, A. S. (1983). Toward a more effective model of research on teaching. *Harvard Educational Review, 53*(3), 294–308. https://doi.org/10.17763/haer.53.3.0105420v41776340

Booth, J., McGinn, K., Barbieri, C., Begolli, K., Chang, B., Miller-Cotto, D., Young, L., & Davenport, J. (2017). Evidence for cognitive science principles that impact learning in mathematics. In: D. Geary, D. Berch, R. Ochsendorf, & K. M. Koepke (Eds.), *Acquisition of complex arithmetic skills and higher-order mathematics concepts* (pp. 297–325) Elsevier Academic Press. https://doi.org/10.1016/B978-0-12-805086-6.00013-8

Brandenburg, D. C., & Ellinger, A. D. (2003). The future: Just-in-time learning expectations and potential implications for human resource development. *Advances in Developing Human Resources, 5*(3), 308–320. https://doi.org/10.1177/1523422303254629

Craig, T., & Marshall, J. (2019). Effect of project-based learning on high school students' state-mandated, standardized math and science exam performance. *Journal of Research in Science Teaching, 56*(10), 1461–1488. https://doi.org/10.1002/tea.21582

Crawford, R. (2016). *History of tertiary education reforms in New Zealand*, New Zealand Productivity Commission, https://www.productivity.govt.nz/assets/Documents/65759a16ed/History-of-tertiary-education-reforms.pdf

Crosier, D., & Parveva, T. (2013). *The Bologna process: its impact in Europe and beyond. Fundamentals of educational planning* (97), UNESCO: International Institute for Educational Planning. ISBN: 978-92-803-1368-0. Retrieved January 1, 2022, from https://unesdoc.unesco.org/ark:/48223/pf0000220649

Darling-Hammond, L. (1999). *Teacher quality and student achievement: A review of state policy evidence*, Center for the Study of Teaching and Policy, http://depts.washington.edu/ctpmail/Publications/PDF_versions/LDH_1999.pdf

Dewey, J. (1897). My pedagogic creed, *The School Journal*, Volume LIV, Number 3 (January 16, 1897), 77–80. http://infed.org/mobi/johndewey-my-pedagogical-creed/

Dolenc, K., Šorgo, A., & PlojVirtič, M. (2021). Signs of a Catastrophe: Predicted Shortage of Teachers of Lower Secondary Science and Technics and Technology in Slovenia. *Journal of Elementary Education, 14*(2), 239–256. https://doi.org/10.18690/rei.14.2.239-256.2021

du Plessis, A. E. (2020). The lived experience of out-of-field STEM teachers: A quandary for strategising quality teaching in STEM? *Research in Science Education, 50*(4), 1465–1499. https://doi.org/10.1007/s11165-018-9740-9

du Plessis, A. E., Gillies, R. M., & Carroll, A. (2014). Out-of-field teaching and professional development: A transnational investigation across Australia and South Africa. *International Journal of Educational Research, 66*, 90–102. https://doi.org/10.1016/j.ijer.2014.03.002

EdWeek Reseach Center. (2022). *1st annual Merrimack College teacher survey: 2022 results.* Merrimack College

Enelund, M., & Henricson Briggs, K. (2020). Tracks for change, flexibility, interdisciplinarity and creativity in engineering education. In J. Malmqvist, J. Bennedsen, K. Edström, N. Kuptasthien, A. Sripakagorn, I. Saemundsdottir, J. Roslöf & M. Siiskonen (Eds.), *Proceedings of the International CDIO Conference* (Vol. 1, pp. 37–47). Chalmers University of Technology.

English, L. D. (2016). STEM education K-12: Perspectives on integration. *International Journal of STEM Education, 3*, 3. https://doi.org/10.1186/s40594-016-0036-1

Erdogan, N., & Stuessy, C. (2016). Examining the role of Inclusive STEM schools in the college and career readiness of students in the United States: A Multi-group analysis on the outcome of student achievement. *Educational Sciences: Theory and Practice, 15*(6), 1517–1529.

European Commission. (n.d.), *The Bologna process and the European higher education area*, Retrieved August 1, 2022, from https://ec.europa.eu/education/policies/higher-education/bologna-process-and-european-higher-education-area_en

European Higher Education Area [EHEA]. (n.d.-a), *Full members*, Retrieved August 1, 2022, from http://ehea.info/page-full_members
European Higher Education Area [EHEA]. (n.d.-b), *Lifelong learning*, Retrieved August 1, 2022, from http://ehea.info/pid34427/lifelong-learning.html
Fulton, K., & Britton, T. (2011). STEM teachers in professional learning communities: From good teachers to great teaching, *National Commission on teaching and America´s future*. https://files.eric.ed.gov/fulltext/ED521328.pdf
Genov, N. (2014). The future of individualization in Europe: Changing configurations in employment and governance. *Eur J Futures Res, 2*, 46. https://doi.org/10.1007/s40309-014-0046-5
Goodlad, J. I. (1983). A study of schooling: Some implications for school improvement. *Phi Delta Kappan, 64*(8), 552–558. https://eric.ed.gov/?id=EJ279532
Graham, R. (2018). *The global state of art in engineering education*, Massachusetts Institute of Technology. ISBN: 9780692089200
Hallinen, J. (2021, November 18). *STEM. Encyclopedia Britannica.* https://www.britannica.com/topic/STEM-education
Hallström, J., & Ankiewicz, P. (2019). Laying down the "T" and "E" in STEM education: Design as the basis of an integrated STEM philosophy. In S. Pulé & M. de Vries (Eds.), *PATT2019: Developing a knowledge economy through technology and engineering education* (pp. 187–194). Msida: Department of Technology and Entrepreneurship Education, University of Malta. ISBN: 9789995714796
Han, S., Capraro, R. & Capraro, M. (2015). how science, technology, engineering, and mathematics (STEM) project-based learning (PBL) affects high, middle, and low achievers differently: The impact of student factors on achievement. *International Journal of Science and Mathematics Education, 13*, 1089–1113. https://doi.org/10.1007/s10763-014-9526-0
Henriksen, D., & Mishra, P. (2020). Of metaphors and molecules: Figurative language bridging STEM and the arts in education. *Leonardo, 53*(3), 316–320. https://doi.org/10.1162/leon_a_01607
Honey, M., Pearson, G., & Schweingruber, A. (2014). *STEM integration in K-12 education: Status, prospects, and an agenda for research.* National Academies Press. ISBN: 978-0-309-29796-7
Huri, N., & Karpudewan, M. (2019). Evaluating the effectiveness of integrated STEM-lab activities in improving secondary school students' understanding of electrolysis. *Chemical Education Research and Practice, 20*, 495–508. https://doi.org/10.1039/C9RP00021F
Jia, Y., Zhou, B., & Zheng, X. (2021). A curriculum integrating STEAM and maker education promotes pupils' learning motivation, self-efficacy, and interdisciplinary knowledge acquisition. *Frontiers in Psychology, 12*, 725525. https://doi.org/10.3389/fpsyg.2021.725525
Kelley, T., & Knowles, G. (2016). A conceptual framework for integrated STEM education. *International Journal of STEM Education, 3*, 11. https://doi.org/10.1186/s40594-016-0046-z
Kirschner, P., Sweller, J., & Clark, R. (2006). Why minimal guidance during instruction does not work: An analysis of the failure of constructivist, discovery, problem-based, experimental, and inquiry-based teaching. *Educational Psychologist, 41*(2), 75–86. https://doi.org/10.1207/s15326985ep4102_1
Krakehl, R., Kelly, A.M., Sheppard, K., & Palermo, M. (2020). Physics teacher isolation, contextual characteristics, and student performance. *Physical Review Physics Education Research, 16(2).* https://doi.org/10.1103/PhysRevPhysEducRes.16.020117
Kwon, H. (2017). Effects of 3D printing and design software on students' overall performance. *Journal of STEM Education: Innovations and Research, 18*(4), 37–42.
Lee, Y., Bicer, A., Kwon, H., Rugh, M., Capraro, R., Capraro, M., & Barroso, L. (2019). Post-secondary ready: Does the STEM curriculum Matter? In: *2019 IEEE Frontiers in Education Conference (FIE) Frontiers in Education Conference (FIE).* https://doi.org/10.1109/FIE43999.2019.9028434
Lipowitz, D., & Berkeihiser, M. (2019). Reengineering the conflict: STEM in the English classroom. *Technology and Engineering Teacher, 79(2)*, 28–33. https://eric.ed.gov/?id=EJ1230951

Lyche, C. (2010). Taking on the completion challenge: A literature review on policies to prevent dropout and early school leaving. In *OECD Education Working Papers*, No. 53, OECD Publishing, Paris. https://doi.org/10.1787/5km4m2t59cmr-en

Marr, B. (2022). The 5 biggest business trends for 2023. *LinkedIn*. Retrieved January 5, 2023, from https://www.linkedin.com/pulse/5-biggest-business-trends-2023-bernard-marr/

Matorevhu, A. (2020). O-level mathematics and science teachers' alignment of STEM skills oriented instructional objectives with assessment in a high stakes examinations environment. *International Journal of Trends in Mathematics Education Research, 3*(1), 1–7. https://doi.org/10.33122/ijtmer.v3i1.165

MetLife. (2012). The MetLife survey of the American teacher: Challenges for school leadership. Metropolitan Life Insurance Company. https://www.metlife.com/content/dam/microsites/about/corporate-profile/MetLife-Teacher-Survey-2012.pdf

Mirel, J., & Goldin, S. (2012, April 17). Alone in the classroom: Why teachers are too isolated, *The Atlantic*. https://www.theatlantic.com/national/archive/2012/04/alone-in-the-classroom-why-teachers-are-too-isolated/255976/?utm_source=copy-link&utm_medium=social&utm_campaign=share

Moore, T., Stohlmann, M., Wang, H., Tank, K., Glancy, A., & Roehrig, G. (2014). Implementation and integration of engineering in K-12 STEM education. In S. Purzer, J. Strobel, & M. Cardella (Eds.), *Engineering in Pre-College Settings: Synthesizing Research, Policy, and Practices* (pp. 35–60). Purdue University Press.

Musselin, C. (2018). New forms of competition in higher education. *Socio-Economic Review, 16*(3), 657–683. https://doi.org/10.1093/ser/mwy033

National Academy of Science, National Academy of Engineering, and Institute of Medicine of the National Academies [NASEM] (2005). *Rising above the gathering storm*, The National Academies Press. https://doi.org/10.17226/11463

National Assessment Governing Board [NAGB]. (2018). *Technology and engineering literacy framework for the 2018 national assessment of educational progress*, National Assessment Governing Board https://www.nagb.gov/content/dam/nagb/en/documents/publications/frameworks/technology/2018-technology-framework.pdf

Nguyen, T., & Redding, C. (2018). Changes in the demographics, qualifications, and turnover of American STEM teachers, 1988–2012 *AERA Open* Volume 4 Issue 3, July-September 2018. https://doi.org/10.1177/2332858418802790

Padwa, L., Kelly, A., & Sheppard, K. (2019). Chemistry teacher isolation, contextual characteristics, and student performance. *Journal of Chemical Education, 96*(11), 2383–2392. https://doi.org/10.1021/acs.jchemed.9b00392

Park, S., Henkin, A. B., & Egley, R. (2005). Teacher team commitment, teamwork and trust: Exploring associations. *Journal of Educational Administration, 43*(5), 462–479. https://doi.org/10.1108/09578230510615233

Partnership for 21st century learning. (n.d.). *P21, Partnership for 21st century learning. A network of Battell for kids*. Retrieved August 1, 2022, from https://www.battelleforkids.org/networks/p21

Pearson, D., Moje, E., & Greenleaf, C. (2010). Literacy and science: Each in the service of the other. *Science, 328*, 459–463. https://doi.org/10.1126/science.1182595

Prince, M., & Felder, R. (2006). Inductive teaching and learning methods: Definitions, comparisons, and research bases. *Journal of Engineering Education, 95*(2), 123–138. https://doi.org/10.1002/j.2168-9830.2006.tb00884.x

Rockoff, J. (2004). The impact of individual teachers on student achievement: Evidence from panel data. *American Economic Review, 94*(2), 247–252. https://doi.org/10.1257/0002828041302244

Sanders, M. (2009). STEM, STEM education, STEMmania. *The Technology Teacher, 68*(4), 20–26. https://www.teachmeteamwork.com/files/sanders.istem.ed.ttt.istem.ed.def.pdf

Shaw, M. (2001). *Engineering problem solving*. William Andrew, ISBN: 9780815514473. https://www.sciencedirect.com/book/9780815514473/engineering-problem-solving#book-description

Shernoff, D. J., Sinha, S., Bressler, D. M., & Ginsburg, L. (2017). Assessing teacher education and professional development needs for the implementation of integrated approaches to STEM education. *International Journal of STEM Education, 4*, 13. https://doi.org/10.1186/s40594-017-0068-1

Sutcher, L., Darling-Hammond, L., & Carver-Thomas, D. (2016). *A coming crisis in teaching? Teacher supply, demand, and shortages in the U.S.*, Learning Policy Institute. https://doi.org/10.54300/247.242

Svärd, J., Schönborn, K., & Hallström, J. (2017). Design of an authentic innovation project in Swedish upper secondary technology education. *Australasian Journal of Technology Education, 4*, 1–15. https://doi.org/10.15663/ajte.v4i1.48

Svard, J., Schönborn, K., & Hallström, J. (2022). Students' perception of authenticity in an upper secondary technology education innovation project. Res. Sci. Technol. Educ. https://doi.org/10.1080/02635143.2022.2116418

Tal, R. T., Dori, Y. J., & Keiny, S. (2001). Assessing conceptual change of teachers involved in STEM education and curriculum development—The STEMS project approach. *International Journal of Science Education, 23*(3), 247–262. https://doi.org/10.1080/09500690175006650l

Thomas, B., & Watters, J. (2015). Perspectives on Australian, Indian and Malaysian approaches to STEM education. *International Journal of Educational Development, 45*(November 2015), 42–53. https://doi.org/10.1016/j.ijedudev.2015.08.002

Tillman, D., An, S., Cohen, J., Kjellstrom, W. & Boren, R. (2014). Exploring wind power: Improving mathematical thinking through digital fabrication. *Journal of Educational Multimedia and Hypermedia, 23*(4), 401–421. Waynesville, NC USA: Association for the Advancement of Computing in Education (AACE).

Trines, S. (2018, December 11). Bologna-type harmonization in Africa: An overview of the common higher education area of the East Africa community. *World Education News + Reviews* https://wenr.wes.org/2018/12/common-higher-area-chea-of-the-east-african-community

U.S. Bureau of Labor Statistics. (2021). *Employment projections, Employment in STEM occupations*, Retrieved January 5, 2022, from https://www.bls.gov/emp/tables/stem-employment.htm

Van Rooij, E. C. M., Fokkens-Bruinsma, M., & Goedhart, M. J. (2020). Identifying potential secondary school teachers among science university students: A latent profile analysis. *Journal of Science Teacher Education, 31*(5), 556–577. https://doi.org/10.1080/1046560X.2020.1729478

Yaki, A. A., Saad, R. M., Renuka, V. S., & Zulnaidi, H. (2019). Enhancing science achievement utilising an integrated stem approach. *Malaysian Journal of Learning and Instruction, 16*, 181–205.

Joachim Svärd is a Swedish certified teacher, MSc, and Ph.D. student at Linköping University, particularly interested in authentic secondary school learning. With an entrepreneurial background, he has studied the relationship between real-world relevant tasks, problem-solving, and perceived authenticity among students in technology education. He is currently working with junior high school youths in a socio-economic deprived suburban area of Göteborg, Sweden

Chapter 17
Makerspaces and the Position of T in STEM

Marc J. De Vries

Abstract This chapter will argue that making activities in Makerspaces have the potential to clarify and enhancing the position of Technology education in STEM Education. Making is at the heart of Technology education with its own pedagogical value, but it also can contribute to the learning of other STEM subjects (and even beyond those). The nature of making and the different types of relations between Makerspaces and schools will be discussed to show that this potential is not realized 'automatically'. This relationship needs to fulfil certain requirements to lead to the realization of that potential.

Keywords Makerspaces · Fablabs · The nature of making · Pedagogy

17.1 Introduction

The relationship between the Maker Movement and Technology Education is somewhat ambivalent. On the one hand, technology teachers applaud the increased attention to making activities, as this is one of the most important types of activities in their school subject, next to designing. Yet, they feel that the popularity of the Maker Movement seems to ignore the fact that making has happened in their subject for as long as it exists. Everyone looks in the direction of the fancy Makerspaces that attract more attention than they ever got with their very similar activities in the context of the school curriculum. What is going on here, they may wonder. Is it the fact that it is often an out-of-school affair? Is it because the makers are not graded for what they do in the Makerspaces? Is it because of the equipment that they could never afford? Is it all of that together? At the same time, there are examples of a strong connection between Makerspaces and the school and also there are Makerspaces which are found within schools. In some cases, the Makerspaces even contribute not just to technology education but to other STEM subjects, and even to subjects other than

M. J. De Vries (✉)
University of Technology Delft, Delft, The Netherlands
e-mail: M.J.deVries@tudelft.nl

W. Fox-Turnbull and P. J. Williams (eds.), *Locating Technology Education in STEM Teaching and Learning*, Contemporary Issues in Technology Education,
https://doi.org/10.1007/978-981-97-1995-2_17

STEM. Every reason to reflect on this ambivalent relationship between Makerspaces and Technology Education. First, we will reflect on the nature of making and its relation to technology. Next, we will see how making can contribute to the education (be it formal, informal or non-formal) of young people. Then we will reflect on the relation of Makerspaces to schools in current practice: What types of relations can be found and how do they play out? Finally, we will discuss the potential of making Makerspaces as a way of accommodating for highlighting the role of the T in STEM. In other words: How can Makerspaces contribute to the position of Technology Education in the curriculum rather than compete with it?

17.2 What Is Making and How Does It Place T in STEM?

One of the unique contributions of Technology in STEM is making things. Making is one of the three main processes in technology and engineering (the other two are designing and using/evaluating). In the making process, more than materials are involved. Energy is needed to transform the material and information is needed to realise a desired transformation. The source of energy and information is different for different types of making processes:

- in manual making both energy and information are delivered by a human being. It is the maker who uses her/his muscle force to process the material and (s)he also directs the tool in such a way that the material gets the desired shape.
- in machine production, the machine delivers the material, but the maker still delivers the information. The drilling machine in the hand of the maker uses electrical energy to impact the material but the maker directs the drilling machine in the right direction.
- in automated making the automated (or robot) delivers not only the energy but also the information. Of course, in the end the information comes from a human being, but it is programmed into the automate beforehand and once this is done the automate can work autonomously and only needs human control because there is always a chance that something goes wrong. In this case, we can speak about AI-powered making, although it can be questioned if the term 'intelligence' can be applied to a process that is not more than data processing, even in its most advanced form.

Doing the making in the school can provide pupils with an understanding of the sort of making processes that exist in the real world and thus contribute to technological literacy. Although it may seem that production today is largely a matter of automated production, this is more of an image that we get from the media. They put a lot of emphasis on advanced technologies, which results in pupils having a distorted view of technology as if it were electronic devices like computers, lasers and robots (De Vries, 2016). A visit to a car factory can help to get more balance in our intuitive image of technology. Even in car factories with a high degree of automation, there are always certain steps in the production process that can best be done manually. Those processes require advanced mental processes that even the most advanced robot cannot perform. Although there is a whole debate about the question of whether humans are or are not fundamentally different from robots, it is still the case that certain actions seem to require a flexible and creative 'mind' that robots perhaps can imitate but not possess. This becomes particularly visible when something unexpected happens that is not in the robot's programme and 'confuses' it, with the possibility of dangerous behaviour. The development that production steps are taken over by robots will continue for years to come, but it is likely that there will always remain a place for human manual interventions.

17.3 The Cognitive and Pedagogical Value of Making

Making played an important role in the evolution of humankind. Stout, Toth, Schick and Chaminade argued that toolmaking caused changes in the human brain. Their experiments with positron emission tomography revealed that with the increased demand for effective visuomotor coordination and hierarchical action organization in tool-making making certain areas of the brain are more activated and therefore more developed (Stout et al., 2008). Their conclusion was that toolmaking was one of the factors for the evolution of humans with their highly developed mental capacities. If this is true, then making can still be seen as a way of stimulating mental development and as such a valuable contribution to the cognitive development of young people. That contribution goes way beyond Technology Education and even beyond school. It would be a deep contribution to the lives of people.

Apart from this, there is a pedagogical value in making. This pedagogical value has been emphasized by scholars like Comenius, Fröbel and Pestalozzi. Comenius saw making skills as very important for education as they connect heart, hand and mind. Using proper tools to make artefacts, beginning with simple ones and then gradually to more complex ones, for him was an important activity in schools. Pestalozzi saw making as a core activity in learning as it does not only contribute to bodily skills but also to mental skills and even to morality. His concern was particularly for less privileged children for whom making was a suitable means of enabling them into valuable members of society. Fröbel turned Pestalozzi's ideas into practice in his Kindergarten (Manning, 2005) in which young children could develop their self-empowerment in a safe environment by playing and tinkering. For him, in particular, making is a contribution to the development of the child's soul. For him, there was a religious motive involved also: making the child become the creative being that God had intended her or him to be. Thus, making is not a 'luxury article' in education but a very basic and necessary element.

Making also appeals to pupils. Primary school teachers know that by experience. Making adds to the fun of education. Completed products can give a feeling of pride and accomplishment. In the line of Pestalozzi: it enables children that have less cognitive ability to show that they have other skills that are equally valuable, but in a different way. Motivation being raised by the fun of making can be an important driving force for all learning, so even if pleasure in school is not seen as a goal (but why not?) it can be seen as a contribution to learning in general.

Making also brings authenticity in schooling. Even in a time when most of the making seems to have been left to machines and automates, there will always be parts of the process that will be done manually because they are, at least for now, too complex for a robot. A visit to a car-producing factory can raise that awareness. Even though at first sight the whole production floor seems to be full of robot arms and huge machines, observing carefully shows that throughout the process there are manipulations that are too complex for a robot to make. In this respect, making allows for authentic learning as making processes are part of the real world and likely will always remain to be so (although in education we find them mostly in somewhat simplified forms because of the limitations of the educational context).

Making activities can stimulate growth in Positive Technological Development, as Strawhacker and Bers showed in a case study with kindergarten children (Strawhacker & Bers, 2018). Positive Technological Development is rooted in Seymour Papert's notion of Constructionism, a theoretical approach that claims metacognitive learning afforded by computer programming. Making can also result in high PTD engagement with children, as Strawhacker and Bers showed. They observed positive behaviour in dealing with (digital) learning experiences in a Makerspace environment.

Another pedagogical value in the making activity is the cultural value of making. Perhaps the most 'extreme' example of a subject in which that dimension is emphasized is Slöyd in the Scandinavian countries. Slöyd has everything to do with learning the culture of the country and feeling connected to its cultural heritage through making. Scholars such as Cygnaeus and Solomon have provided a theoretical foundation for that. Making artefacts in Slöyd enables getting to know the cultural history of one's country and thus contributes to children becoming citizens with an appreciation for the cultural accomplishments of the country or region in which they live. As Hallström argued in his chapter in the International Handbook of Technology Education (Hallström, 2018), this cultural value, but also the above-mentioned value for personal development as emphasized by Fröbel and Pestalozzi, among others, gets lost in the form of Technology Education that shifts from a craft orientation to a more digital production-oriented curriculum. That is the reason why in Nordic countries often Slöyd is a school subject that exists next to Technology Education (Hallström, 2018).

In summary: making has an educational value that goes way beyond Technology Education and therefore deserves to be in the school curriculum. But what is the current practice of making in the Maker Movement? That will be the focus of the next section.

17.4 Makerspaces in Practice

Currently, a lot of making in education takes place in so-called makerspaces. We also find the term Fablabs for those facilities. Obviously, that term is a contraction of Fabrication and Laboratory. Makerspaces and Fablabs are dedicated rooms where making equipment is available for anyone who likes to spend some time there and make something. Many makerspaces often function in the context of non-formal and informal education. They are then not necessarily connected to school curricula. They are similar to science museums but then focus on making rather than science. Makerspaces have gone through rapid development and have quickly gained popularity.

Makerspaces vary in size and equipment but often there are at least some high-tech devices like 3D printers and laser cutters. The presence (and popularity) of 3D printers in particular raise the question of whether or not the pedagogical value of making is still realised when pupils do not seem to do much more than downloading predesigned drawings and pushing buttons to initiate the making process. The relation with designing gets lost this way. I will come back to this in the next section. Also, the making itself becomes a very shallow activity because the bodily dimension is reduced to practically nothing. No hand–eye coordination, for instance, is needed (or hardly any) when pushing a button.

The presence and popularity of makerspaces outside schools have been received with mixed feelings by technology teachers. On the one hand, they were happy to see that making was appreciated as a worthwhile activity. On the other hand, it annoyed them as making had been promoted for years in technology education and mostly still had a low profile there, certainly compared to other school subjects. Makerspaces seemed not to need technology education for their success. What technology education had been struggling with for many years, makerspaces suddenly seemed to accomplish in just a couple of years. Technology teachers have worked in makerspaces ever since technology education existed, but the term 'makerspace' was never used for those rooms and now suddenly this term seems to have a magical effect on the 'outside world'. Technology labs increasingly have advanced equipment also but for some reason, the same devices in an out-of-school room are more appealing than the school lab.

There are, however, more and more efforts to link makerspaces to formal education. Teachers from school can go to makerspaces to do things that they normally cannot do in their own school. Makerspaces, for instance, are usually equipped with more 3D printers and other expensive equipment than schools can afford to buy. This can be the beginning of a broader movement towards connecting makerspaces to schools. There are also schools that open up their technology lab as a place where children from outside can come and make things. Thus, the makerspace becomes an outreach activity for the school.

17.5 Making and the T in STEM

In this book, the specific focus is on T in the context of STEM. How does that play out for making processes? In the first place, there is a relation with science because the maker has to take into account material properties in the making. These can be explored by trial and error, which is typically what happens in traditional craftsmanship. Sometimes, however, science is needed to get more in-depth knowledge of material properties and then that knowledge can be used to inform the making process. In making, energy transformations take place. These, too, can be explored and learnt in an intuitive way, but here, too, science can inform about, for instance, the efficiency of the energy transformation and how to improve that. Thus there is a clear relation between S and T in the making process related to the material and the energy dimension. In the information dimension, there is a clear relation with mathematics. Mathematical modelling can be used to inform the making process and play a role in the information processing that has to take place in the making process. In that case, the E in STEM also becomes visible. The need to make mathematics more explicit grows when we move from manual to machine making and even more when the step towards automated making is made. Mathematical modelling plays a vital role in engineering from its earliest years and still today. In Fablabs, mathematical models are used as designs that can be fed into a 3D printer. That is by no means a typical educational situation as it happens in real engineering also. Even such large constructions as bridges are now engineered for production by 3D printers. Even food and tissue can be engineered for this type of production. Thus, making in education can illustrate how mathematics and engineering are connected, which is one of the aims of (integrated) STEM Education.

Subjects like Sloyd and craft, however, were often quite remote from science and mathematics education and so are most makerspaces. That is changing, however. For some decades now there has been an interest in so-called design-based learning. That means that design activities are not only used as a means to learn design but also to learn other things, for instance understanding concepts. Those concepts can be technological concepts, such as systems, but also concepts in science and math. Even without making the designed products, there is a learning effect in terms of concept learning, even if it is not optimal (see later on in this chapter). The combination of designing and making is therefore potentially a powerful strategy for learning concepts in science, technology, engineering and mathematics. There are examples of makerspaces that let pupils design a regular polygon and making it out of matter seems to add to their understanding of the nature of that figure. I use the term ‘potential’ as we know from research that this does not happen without the teacher taking a role as a guide in this process. Without teachers stimulating learners to relate the design challenge to certain concepts that can help analyse the problem or find possible solutions, pupils by themselves will just focus on the practical problem and try to solve it in a ‘trial-and-error’ mode. Kolodner and others have shown that the teacher is indispensable for concept learning to take place.

17.5.1 *Embedding Making in the School Curriculum Through Makerspaces*

Linking the making activities to the school curriculum would have consequences for makerspaces. The activities in those rooms can then no longer be limited to downloading an existing design from the internet, uploading it into the 3D printer and pushing the make button in a place that is unrelated to the school. This separation of designing and making is not good for maker education anyway Nemorin (2017). Nemorin uses fairly strong expressions for the frustrations this can lead to a lack of pragmatic engagement (and a loss of fun along with that), affective labour of failing (because the 3D printers do not fulfil the promises they make due to their complexity) and 'mediated alienation' (due to detachment of the product from my own creativity). Fasso and Knight have therefore argued for design as the process-base of a Makerspace, both in formal and informal/non-formal education (Fasso and Knight 2019). They show examples of completely predesigned commercially purchasable kits for making artefacts, in which there is no space for creative designerly input from people. According to them, this is a reductionist approach that greatly reduces the educational value of making. Worsley and Blikstein claim that making can contribute to the learning of reasoning strategies (Worsley & Blikstein, 2017). Materials-based reasoning was one of the four reasoning types they found in a study among 9th–12th grade students (the other three were unexplained reasoning, example-based reasoning and principle-based reasoning). It is obvious that materials-based reasoning can be directly related to making activities. This reasoning is needed in design activities when decisions need to be taken about the material in which the design is to be realized. Gravel, Tucker-Raymond, Kohlberger and Browne found that experienced makers claim that they gained STEM-related design literacies such as identifying, organizing and integrating information through their making experience (gravel et al., 2018). Designing and making therefore should not be separated in order to make Makerspaces fruitful for learning at school.

Makerspaces would then become part of the school curriculum, including its learning goals (and still keep their public function). This symbiosis looks promising and future years will show if it can fulfil its potential. We have already seen an example of how making can contribute to mathematics education and making can be connected and other examples are known from the literature (Doorman et al., 2019). This can go way beyond STEM. In a subject like history, artefacts play an important role (although this is not always quite visible in the history education curriculum, unfortunately). Pupils can get an understanding of the ingenuity of people in the past, even in prehistoric eras, by making copies of the devices they made for daily life and experiencing how difficult it is to make them in such a way that they realize the desired function. In musical education, making an instrument can provide insights into its nature and possibilities for producing different sounds. In economics education, making products in class can develop an understanding of what determines the price of products on the market. This can also be realised by turning the already existing technology education lab into a place where activities take place that connect to the whole school curriculum. This would make the T not only visible in STEM but in the whole school. It would imply a design-and-make-based education, not just for learning concepts in technology and other STEM subjects (see above), but also for other school subjects. The term design-based learning is well known and its value for concept learning has been well established by educational research (Henze and De Vries 2021). Perhaps it would be better to use the term design-and-make-based education to emphasize the desirable close relation between designing and making in concept learning.

17.6 Is Making a Necessity or a Luxury?

I have argued for the value of making in positioning Technology Education in the school curriculum in particular in relation to STEM. But is it really necessary? Is there, not the option of seeing design as the core of technology rather than making? To answer that question we have to look at the nature of design. It may seem that designing is a purely mental activity. If at all there is a component that needs material next to thinking, it is drawing. But that is not the same as making. The drawing certainly seems a necessary component in designing. Probably most designers will tell us that creating a design only in their minds does not work. “I think with my pencil”, is an expression that I once heard from a professional designer. Trying to keep the whole design only in one’s mind would probably lead to ‘cognitive overload’. The paper serves as an ‘extended mind’. But, in a similar way making could very well also be an important non-mental element in designing, as it enables one to explore the design in a way that the mind only does not enable. Another expression that I heard from a professional designer: “I cannot make a design to be realized in a certain material if I did not have that material in my hands, allowing me to touch its surface, bend it, stretch it, etcetera”. The design is not created to stay in the mind. It is meant to be realized in materials. In the context of technology, one could question if a design is a design at all if it is not there to be realized in materials. Apart from the fact that the intrinsic pedagogical value of making would get lost if we expel it from Technology Education, and from the curriculum at large, also something of the essence of designing would get lost. Where in the previous section I argued that the value of making in education is lost when it is not combined with designing, here I argue that it also holds the other way round: without realising the design in materials, the value of designing as a learning activity is very much diminished, as something of the nature of design gets lost then.

17.7 Conclusion

At the moment Maker Education is only partially connected to the school curriculum and to STEM. However, there is a clear potential for making projects become an activity that pervades the whole curriculum. This can be done both by using out-of-school Makerspaces and also with inside-school Makerspaces, which currently are mainly used in the context of Technology Education. But they can serve as a bridge between Technology Education and other STEM subjects, and even beyond that to other school subjects such as history and economics. It needs further exploration of what this would mean for those Makerspaces. Clearly, it is not enough to have fancy machines with a certain magic that appeals to pupils. Martin (2015) mentions that digital tools as only one of three elements that are needed to use Makerspaces for enhancing engagement in the practice of Technology education and STEM education. The other two are a community structure and a maker mindset. The community comprises both relations between teachers and pupils and between teachers and teachers. A good cooperation between teachers of different school subjects is necessary to make this work. This means that an understanding of each other's disciplines and the way they are taught should be part of teachers' pedagogical content knowledge (PCK). Fan showed that the capacity to build interdisciplinary connections was a necessary condition for implementing maker education by teachers (Fan, 2022). This has implications for teacher education programmes. The equipment needs to allow for a link with design activities and for having a physical experience with the materials and their properties. This is needed for the maker mindset. The potential of Makerspaces to serve as a means through which Technology Education can get a good position in STEM and in the whole school curriculum only works when conditions like these are fulfilled. This can take Makerspaces in schools to the next level. The future will show if schools will be able to meet this challenge.

References

de Vries, M. J. (2016). *Teaching about technology*. Springer.

Doorman, M., Bos, R., de Haan, D., Jonker, V., Mol, A., & Wijers, M. (2019). Making and implementing a mathematics day challenge as a makerspace for teams of students. *International Journal of Science and Mathematics Education, 17*(Suppl 1), 149–165.

Fan. (2022). An importance-performance analysis (IPA) of teachers' core competencies for implementing maker education in primary and secondary schools. *International Journal of Technology and Design Education, 32*, 943–969.

Fasso, W., & Knight, B. A. (2019). Identity development in school makerspaces: Intentional design. *International Journal of Technology and Design Education, 30*, 275–294.

Gravel, F. G., Tucker-Raymond, E., Kohlberger, K., & Browne, K. (2018). Navigating worlds of information: STEM literacy pravtices of experienced makers. *International Journal of Technology and Design Education, 28*, 921–938.

Hallström, J. (2018). Exploring the relationship between technology education and educational Sloyd. In: Vries, M.J. de (Ed.), *Handbook of technology education* (pp. 73–84). Springer.

Manning, J. P. (2005). Rediscovering Froebel: A call to re-examine his life and Gifts. *Early Childhood Education Journal, 32*(6), 371–376.
Martin, L. (2015). The promise of the maker movement for education. *Journal of Pre-College Engineering Education, 5*(1), 30–39.
Nemorin, S. (2017). The frustrations of digital fabrication: An auto/ethnographic exploration of '3D Making' in school. *International Journal of Technology and Design Education, 27*, 517–535.
Stout, D., Toth, N., Schick, K., & Chaminade, T. (2008). Neural correlates of early stone age toolmaking: Technology, language and cognition in human development. *Philosophical Transactions of the Royal Society, 363*, 1939–1949.
Strawhacker, A., & Bers, M. U. (2018). Promoting positive technological development in a kindergarten makerspace: A qualitative case study. *European Journal of STEM Education, 3*(3), 09.
Worsley, M., & Blikstein, P. (2017). Reasoning strategies in the context of engineering design with everyday materials. *Journal of Pre-College Engineering Education, 6*(2), 58–74.

Marc J. De Vries is full professor of Science and Technology Education and professor of Christian Philosophy of Technology at Delft University of Technology, the Netherlands. He currently serves as the editor-in-chief of the International Journal of Technology and Design Education. He wrote a monograph on philosophy of technology for technology educators and a monograph on the history of the Philips Research corporate industrial laboratory. He coordinates the series of international Pupils' Attitude Towards Technology (PATT) conferences.

Chapter 18
'We Treat Everyone Equally, but…'—Gendered Attitudes and Perceptions in STEM

Sonja Niiranen, Johanna Lätti, and Sini Teräsahde

Abstract Gender segregation, both horizontal and vertical, in the fields of technology and engineering, remains persistent. In many countries, the percentage of women studying STEM in higher education is around 20%, and this segregation continues in working life as differences in career paths, salaries, jobs, and positions. To facilitate a more equitable gender balance, particularly in higher education technology-oriented fields, it is important to expand our knowledge and awareness of the mechanisms that perpetuate gender segregation, such as gender-related attitudes, stereotypes, and cultural beliefs. This chapter discusses gender-related topics in STEM education and in technology-oriented careers based on two types of data collected during 2020–2021. Interview data were collected through 14 thematic semi-structured interviews with academic staff members in engineering higher education. Survey data were collected from 128 answers of female alumni who had graduated in technological fields from Tampere University or the former Tampere University of Technology. Our findings reveal that gender equality issues are not acknowledged in engineering higher education. Most of the interviewed staff members felt that their guidance practices were gender-neutral and that gender was an irrelevant factor in guidance. At the same time, they identified some gendered patterns in students' performance and orientation, stereotypical characteristics associated with female and male students, and different expectations of technical skills, which corresponded with the alumni experiences of gender-differentiated treatment. The paper concludes by providing suggestions and further directions to support gender-aware approaches in technology education.

Keywords Gender equality · Unconscious bias · Gender awareness · STEM

S. Niiranen (✉) · J. Lätti · S. Teräsahde
Tampere University, Tampere, Finland
e-mail: sonja.niiranen@tuni.fi

W. Fox-Turnbull and P. J. Williams (eds.), *Locating Technology Education in STEM Teaching and Learning*, Contemporary Issues in Technology Education,
https://doi.org/10.1007/978-981-97-1995-2_18

18.1 Introduction

STEM continues to gain prominence in many countries, and it is likely that it will remain a significant area of research in technology education, as it is one of the subject areas included in the STEM reform movement (Fan & Yu, 2017; Williams, 2016; Xu et al., 2020). One remarkable challenge related to STEM fields is gender segregation, which remains persistent. There are still relatively few women who enter occupations in the natural sciences, engineering, and technology fields. In many countries, only around 15–20% of women study and work in STEM-related fields. On average, the share of women employed in technology-related fields globally does not exceed 25%, even though EU policymakers and technology actors share the aim to achieve gender balance and increase diversity (see European Commission, 2016, 2018). However, it should be noted that there are contextual differences between countries, both in Europe and globally, and in some countries, the number of women in STEM fields is relatively higher.

In recent decades, many international projects have been introduced with the goal of inspiring women and girls to become more engaged in STEM. Considering the importance of technology in our daily lives and in building future societies, it is important to understand the gendered bias and inequalities and to promote more diverse education in related fields. Thus, the gender imbalance in technology can be seen as an educational challenge as well as an issue related to ethics and responsibility in teachers' professional competence. Striving for more equality and less segregation in technology calls for recognising the key problems regarding gendered stereotypes and attitudes in STEM and technology education. Despite various efforts to promote equality and to balance the gender distribution in technology, obstacles still hamper the access and involvement of girls and women. While gender issues in education are a well-established field of research and have received considerable international research attention, more knowledge is needed to understand the persistent male dominance and its implications in STEM and technology fields.

In this chapter, we will identify gendered perceptions and experiences of the academic staff and engineering higher education students at Tampere University, Finland. All data were collected as part of the project 'Equal career paths for women—NOW!', which was funded by the European Social Fund (2020–2022). The project produced topical knowledge about the causes and consequences of gender segregation in Finnish higher education and working life and created, piloted, and consolidated new gender-conscious models of study and career counselling, career services, and entrepreneurial support for female students and professionals. Our aim was to examine the inequality and gendered processes that women may experience when studying in engineering higher education. We analysed interview data collected from the university staff members ($N = 14$) and investigated (1) what kind of gender equality issues are attached to student guidance practices in engineering higher education, (2) whether staff members recognise differences in male and female students' orientations and career expectations, and (3) what kind of gendered stereotypes are embedded in their descriptions. Further, based on the survey data ($N = 128$), we examined the experiences of female alumni from the technology and engineering fields and the implications of gender in study and work life.

Through the data, we will shed light on the gender concerns in STEM and technology education, adding to the knowledge about perceived gendered differences in studies, gendered attitudes about female and male students, and images created about working life during studies. Although our study is focused on higher education, similar issues can be observed at lower levels of education. We will also provide suggestions for ways to increase gender awareness in STEM, particularly in technology education, with a view to developing strategies to reduce gender segregation in technological disciplines.

18.2 Gendered Attitudes and Perceptions in STEM

There is a high level of occupational segregation (horizontal segregation) in the Finnish labour market compared to other European countries (European Commission, 2009). Additionally, the fields of education are among the most gender-segregated in Western countries. The overall proportion of women studying in technology fields is 19%, but there are also gender divisions within these fields (Keski-Petäjä & Witting, 2018; Naukkarinen & Bairoh, 2020). Women comprise one-fifth (22%) of Master of Science in Technology students and employers in technology companies. Within these fields, both *horizontal* and *vertical divisions* are remarkable. Male-dominated subjects in universities consist of mechanical, construction and electrical engineering, automation technology, physics, and ICT (10–25% are women), while relatively more women study environmental and industrial engineering, material technology, chemistry, architecture and biotechnology, and biomedical engineering (30–80% are women). This segregation persists in working life. Even if men and women have the same educational background in technology, there is divergence between genders upon labour market entry, and this manifests throughout working life as differences in occupations, wages, and career paths (Naukkarinen, Bairoh & Putila, 2021) and as a trend of men inhabiting most of the leadership positions (vertical segregation) (Bairoh, 2019). Women's career advancement is somewhat slower compared to men's, and men dominate leadership positions in the private sector and listed companies. Further, the median salary for women with a university degree in technology or engineering is 87% that of men's salary, mainly because of the different sectors and positions they hold. However, five per cent of the wage gap is unexplained (Bairoh, 2019). Pay gaps have been found to emerge early, as they exist between men and women who are employed and work in engineering jobs during their studies (TEK, 2020). Thus, it is important to identify possible gender bias in STEM and technology education.

Technology (education) has a deeply gendered history and traditions. The concepts of technology and technological activity remain much more strongly associated with masculinity than with femininity (Murphy, 2006). This automatic and unconscious bias about the interests and competencies of men and women is deeply rooted in cultural beliefs. Gendered perceptions occur at an early stage, as children and young people already associate intelligence as well as mathematical and technical skills with men rather than women (see Diez et al., 2022; Hallström et al., 2015). This phenomenon is also evident in engineering higher education, where teachers report similar stereotypical assumptions (Lätti, Niiranen & Teräsahde, 2021). Gendered images may also influence young people's attitudes towards the field (Ardies, 2015; Ardies et al., 2015; Moshe & Khaled, 2012). Research shows that gender differences in choice of studies emerge in adolescence (Mercedes López-Sáez et al., 2011).

Researchers have pointed out that although girls and boys perform equally in STEM-related fields, girls and women who are talented in mathematical subjects choose, for example, medicine or commercial rather than technical studies (Mercedes López-Sáez et al., 2011). Implicit attitudes are more favourable towards women studying medicine and men studying engineering, and girls who choose technology are viewed more negatively than girls who choose other fields of study (Mercedes López-Sáez et al., 2011). Programme for International Student Assessment (PISA) studies indicate that the differences between boys and girls in mathematical abilities are not big, and in many countries, such as Finland, girls even outperform boys, contrary to the stereotype (OECD, 2020). However, girls' perceptions of their weaker mathematical and technical skills develop early. Female students express less confidence in their competencies and have lower self-efficacy than male students in various STEM domains. The same phenomenon is recognised among higher education students in the field of technology (Lätti et al., 2021; Marshman et al., 2018). Furthermore, although the number of boys and girls who choose STEM-related subjects in lower school levels is almost equal in many countries, the relative number of women decreases at higher levels of education.

18.3 Data and Respondents

18.3.1 Data Collection

As a source of discussion in this chapter, we have used two types of data collected during the NAU project in 2020–2021. The interview data consist of 14 thematic semi-structured interviews, in which academic staff members in engineering higher education describe the gender differences and stereotypical attitudes they experience in their work. Interviewees were selected based on their various experiences and responsibilities in counselling, guiding, and/or supervising duties from all four faculties, which offer Master of Science in technology degrees in the studied university. Faculties and degree programmes include highly male-dominated fields (e.g. automation technology and mechanical engineering) but also fields with relatively high numbers of women (e.g. environmental engineering and health technology). Interviews were carried out using Zoom, and they focussed on three main themes: equality issues in faculties, perceived gender differences, and (in)equal practices in guidance and counselling. Interviewees were asked to describe the visibility of equality issues in their units, the study culture for minorities, whether they have perceived any differences between male and female students, and whether there are some key issues requiring development with regard to equality in guidance and counselling.

Additional data are from an alumni survey conducted in March 2021. An invitation to respond to the survey was sent to all females who had graduated from technological fields at Tampere University or the former Tampere University of Technology before the merger in 2019. The survey included statements and questions that addressed the significance of gender in studies and work. However, only the findings concerning experiences during studies were scrutinised for the purposes of this chapter. The statements were answered by choosing the option closest to one's experience from the five-point Likert scale. Unlimited space was provided to answer the open-ended questions.

18.3.2 Analysis of the Data

All interview data and the responses to the open-ended survey questions were analysed using qualitative, thematic content analysis, and quantification (Schreier, 2014). The survey statements were analysed with descriptive statistics (May, 2017).

In the first phase of the interview data analysis, relevant themes and subcategories for each research question were identified, and the data were coded under each category. In the second phase, identified themes were classified based on the frequency of their occurrence. Finally, general descriptions of each theme were created with illustrations from the original data. In the analysis of the open-ended questions of the survey, the responses were classified based on their contextual similarities, and the cases in each class were quantified.

The interviews were treated completely anonymously to ensure that individual interviewees could not be identified. It should be noted that the group of interviewees was rather selective, as it consisted of academic staff members who considered the topic important in their own work and thus were more likely to agree to the interview. In addition, the interviews were conversational, and the interviewer may have asked more specific follow-up questions and thus guided the course of the interview.

18.3.3 Respondents

The interviewees held different academic positions, with the titles of researcher, university teacher, or professor, and their experience of active guidance varied from two to 20 years. The guidance was understood in a broad sense as involving supervising bachelor, master, and Ph.D. thesis (often done in cooperation with companies), different individual and group exercises and laboratory experiments, tutoring students, and following their practical training and internship in working life. Six of the interviewees were women, and eight were men.

The survey respondents ($N = 128$) graduated between the years 1987–2021. The number of responses varied between the disciplines. The highest share of responses was received from the respondents with backgrounds in environmental and energy engineering ($n = 37$) and industrial engineering and management ($n = 20$). More than ten responses were received from electrical engineering ($n = 14$), engineering and natural sciences ($n = 13$), information technology ($n = 12$), and engineering materials science ($n = 11$). Less than 10 responses were received from automation technology ($n = 8$), biotechnology and biomedical technology ($n = 4$), civil engineering ($n = 3$), mechanical engineering ($n = 2$), and other subjects ($n = 4$). The share of respondents between subjects corresponded with the segregated gender distribution between the subjects and study lines of technology, according to the equality statistics of the former Tampere University of Technology in 2017–2018 (TTY, 2018).

18.4 Findings

18.4.1 Gendered Perceptions and Attitudes Regarding Male and Female Students

The interviewed engineering higher education staff members identified some gender differences in students' performance and orientation towards studies. They described female students, more often than male students, as performing high-quality work and succeeding in their studies. They also viewed them as motivated, demanding, and proactive in asking for instructions. Male students, in contrast, were seen to represent more variety, with a higher number showing less ambition to achieve excellent grades.

> On average, the intelligence and interest that these female students have, the material is just of higher quality among females who come here. And then [the starting level of] the male ones is more from side to side. (Male, automation and mechanical engineering)

> Female students are perhaps a little more willing to ask if something is unclear. (Male, industrial engineering and management)

> I'm somehow used to the fact that girls are kind and perform and then if they come with that they can't, don't want to do something, it's really difficult for me to accept. (Female, civil engineering)

At the same time, staff members described female students as insecure about their own skills and competencies and in need of more support. Female students were also considered to be more likely to be perfectionists and to want to know everything, whereas male students were assumed to be confident about their interests. A few staff members also pointed out that women, due to their minority status, may be better remembered and given more attention and offered more help. Such perceptions may be reflected in women's insecurity about their own skills as technology students and experts. Many of the interviewees mentioned that female students often need to be encouraged to trust their own skills.

> With women, it's very strong that they get really caught up in the details and feel like they should always understand everything. (Female, ICT)

> I feel that women perhaps need more belief in their own doing and encouragement, than the male students. (Female, civil engineering)

> As a woman, it is easier to ask for help, because as a woman I feel that help has been pushed there all the time. (Female, ICT).

The interview data revealed stereotypical characteristics and different expectations about the technical skills of female and male engineering students. The most frequently mentioned stereotypical perceptions were related to mathematical competence and technical knowledge, where women are considered weaker. The survey data also revealed that almost half (12% fully agree, 28% partly agree) of the female respondents had encountered questioning of interest in their field because of their gender (Fig. 18.1, statement 1).

> But the difference in skills is that female students have poorer mathematical skills. (Male, materials science and environmental engineering)

> So quite often it seems that there perhaps were some kind of preconceptions that women are weaker in terms of mathematical skills compared to men, so in a way it was assumed that if a woman gets a good number on the exam, she will do well because she is hardworking, rather than the fact that she would know the subject. (Female, ICT)

There was also a common perception among the interviewees that women's success can be explained by their hard work and diligence rather than by innate talent. The interviewees also communicated assumptions that female students had less previous experience in relation to their technical skills. Women were not necessarily expected to have the same technical background and hobbies as male students, for whom technical competence and knowledge were thought to be innate. For instance, the image of a nerd is inherently masculine, to the exclusion of women.

> Women are treated a bit like 'women don't know enough about technology, or not in the same way' than these super nerds who are born with the keyboard in hand'. (Male, ICT).

At the same time, female students were described as having the qualities of conscientiousness, accuracy, and diligence following instructions and keeping to schedules. They asked more questions and demanded a lot from both their studies and themselves. In comparison, male students were viewed as adopting a more nonchalant attitude towards their studies.

> Well, in my opinion, girls are clearly more conscientious and perhaps follow schedules better than boys. (Female, civil engineering)

> They [female students] tend to have a higher bar, while boys are usually happy with themselves when they just do something. (Female, civil engineering)

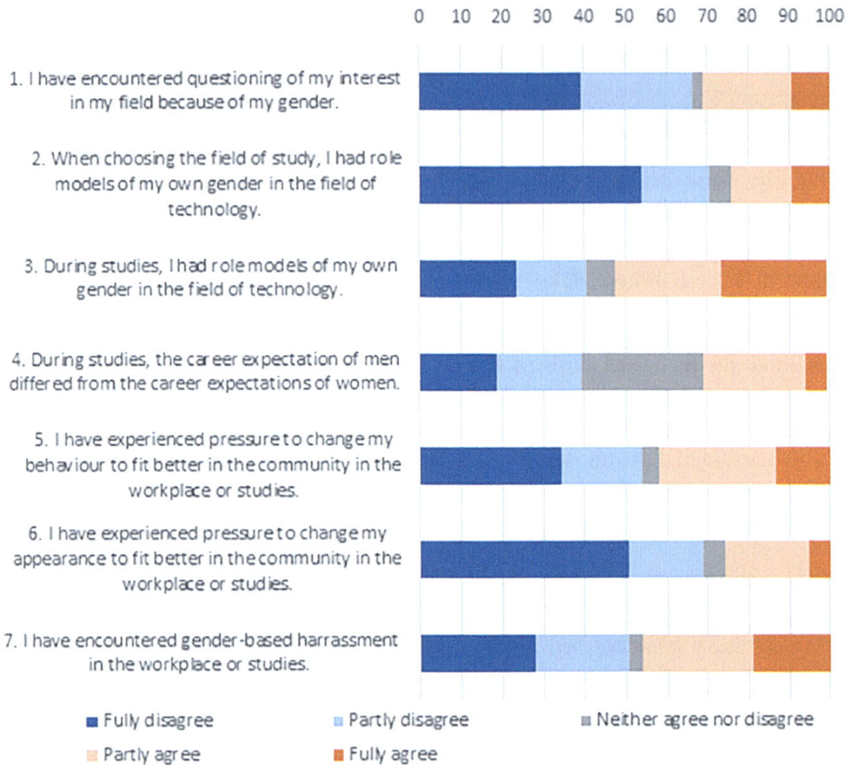

Fig. 18.1 Female respondents' perceptions of the significance of gender in technology studies

In addition, female students' tendency to take their performance more seriously was related to their ways of showing emotions, for example, through crying. Such behaviour was not regarded as appropriate for male students, or it would be less expected. Women asking for and receiving help could be considered as over-reliance on teachers, while some sort of 'self-survival ethos' was associated with men.

> Male students may be a bit more like that, that you have to cope with it yourself, and not go to the teacher's lap and cry, but you have to get things done. (Male, industrial engineering and management)

> And it might have been surprising if some boy had come to cry to me about his grade. I don't know if it's also related to this crying and this kind of behaviour, that it is more acceptable for girls and then one gives more easily like freedoms because of the gender in this situation. (Male, ICT)

One-third of the survey respondents (34%) had not experienced gender-related attitudes or assumptions during their studies, which was interpreted based on three types of answers: simple 'no' or 'I did not experience' replies, expressions of perceived equal treatment, and expressions of some experiences of gendered attitudes and assumptions that were not considered to be problematic. A quarter of the respondents (24%) did not answer this question at all, suggesting that the question might have been regarded as controversial. However, almost half (42%) of the female respondents indicated that they had experienced gendered speech involving attitudes and assumptions during their studies. This included, for example, professors' and teachers' sexually toned, chauvinistic, and discriminative comments. They also indicated that their male student colleagues had stereotypical attitudes and assumptions that underrated female students' skills in math, physics, and coding, which appeared, among other things, while doing group work assignments together. Male students had expressed 'that the so-called soft fields of technology, like environmental technology or chemistry, were better suited to women', and they had looked down on female-dominated fields, such as humanities.

18.4.2 *Images of Work Life Created in Studies*

Some interviewed engineering higher education teachers mentioned observable gender divisions among faculty staff members and the small number of female students in male-dominated fields. In most studied fields, the number of women professors, tenure-track position holders, and researchers is relatively low, while the share of female teachers is relatively higher. In addition, the interviewees pointed out that visiting lecturers, who are often high-level managers in companies, are mainly men. This imbalance was considered problematic from the perspective of role models visible to students, and some interviewees invited female experts to highlight diversity in working life. At the same time, this gender imbalance is something people are used to and do not necessarily pay attention to.

> I'm just used to the fact that our unit, too, in practice, it goes like this: all the professors are men and the administrative staff are women. And then the majority of the people at the coffee table were boys. (Female, civil engineering)

The survey results indicated that most (70%) of the female respondents did not have a role model of their own gender at the time they selected their field of study (statement 2), but approximately half of the respondents (52%) had found a role model of their gender during their studies (statement 3). It is worthwhile to note, however, that two out of five respondents (41%) had experienced a lack of role models during their studies as well. The finding suggests that these young adults apply to higher education and make their study choices mainly based on the stereotypical masculine representations of technology disciplines. Male role models in the highest positions, such as professors or business managers, may contribute to weakening women's identification with these positions and the qualities they require.

In the interviews, a few respondents described male students as having more straightforward and clearer career goals. Men were pictured as more determined to become managers and focussed on more demanding positions at an early stage of their studies, whereas women's orientation was seen to develop later, and their goals were considered somewhat less clear.

> Women have to be encouraged a little bit to find out what kind of things they are interested in and then, little by little, during the conversation, it often becomes clear what they want to aim for. (Male, materials science and environmental engineering)

Although there are some typical gender choices that guide both female and male students' orientations, most of the engineering higher education staff members did not observe any differences in career goals or expectations among the students. Some interviewees pointed out that women apply to slightly different degree programmes, such as industrial engineering, design, or building protection in architecture, and a few of them assumed women were less interested in male-dominated environments, such as construction sites. Teachers also recognised some stereotypical attitudes of what jobs and skills are suitable for men and women.

> I think this is very strongly reflected in the professors' ideas that which student is taken for something, which is taken for summer work, so yes there are boys prioritised because they have the strength and they are able to do the jobs, more willing to do physical work. [...] If you're a very feminine woman, then you'll choose a major in economics or traffic and transport engineering. (Female, civil engineering)

Moreover, women were viewed as more likely to apply for pedagogical studies. A proportionally high number of female students orientate themselves towards becoming qualified as teachers, whereas programming, for instance, attracts more male students. Staff members' descriptions of study choices also revealed some implicit valuation given to different competencies as well as attitudes about what constitutes real and valuable technical work. For example, pedagogical knowledge was assumed to be of interest to female students but also viewed as less important than technical work and professionalism.

> Perhaps in our unit, the emphasis is on the fact that we also have those student teachers who do their pedagogical studies at the same time. So this has been talked about a lot recently, whether you are automatically a worse mathematician if you take pedagogical studies (Female, ICT)

The views of the survey respondents diverged regarding whether the career expectations of men differed from those of women (statement 4), as a little over one-third of respondents disagreed (partly 20%, strongly 19%), while nearly one-third of them agreed (partly 26%, strongly 6%) or were unsure (30%). Although the data do not reveal clear differences in career expectations between female and male students, the relative number of women still decreases at higher educational levels, (e.g. postgraduate students and researchers), and the jobs and career development of men and women diverge in working life.

18.4.3 Study Environment and Fitness to the Community

Half of the interviewed engineering higher education teachers ($n = 7$) indicated that in the study environment, gender is not relevant or they have not paid attention to the matter. Most of them considered study practices to be gender-neutral and did not observe significant equality problems or different treatment between female and male students. Regarding their own guidance, counselling, and tutoring duties, they emphasised that the gender of their students did not matter, and they treated everyone similarly. The interviewees described the study environment as a good place where everyone can be themselves. A male-dominated field may also be seen as a place for some types of women who fit in.

> At least I haven't noticed that there has been any conscious difference in attitude. So in my opinion, everyone is treated in the same way. (Male, physics)

> Personally I haven't noticed any differences in treatment based on, for example, a person's gender. (Male, electrical engineering)

> I don't feel that there is any difference, that the attitude is the same regardless of gender. The content is what we are talking about, so I don't feel that it would have an impact on anything in particular. (Male, industrial engineering and management)

> They kind of know that there will not be a lot of women [...] the women who are there, they do blend in. (Female, civil engineering).

The experiences of the study environment by alumni were somewhat contradictory to the views of the staff. First, almost half of the survey respondents had experienced pressure to change their behaviour (statement 5, 17% fully agree, 37% partly agree), and one-third had experienced pressure to change their appearance (statement 6, 7% fully agree, 27% partly agree) to fit better in the workplace community or during studies. Second, the open answers revealed that several respondents had experienced joking about women related to evaluative comments about female students' appearance in students' free-time activities. Third, almost half of the respondents (46%) had encountered gender-based harassment in the workplace or during their studies (statement 7, 19% fully agree, 27% partly agree). Being harassed appeared to be a common experience, indicating that the working environments of the respondents lacked physical and/or psychological safety.

The data also showed that the study environment, including free-time activities, tended to be more favourable to men in terms of creating networks that would be essential later in their work. Even though most female respondents indicated that they had managed to network well or excellently during their studies, one-third (35%) of the respondents perceived that men had wider networks and better opportunities to network than women. Seven respondents specified that the 'sauna culture' favoured networking between men and created a potentially sexist environment for women.

18.5 Towards Gender Awareness in STEM Teaching

It is still evident that an increase in the number of women in technical careers has not yet been achieved in EU countries, and the reluctance of women to enter occupations in the natural sciences or technology-related fields is still a challenge many educators face worldwide (e.g. European Commission, 2016, 2018). In working life, women remain in the areas that are traditionally considered feminine, such as health, education, and services, while men take positions of a technical nature, such as engineering, science, and mathematics (Diez et al., 2022). If gender-related segregation is not actively dismantled in education, this gender imbalance of women entering technical fields will continue and women will have fewer possibilities to advance in their careers. To support a more equitable gender balance, it is important to expand our knowledge and awareness of the mechanisms that uphold gender segregation. Our findings reveal that gender equality issues are not acknowledged in engineering higher education. Most staff members did not pay much attention to gender issues, and they emphasised that all students are treated equally. At the same time, they identified gendered patterns in students' performance and orientation, stereotypical characteristics attached to female and male students, and different expectations of technical skills based on gender.

When considering gendered perceptions in technology field studies in more detail, the staff members described female students as motivated, active, demanding, and as those who do high-quality work and succeed in their studies. However, at the same time, they viewed females as more likely to be insecure about their own skills and competencies and to require support. Based on the survey, the female respondents had encountered questioning about their interest during their studies because of their gender. Despite boys' and girls' equal or near equal achievement in mathematics, researchers have reported that girls tend to have lower self-efficacy and self-confidence in relation to science and technology in primary school (Endepohls-Ulpe et al., 2012; Hallström et al., 2015; Murphy, 2007). Moreover, it has been noted that in technology education, girls need to receive more encouragement and support for their competence from their teachers (Virtanen et al., 2015). This is consistent with a study on teachers' influence on students' self-efficacy and attitudes towards exact science studies, which showed that perceived attention from a teacher was more closely related to self-efficacy in girls than in boys (de Weerd & Rommes, 2012). In contrast, compared to girls, boys feel more self-confident and are more eager to test and try something new, especially in relation to technology activities (Hallström et al., 2015; Virtanen et al., 2015). Teachers should be aware of how their behaviour can influence specific activities and students' self-efficacy beliefs, particularly their ability to help strengthen girls' self-efficacy (Güdel et al., 2018).

The interviewees pointed out some gendered perceptions related to mathematical and technical competencies and expectations as well as typical and suitable behaviours for male and female students. They recognised cultural beliefs, such as women's weaker mathematical competence and men's natural technical knowledge. In addition, female respondents had experienced gendered speech, such as professors' and teachers' sexual, chauvinistic, and discriminative comments and male colleagues' stereotypical attitudes and assumptions, which underestimated female students' skills in math, physics, and coding. These findings reveal that stereotypical perceptions and attitudes are still present behind gender-neutral ideals in engineering higher education. From the perspective of everyday practices, gendered attitudes remain hidden and unidentified. To identify the mechanisms that might hamper the equal participation and advancement of women in the field of technology, it is necessary to recognise that unconscious gender biases might already exist in the early years of STEM and technology education. Thus, we must all, as a technology and engineering education community, reflect on our attitudes regarding gender and how they affect our speech, gestures, actions, and behaviours. Negative communication and encounters should be replaced with positive and encouraging communication that supports the growth and development of students' identity and self-esteem. To achieve this, we should understand that there are individual differences as well as group differences between the needs, behaviours, and attitudes of females and males. Further, attention should be placed on dismantling assumptions about what girls and boys can and want to do, and students should be offered the support needed to develop new learning habits (Murphy, 2007).

When examining what kind of images students have and receive of engineering work life, it became evident that most of the female respondents had not had a role model of their own gender at the time they chose their field of study, and only half of the respondents had found a role model of the same gender during their studies. Additionally, in most technical fields, the number of women in universities is relatively low, as professors, tenure-track position holders, researchers and visiting lecturers (i.e. high-level managers in companies) are mainly men. This imbalance was considered problematic from the perspective of role models visible to students. The low number of women deprives them of the opportunity to contribute to research and innovation on an equal footing and given the different perspectives that women bring, the quality of research and innovation suffers as well (European Commission, 2016, 2018). Hence, it is important to develop interventions to create equal opportunities for all to enter technology fields of work. Technology education can have an important impact on society by providing all children with possibilities and opportunities to participate in modern technologies and to acquire the knowledge necessary to perform a wide variety of jobs. Education and student guidance are important in providing positive images of technology, its professions, and competencies. If stereotypical attitudes and images of technology remain unconscious, they may further reinforce gendered segregation. Thus, it is crucial for schools to accept more responsibility for providing students with greater opportunities to obtain practical experience and information about career options in technology-oriented fields. Also, experienced teachers of technology-related subjects are important in supporting the development of the technology skills and knowledge that students will require in their working lives.

Teachers play a key role in dismantling gendered practices and renewing the image of technology education because they are well-placed to alter students' perceptions and identities (Murphy, 2007). Therefore, teachers should adopt language that avoids the reproduction of gender norms, stereotypes, and hierarchies and of the existence of unconscious association barriers (i.e. unconscious bias). Unconscious bias is something that one is unaware of, and it occurs outside of one's control. It happens automatically and results in quick judgments and assessments of people and situations, influenced by one's background, cultural environment, and personal experiences (Equality Challenge Unit, 2013). These associations subconsciously influence ideas about which fields and jobs are appropriate for which gender and this can have an impact on whether someone applies for and obtains certain jobs. The impact of biases can be reduced by increasing teachers' and study counsellors' awareness of gender equality and the impact of gender on careers. A teacher or study counsellor who acts in a gender-sensitive way can instil hope in students (e.g. Yoon et al., 2019), helping them to discover and develop their own strengths and encouraging them to make atypical choices (European Commission, 2005).

Gender-aware teachers recognise that gender affects careers and that segregation narrows students' perceptions of a field. They systematically take gender into account in their teaching and guidance, for example, by introducing career options that are considered atypical for the student's gender and offering a wide variety of role models and support (European Commission, 2005; Juutilainen, 2003, p. 63; Niles, 2011). Gender awareness also refers to awareness of the existence of unconscious biases (Equality Challenge Unit, 2013). These biases may unconsciously affect perceptions about which jobs are suitable for different genders, which in turn can influence the choice to apply for positions or to pursue career advancement. Conscious action requires constant reflection from teachers and a desire to develop and improve their methods. For instance, by presenting a variety of role models, teachers have the opportunity to diversify the images associated with technical fields.

The findings also shed light on the controversy related to the significance of gender in the study environment and the importance of students' free-time activities to community building and networking. Although the staff considered the environment to be safe and equal, the many student experiences of difficulty fitting into the community, gender-based harassment, and unequal opportunities for networking cannot be ignored. These findings urge educators to pay attention to safety in study environments. Indeed, arranging study environments as a 'safe community' in which students are encouraged to ask questions and get to know their peers is an effective way to support the learning of both girls and boys and could possibly be an asset in promoting diversity in technology education (Kangas et al., 2017). Moreover, the concept of a 'safe space', which originated from minority cultures and includes both separatist and inclusive safe spaces, could be useful in advancing equality in technology education (The Roestone Collective, 2014). In basic education schools, a separate safe space could mean, for example, organising technology clubs for girls only, thus providing them with encouraging circumstances to test and play with technological devices free from assumptions about girls' weaker technological skills. Creating inclusive safe classrooms is already considered a pedagogical requirement and an ethical obligation (The Roestone Collective, 2014). Still, educators must work hard and need institutional support to create an atmosphere in which everyone is equally encouraged to participate, there is zero tolerance for bullying and harassment, and the measures stated in the schools' equality plans are implemented (Lahelma, 2021). This is crucial because, as our results show, students may experience gender-based harassment and problems related to community fitness and building.

18.6 Conclusion

This study gave a voice to female engineering graduates and examined academic staff members' perceptions of gender differences in engineering higher education. It seems that in engineering higher education, gender is not the subject of much debate—or it is only discussed in very limited contexts—and so the real problems may remain hidden. Teachers are more likely to emphasise the gender neutrality of teaching, guidance, and the study environment, even though reducing inequalities requires a conscious approach to detect gendered practices and attitudes and to actively strive to change them. As teachers and educators play a key role in dismantling gendered practices, they should be aware of how their behaviour can influence specific activities and students' perceptions. Thus, it is crucial to provide teachers with information and training on gender awareness in pedagogy. We suggest that instead of emphasising gender neutrality ('gender does not matter'), more gender-aware teaching and guidance could be promoted in STEM. This need for gender-aware teaching is based on the strong gender-based segregation, hidden biases, and attitudes that persist in the field of technology.

Our findings reveal that some women who have studied in technical fields experienced gendered treatment, attitudes, and harassment both in their studies and at work. More attention should be paid to gender and its mechanisms and to the pedagogical questions of implementing gender-aware teaching and guidance at lower levels of education. In practice, gender-aware teaching requires more knowledge and consciousness about gendered attitudes and practices, the will to act equally, and effective strategies for diversifying the images related to technical fields and professions. Our research suggests that while teachers may have a willingness to act equally, they often lack adequate tools and support. Thus, educational institutions should take responsibility for creating an equal learning culture and reducing stereotypical assumptions. Hopefully, our research will enrich the ongoing discussions on gender concerns in STEM and technology education.

References

Ardies, J. (2015). *Students' attitudes towards technology. A cross-sectional and longitudinal study in secondary education*. (Ph.D. dissertation). Universiteit Antwerpen, Antwerpen.

Ardies, J., De Maeyer, S., & Gijbels, D. (2015). A longitudinal study on boys' and girls' career aspirations and interest in technology. *Research in Science and Technological Education, 33*(3), 366–386.

Bairoh, S. (2019). Valitaanko pätevä vai nainen? Sukupuolten tasa-arvo ja johtotehtäviin eteneminen tekniikan korkeakoulutettujen työpaikalla. [Shall we select a qualified or a woman? Gender equality and the advancement to the leading positions in the workplaces of people having a higher education degree in technology]. Diversiteetti tekniikan alalla -hankkeen tuloksia. [Results from the project Diversity in the Field of Technology]. Helsinki: Academic Engineers and Architects in Finland TEK. https://www.tek.fi/sites/default/files/attachments/tek_pateva-vai-nainen_raportti_2019.pdf de Weerd, J., & Rommes, E. (2012). To beta or not to beta? The role of teachers in the gendered choice of science and technology by secondary school students. In C. Quaiser-Pohl & M. Endepohls-Ulpe (Eds.), *Women's choices in Europe: Influence of gender on education, occupational career and family development* (pp. 63–78). Münster: Waxmann.

Diez, J. -L., Ramos, A., & Candela, C. (2022). Static and dynamic assessment of STEM gender stereotypes in secondary education using a novel cluster-based analysis. *International Journal of Technology and Design Education*.https://doi.org/10.1007/s10798-022-09746-1

Endepohls-Ulpe, M., Ebach, J., Seiter, J., & Kaul, N. (2012). Barriers and motivational factors for taking up a career in a technological field in Germany and Austria. In C. Quaiser-Pohl & M. Endepohls-Ulpe (Eds.), *Women's choices in Europe: Influence of gender on education, occupational career and family development* (pp. 79–93). Münster: Waxmann.

Equality Challenge Unit. (2013). *Unconscious bias and higher education*. London. Retrieved December 20, 2022, from https://www.google.com/url?sa=t&rct=j&q=&esrc=s&source=web&cd=&cad=rja&uact=8&ved=2ahUKEwig4HkoLX7AhVskosKHVWsD5cQFnoECAwQAQ&url=https%3A%2F%2Fdiversity.caltech.edu%2Fdocuments%2F19785%2Funconscious-bias-and-higher-education-compressed.pdf&usg=AOvVaw1Turj2ZoKVRtKeY-T_ym0u

European Commission. (2005). *Overcoming gender blindness in careers guidance. EQUAL success stories: Development Partnerships working against discrimination and inequality in Europe.* Directorate-General for Employment, Social Affairs and Equal Opportunities. Retrieved July 11, 2022, from https://ec.europa.eu/employment_social/equal_consolidated/about/key-doc.html

European Commission, Directorate-General for Employment. Social Affairs and Inclusion, Verashchagina, A., & Bettio, F. (2009). *Gender segregation in the labour market: Root causes, implications and policy responses in the EU*. https://doi.org/10.2767/1063

European Commission. (2016). *She figures 2015—gender in research and innovation*. Publications Office of the European Union.

European Commission. (2018). *She figures 2018*. https://ec.europa.eu/info/publications/she-figures-2018_en

Fan, S.-C., & Yu, K.-C. (2017). How an integrative STEM curriculum can benefit students in engineering design practices. *International Journal of Technology and Design Education, 27*, 107–129.

Güdel, K., Heitzmann, A., & Müller, A. (2018). Self-efficacy and (vocational) interest in technology and design: An empirical study in seventh and eighth-grade classrooms. *International Journal of Technology and Design Education, 29*, 1053–1081. https://doi.org/10.1007/s10798-018-9475-y

Hallström, J., Elvstrand, H., & Hellberg, K. (2015). Gender and technology in free play in Swedish early childhood education. *International Journal of Technology and Design Education, 25*(2), 137–149. https://doi.org/10.1007/s10798-014-9274-z

Juutilainen, P. -K. (2003). Elämään vai sukupuoleen ohjausta? Tutkimus opinto-ohjauskeskustelun rakentumisesta prosessina. [Counselling towards life or towards gender? Process research of career councelling conversations]. (Ph.D. dissertation). Joensuu: Joensuun yliopisto. http://urn.fi/URN:ISBN:952-458-330-5

Kangas, J., Rantanen, E., & Kettunen, L. (2017). How to facilitate freshmen learning and support their transition to a university study environment. *European Journal of Engineering Education, 42*(6), 668–683. https://doi.org/10.1080/03043797.2016.1214818

Keski-Petäjä, M., & Witting, M. (2018). *Alle viidennes opiskelijoista opinnoissa joissa tasaisesti naisia ja miehiä—koulutusalojen eriytyminen jatkuu* [Less than a fifth of student's study in disciplines with even numbers or women and men—the gender segregation of educational fields continues]. Statistics Finland. https://www.stat.fi/tietotrendit/artikkelit/2018/alle-viidennes-opiskelijoista-opinnoissa-joissa-tasaisesti-naisia-ja-miehia-koulutusalojen-eriytyminen-jatkuu/

Lahelma, E. (2021). Revisiting lower secondary schools in the 1990s: Reflections on and interpretations of sexual and gender-based harassment. In Y. Odenbring & T. Johansson (Eds.), *Violence, victimisation and young people: Education and safe learning environments* (pp. 15–29). Springer International Publishing AG.

Lätti, J., Niiranen, S., & Teräsahde, S. (2021). We treat everyone equally, but…. Finnish engineering teachers' perceptions of gendered differences in guidance and counselling. In *SEFI Conference Proceedings*.

Marshman, E., Kalender, Z., Nokes-Malach, T., Schunn, C., & Singh, C. (2018). Female students with A's have similar physics self-efficacy as male students with C's in introductory courses: A cause for alarm? *Physical Review Physics Education Research, 14*, 020123. https://doi.org/10.1103/PhysRevPhysEducRes.14.020123

May, A. (2017). Simple descriptive statistics. In M. Allen (Ed.), *The SAGE encyclopedia of communication research methods* (pp. 2–8). Sage Publications Ltd.

Mercedes López-Sáez, M., Puertas, S., & Milagros Sáinz, M. (2011). Why don't girls choose technological studies? Adolescents' stereotypes and attitudes towards studies related to medicine or engineering. *The Spanish Journal of Psychology, 14*(1), 74–87.

Moshe, B., & Khaled, A. (2012). Teaching image-processing concepts in junior high school: Boys' and girls' achievements and attitudes towards technology. *Research in Science and Technological Education, 30*(1), 81–105.

Murphy, P. (2006). Gender and technology. Gender mediation in school knowledge construction. In J. R. Dakers (Ed.), *Defining technological literacy: Towards an epistemological framework* (pp. 219–237). New York: Palgrave MacMillan.

Murphy, P. (2007). Gender and pedagogy. In D. Barlex (Ed.), *Design and technology: For the next generation* (pp. 236–251). Shropshire: Cliffeco Communications.

Naukkarinen, J. K., & Bairoh, S. (2020). STEM: A help or a hinderance in attracting more girls to engineering? *Journal of Engineering Education, 109*, 177–193. https://doi.org/10.1002/jee.20320

Naukkarinen, J. K., Bairoh, S., & Putila, S. (2021, July). *Gender segregation in the occupations of Finnish engineers*. Paper presented at 2021 ASEE Virtual Annual Conference Content Access, Virtual Conference. https://peer.asee.org/37221

Niles, S. G. (2011). Career flow: A hope-centered model of career development. *Journal of Employment Counseling, 48*(4), 173–175. https://doi.org/10.1002/j.2161-1920.2011.tb01107.x

OECD. (2020). Girls' and boys' performance in PISA. In *PISA 2018 results (Volume II): Where all students can succeed*, PISA. Paris: OECD Publishing.https://doi.org/10.1787/b5fd1b8f-en

Schreier, M. (2014). Qualitative content analysis. In U. Flick (Ed.), *The SAGE handbook of qualitative data analysis* (pp. 170–183). Sage Publication Ltd.

TEK (2020). *Student survey 2020*. Academic Engineers and Architects in Finland TEK.

The Roestone Collective. (2014). Safe space: Towards a reconceptualization. *Antipode, 46*(5), 1346–1365.https://doi.org/10.1111/anti.12089

TTY (2018). *Tampereen teknillisen yliopiston tasa-arvo- ja yhdenvertaisuustyön raportointi* [The report of the equality and diversity work of Tampere University of Technology]. Retrieved December 28, 2021, from https://www.google.com/url?sa=t&rct=j&q=&esrc=s&source=web&cd=&cad=rja&uact=8&ved=2ahUKEwi8wJOg_oX1AhVCs4sKHf96AtkQFnoECAcQAQ&url=https%3A%2F%2Fcontent-webapi.tuni.fi%2Fproxy%2Fpublic%2F2019-10%2Ftampereen-teknillisen-yliopiston-raportointi.pdf&usg=AOvVaw0cGuwYMuOKc8CNa42wADH3

Virtanen, S., Räikkönen, E., & Ikonen, P. (2015). Gender-based motivational differences in technology education. *International Journal of Technology and Design Education, 25*(2), 197–211. https://doi.org/10.1007/s10798-014-9278-8

Williams, P. J. (2016). Research in technology education: Looking back to move forward … again. *International Journal of Technology and Design Education, 26*, 149–157.

Xu, M., Williams, J. P., Gu, J., & Zhang, H. (2020). Hotspots and trends of technology education in the International Journal of Technology and Design Education: 2000–2018. *International Journal of Technology and Design Education, 30*, 207–224.

Yoon, H. J., Bailey, N., Amundson, N., & Niles, S. (2019). The effect of a career development programme based on the hope-action theory: Hope to work for refugees in British Columbia.

British Journal of Guidance & Counselling, 47(1), 6–19. https://doi.org/10.1080/03069885.2018.1544827

Sonja Niiranen is a University Lecturer in Engineering Pedagogy. She received her master's and doctoral degree in education in 2006 and in 2016, respectively. She worked 13 years as a technical craft and technology education teacher in the Department of Teacher Education at the University of Jyväskylä. She is currently working in the Faculty of Education and Culture at Tampere University as part of a Professional Growth and Learning (PGL) research group and delivering courses on university pedagogy for the academic staff members at the university.

Johanna Lätti is a Doctoral Researcher in the Faculty of Education and Culture at Tampere University. She graduated with a Master of Arts (Education) degree in 2012. Her research covers gender equality politics within higher education institutions and academic work. She has previously worked as a researcher in the research and development project 'Equal Career Paths for Women—NOW!', which is funded by European Social Fund and which observes gender equality and women's position in the field of technology.

Sini Teräsahde is a Doctoral Researcher in the Faculty of Education and Culture at Tampere University. She graduated from the faculty's Lifelong Learning and Education Programme with a Master of Arts (Education) degree in 2010. She has worked as a project researcher and responsible expert in the PGL research group's project 'Equal Career Paths for Women—NOW!' which is funded by the European Social Fund, and which observes gender equality and women's position in the field of technology.

Chapter 19
Conceptual PlayWorlds: Bringing Together Imaginary Play and Imagination in Design to Amplify Technological Learning

Marilyn Fleer

Abstract The notion of play as a foundation for design-based learning is increasingly recognised. Children imagine in play, but they also imagine their designed solutions in technology education. How play creates the learning conditions for STEM has been the focus of research in the Conceptual PlayLab. Through a series of studies into imagination in play and imagination in STEM, a model of practice has been developed. Called a Conceptual PlayWorld, this model of STEM teaching is for children from early childhood through to the end of primary school. We followed 13 children (3.0–5.8 years, mean age of 4.8 years) and two teachers engaged in a Conceptual PlayWorld of Robin Hood in an integrated STEM unit, where technological activity was amplified and design-based learning supported. Results show how over three periods, children's designs developed from collective problem formulation to collective design solutions, and how the play narrative amplified design in technology education. Drawing on a cultural-historical conception of imagination and creativity, this chapter outlines the nature and importance of play in bringing forward design in technology education as a unique characteristic of STEM.

Keywords Play · Conceptual PlayWorlds · Technology · Engineering · Technology

19.1 Introduction

Design-based learning is a form of inquiry-based pedagogy where design thinking and design processes are developed. Whilst design-based learning is core to interior design, graphic design, engineering, and architecture, in this chapter the focus is on

M. Fleer (✉)
Monash University, Peninsula Campus, Australia
e-mail: marilyn.fleer@monash.edu

W. Fox-Turnbull and P. J. Williams (eds.), *Locating Technology Education in STEM Teaching and Learning*, Contemporary Issues in Technology Education,
https://doi.org/10.1007/978-981-97-1995-2_19

design in technology education for children aged three to eight years. In this under-researched area, scholars have identified studies into design cognition (Crismond & Adams, 2012), studies into children's drawings (Kelley & Sung, 2017) and studies of learning through design (English et al., 2020). However, very little is known about how imaginary play and imagination in design can work together for design-based learning to support technology education. In this chapter, special attention is directed to considering how technology becomes a central tenet of children's play in classrooms. The central problem addressed in this chapter is, 'How does imagination in play within a STEM Conceptual PlayWorld create conditions for design-based learning in technology?'

To answer this question, this chapter begins with a theoretical discussion of imagination in play and design, followed by details of a Conceptual PlayWorld model of practice, where the concept of the zones of development in assessing children's learning is considered. The chapter concludes by theorising the relations between collective problem formulation and collective design solutions in a Conceptual PlayWorld where the development of children's designs deepens their play and amplifies their learning. Amplification is conceptualised in this chapter as the activities of teachers and the heightened responses of children that collectively lead to increased engagement, motivation, and learning in technology education. The implications for foregrounding play beyond the early years of school are also considered.

19.2 Imagination in Play

The theoretical foundation of this chapter has its roots in Vygotsky's (1966) conception of play, where children create an imaginary situation, change what they see (visual field), and imagine something new (imaginary field). This imaginary situation affords new actions and possibilities for children. It is through play, that a child can explore their everyday world by making conscious the rules of how things work, and the roles of people within their community. For example, imagining being a civil engineer and creating a dam in the sandpit, provides experiences of force, structure, and materials. But with support, these play narratives and embodied explorations, become designs and structures to test out the new technological solutions imagined. But how does this kind of imaginary play afford learning in design within a technology curriculum context? To understand this problem, we need to amplify imagination in design within play situations of a Conceptual PlayWorld.

19.3 A Conceptual PlayWorld

The research underpinning a Conceptual PlayWorld teaching model is based on ten years of foundational research into STEM and play (Fleer, 2018), and recently two years of research explicitly examining imagination in engineering play as part of a

Laureate programmatic study funded by the Australian Research Council. From this research into STEM learning, a series of Conceptual PlayWorlds were developed—a Scientific PlayWorld (Fleer, 2017a), an Engineering PlayWorld (Fleer, 2020), and a Digital PlayWorld (Fleer, 2017b).

A Conceptual PlayWorld model of STEM learning consists of five characteristics, where children in STEM imagine their designed solutions in engineering education (Fleer, 2020), wonder and inquire in a Scientific PlayWorld (Fleer, 2017a), and engage in meta-imaginary situations in a Digital PlayWorld (Fleer, 2017b). In this chapter, the focus is on the design and technology Conceptual PlayWorld. In a design and technology Conceptual PlayWorld, children play in an imaginary situation, and teachers take an active role in orienting children towards the learning of concepts in design and technology. The teachers plan their program in relation to the characteristics of a Conceptual PlayWorld (Table 19.1).

Table 19.1 shows the planning proforma for a Conceptual PlayWorld that brings forward design in technology for what follows. The characteristics (Column 1) and play narrative of design-based learning are shown (Column 2). The story of Robin Hood acts as the basis for the emergent Conceptual PlayWorld, and this example is used to showcase the essence of the teaching practices adopted by the teachers who participated in the research. This particular Conceptual PlayWorld is oriented to the technology curriculum because children design and prototype solutions. But what is different to other models of teaching technology (see Fig. 19.7 later) is that the design, make, and appraise processes that are characteristic of technology education sit within children's imaginary play. It is the imaginary play narrative that drives the children's need for learning and engaging with the design, make and appraise processes.

The Conceptual PlayWorld of the story of Robin Hood that is the focus of this chapter is an integrated STEM unit where teachers plan (Table 19.1) a play narrative

Table 19.1 STEM PlayWorld to amplify design and technological play

Pedagogical characteristics	Play narrative
Selecting a story	The story of Robin Hood brings drama and socially oriented play into classrooms
Designing an imaginary PlayWorld space	The outdoor fort becomes a time machine, and the outdoor play area becomes Sherwood Forest
Entering and exiting the Conceptual PlayWorld	Children go back in time through the time machine to meet the castle engineer
A problem to be solved	How to get the treasure out of the castle in order to redistribute the wealth back to the villagers. A machine needs to be designed and made
Role of the teacher	Teachers are characters in the story and take on a play role in support of collective imaginary play

that can support children's learning and play in technology. In its implementation over two 90 min periods each week for a term of schooling, children listen to the story reading of Robin Hood (chapters) followed by jumping into the story in character as they enter a time machine (children's fort) to visit Sherwood Forest so they can help the villagers. The social problem in the children's play drives the design and technology solutions needed to retrieve the treasure, to undertake in situ research (meeting the Castle Engineer) and return to the classroom through the time machine for further research using YouTube, excursions, books, etc., in order to solve the problems they meet in their play. For example, how to enter the castle without being noticed. This means 1. learning about a pulley system to lift the drawbridge, 2. studying and understanding the routines of the castle, so as to know when is the best time to go to the treasure room, 3. using Google Earth to study a plan view orientation and to read cross-section books, in order to effectively draw an escape plan to get safely out of the castle.

As is characteristic of a Conceptual PlayWorld, the children's play develops over time. The children meet further problems when they visit the Castle, such as, how to lift the treasure out of the dungeon. This means inventing a simple machine. Children consider what are the designed solutions to these problems which they can take back in time to Sherwood Forest, where they need to use their prototypes and designed solutions to retrieve the treasure (the outdoor play area is imagined as a Castle, with a pulley system attached to play structures).

In the classroom teachers use mind maps to record on large sheets of paper the children's thinking about the problems they meet in their Conceptual PlayWorld. The mind maps also capture children's ideas about how to solve the problem and to make visible to the group the collective problem(s) and solution(s) to take back into their Conceptual PlayWorld. The collective designing, prototyping, and researching in the classroom are mirrored and driven by the role-playing in Sherwood Forest. The reading of a chapter in the book each week brings new adventures to take into the Conceptual PlayWorlds. Additional amplification of the emotional and social motive to help is done by teachers in their planning, such as letters from Friar Tuck (teacher in character) to the children asking for help. These go beyond the story narrative and bring forward the design and technology needed, where they seek out engineering solutions, and the learning of scientific concepts associated with force. The concepts act in service of the children's play so that they can solve the social problem of helping the starving villagers by retrieving the treasure using their designed and prototyped simple machine.

19.3.1 Following Children and Teachers in a Conceptual PlayWorld

In the overarching study reported elsewhere (Fleer, 2021), the Conceptual PlayWorld that is described in practice above began with an intervention (Table 19.1) into teacher practices for promoting STEM learning of children aged three to seven years. In this chapter, the focus is on understanding the nature of design-based learning as it uniquely emerged in a Conceptual PlayWorld under the lens of technology education. Building on earlier research (Fleer, 2021), this chapter asks: How does imagination in play within a STEM Conceptual PlayWorld create conditions for design-based learning in technology education?

A total of two teachers, a technical assistant and 13 children from two classes (5 from preschool and 8 from school) came together regularly over two terms to participate in a STEM Conceptual PlayWorld featuring the story of Robin Hood. The children were aged 3.0–5.8 years (mean age of 4.8 years). Ten children were boys and three were girls. The children and the teachers have European heritage backgrounds.

The teachers (Olivia and Ruth) were degree-qualified and had over ten years of teaching experience. The technical assistant (Bob) helped classroom teachers with digital support, but for the Conceptual PlayWorld program, he acted as a character in the story (Castle Engineer) as well as providing support with the building of structures that the children had designed. Ethics approval for undertaking research was granted through Monash University and the Department of Education and Training. Pseudonyms have been used throughout this chapter.

19.4 What We Did

Digital video observations, the gathering of children's designs, and undertaking photographic documentation of design-based learning, constituted the data collection method. Teacher interviews and weekly planning sessions were also digitally recorded and 100 h of field notes were made. A total of 153.3 h of digital documentation (including teacher interviews in situ or during the seven teacher planning sessions) and 1724 photographic images and drawings were gathered.

By using two digital cameras, it was possible to video record the actions of children within the classroom by following them and documenting their activities (inside and outside). The second camera was put on a tripod and captured the whole classroom.

19.4.1 How Concepts Helped Us

The dialectical concept of imagination and creativity (Vygotsky, 2004) was used to understand the relations between imagination in play and imagination in design in the context of a technology education program of a Conceptual PlayWorld.

A conceptual synergy exists between a cultural-historical conception of play (Vygotsky, 1966) and imagination and creativity (Vygotsky, 2004). Vygotsky's system of concepts is vast (see six volumes of the collected works) and the selection of concepts for analysis must always be in relation to the core problem under study.

In the context of studying development as a process, rather than only an end product (Vygotsky, 1997), imagination is conceptualised as dynamic and in relation to creativity. Imagining something that does not yet exist, when created brings forward new possibilities for action. The process and the creations can be emotionally charged and experienced dynamically in play.

It is through bringing aspects of different experiences or solutions together in new ways (creativity), that the technologist, engineer, or architect imagines and generates new designs (new imaginings). Studying how children do this in design-based learning as they engage (or not) in the process of new creations gives insights into their imagining in the present moment, their possible past experiences, and future solutions to technological problems. This constitutes a cultural-historical conception of the genesis and developmental processes of design-based education under the conditions of a Conceptual PlayWorld (Fleer, 2021).

In this chapter, the children's designs over two terms were conceptually analysed in relation to imagination and creativity (Vygotsky, 2004) and imagination in play (Vygotsky, 1966) where special attention was paid to possible design strategies (e.g. understanding the challenge, represents ideas, revises design solutions) emerging during the transformation of beginning designers (e.g. linear design steps) and informed designers (e.g. back and forth iterative feedback cycle) (Crismond & Adams, 2012) when under the conditions of a Conceptual PlayWorld (Fleer, 2018).

19.5 What We Learned

It was found that a Conceptual PlayWorld amplified imaginary play and this in turn drove the design process for a technological solution. The results are showcased through first, typical examples of children's designs, and second, through a summary narrative of the digital observations of how the design challenge progressively developed in relation to the children's play actions and design solutions in a context of a technological creation.

19.5.1 Children's Designs in a STEM Conceptual PlayWorld

Three periods of design emerged. They were:

1. Collective problem formulation.
2. Researching and developing designs to deepen play.
3. Collective design solutions.

Fig. 19.1 Typical example of a design in the first period of a Conceptual PlayWorld—front-view perspective

19.5.2 Collective Problem Formulation

The collective problem of helping the poor villagers in the Conceptual PlayWorld of Robin Hood resulted in the children wanting to retrieve the treasure from the castle. This generated a series of designed solutions to support the problem of how to enter and exit the castle without being caught. Beginner designer solutions were from a front-view perspective (Fig. 19.1), and this did not give enough detail of an escape route. Conceptualising the inside of the castle as an unfolded box as the beginning of a plan view perspective (Fig. 19.2) demanded a higher level of graphicacy skills than the children had. More developed design solutions emerged as the play imperative intensified—wanting to design a simple machine to retrieve the treasure (see further below). This gave rise to cross-sectional drawings (Fig. 19.3) as children began to study castles in books and build models (Fig. 19.4).

19.5.3 Researching and Developing Designs to Deepen Their Play

The development of children's design capabilities was driven by the collective problem and the realisation that the early forms of the designs were not helpful for their play inside of the imaginary situation of a Conceptual PlayWorld of Robin Hood (Fig. 19.1). The collective problem was emotionally experienced as part of the children's play, and the development of the children's design skills (Fig. 19.4) appeared

Fig. 19.2 Typical example of a design in the second period of a Conceptual PlayWorld—plan view perspective

Fig. 19.3 Typical example of a design in the final period of a Conceptual PlayWorld—cross-sectional perspective

to be driven by the strong motive to help the villagers, whilst also keeping themselves safe when sneaking into the castle. Designs were being iteratively realised within the imaginary play frame and authentically evaluated within the narrative of the children's play. This finding is in keeping with the developmental period of children in

Fig. 19.4 Researching and developing designs and solutions to deepen play

early childhood and early primary school to want to play. It is also in keeping with the processes of designing, making, and appraising in technology education. What is new is how the collective problem that the children were seeking to solve, created the need for quality design solutions. This adds to what is known about how to set up the pedagogical conditions for the development of informed designers (Crismond & Adams, 2012) in design-based education to support technological thinking.

19.5.4 Collective Design Solutions

Informed designers (Crismond & Adams, 2012) brought to their technological problem of how to get the treasure out of the castle the solution of designing a 'grabby' hand machine (Fig. 19.5). Whilst children explored different dimensions of how to do this through researching simple and complex machines using YouTubes, studying castle designs to understand the context (Fig. 19.4), and using Google Earth to view real castles, they ultimately went from the collective problem to a collectively designed solution (Fig. 19.5). The designs (Figs. 19.1, 19.2, 19.3, 19.4 and 19.5) are illustrative of how beginning designers show evidence of becoming

Fig. 19.5 Collective design solutions

informed designers. The development was from individual drawings where graphicacy skills and perspectives were difficult, through to collectively designing solutions to the shared technological problem. Solving the problem and designing the solution appeared in relation to the narrative emerging in the children's play.

19.5.5 How the Play Narrative Amplifies Designs in Technology Education

It was found that a sense of collectiveness emerged through design when formulating the problem in children's play in a Conceptual PlayWorld. Having a collective problem to solve in their play meant the children developed the play narrative as a whole and at the same time developed new design solutions that they wished to create and use in their play. Their motive to play and be in the imaginary Conceptual PlayWorld of Robin Hood created a sense of togetherness and a keen interest in each other's design ideas. The imaginary play drove the design process, and they in turn enriched the children's play—acting as an emotionally charged and motivating force for technological learning and design-based thinking. For example, the teachers and children are seated on the mat in a circle. The first chapter of the story of Robin Hood has been read. The children discuss the story and many comments are made about the villagers being poor and needing help. A collective problem is brainstormed by the children and scribed by Olivia on a large sheet of paper that she has placed on the floor in the middle of the circle of children. She writes down the children's comments:

> [How to] Get the treasure back.

We need the plans of the castle.

Where is the treasure room?

Tools to make?

Materials [needed]?

Drawing models and testing [out ideas and designing an escape plan]

(Digital Observations from Term 3)

It was found that the design ideas that emerged were in relation to the *play problems* that were created by either the teachers or the children. These problems motivated the children to want to know more for solving the *play problem* they were encouraging, such as, how a drawbridge works so that you can get into the castle and what is the layout of a castle (and the guard's routines) when finding the treasure or planning the escape route. This in turn drove their need to research and develop their *technological thinking* about simple machines, and how ropes and chains acted in pulley systems, and *design-based thinking*, such as when considering cross sections and plan views in castle designs (see Figs. 19.1, 19.2 and 19.3).

Teacher planning always began with the play narrative, for example:

> The sheriff has changed the security system and Robin and the treasure are trapped in the castle. The poor villagers need their taxes returned to them [planning document: Robin Hood Storyline].

To deepen the problem and generate the need for more designed solutions, the play narrative also needed to be developed, for example:

> Visits to Bob in the castle in Sherwood Forest to find out about castle engineering [planning document: Imagination space].

Technological investigations were planned within the narrative of the play problem, as shown in Table 19.2 for Term 3 (Column 1) and Term 4 (Column 2):

Table 19.2 Extract from planning document for the Conceptual PlayWorld of Robin Hood

Term 3 planning	Term 4 planning:
How simple machines like pulleys work?	Imagining and designing a complex machine using what we know about simple machines (using plan and side views in our design)
Where can we find pulleys? Wedges? Incline planes? etc.	
Ropes and chains: How do they work in the pulley system? How do we use them in our schoolyard? How were they used in mediaeval times?	

19.5.6 Play Creates the Zone of Proximal Development for Amplifying Children's Technological Thinking and Design Imagining

We also learned that technological learning became visible in the play actions of the children in the imaginary situation of a conceptual PlayWorld. Vygotsky (1966) argued that play creates the zone of proximal development. He said that a child is a head taller than him or herself when in play. In this classroom, the children's motivation for technological learning was high—they wanted to solve the social problem in the story. Their need to find solutions to the social problem drove their engagement in design and technology. Through building competencies in design and technology by being introduced to mature forms of the concepts that were needed to progress the play, children could take their designed solutions back into their play. It was in the play that the children's real form of development could be seen. The children's thinking could be seen in the interactions and in their designed solutions being played out in imaginary situations.

It was in the play that the children met further problems—that is, there was a level of authentic exploration of their designed solutions—did their solutions meet their needs, were their solutions robust and fit for purpose? The children's motivation to fine-tune their prototype or refine their escape plans was exceedingly high and meaningful when in the context of play. This is suggestive of an authentic accessible moment (Fleer, 2015) in design and technology, where the play's actions give insights into what children can do either individually or collectively under the conditions of being motivated to do the best they can. In this zone of proximal development, it is possible to identify more authentically how children are learning, what they are learning, and what further technological challenges could be planned to stretch and develop them further.

We also learned that the mind maps acted as collective placeholders of their emergent technological thinking that they could progressively add to over time. Children's engagement with the social problem and designed solution acted out in their play, made visible how they were individually and collectively thinking in situations of meeting problems that needed newly designed solutions. Consequently, the context of a Conceptual PlayWorld created the zones of development for authentic assessment of children. In line with Vygotsky's conception of play (Vygotsky, 1966) and his conception of the zones of actual, proximal, and potential development (Vygotsky, 1997, 1998), it is possible to conceptualise assessment in a Conceptual PlayWorld as shown in Fig. 19.6.

In summary, a cultural-historical conception of assessment brings forward a form of actual assessment (what children do independently), proximal assessment (that which children successfully do with support from a more competent other), and potentive assessment (mature forms of design and technology are accessible in the child's environment).

In Fig. 19.6, we can theorise that assessing children's thinking in technology can be individual when at the level of a particular child's abilities. That is, to capture

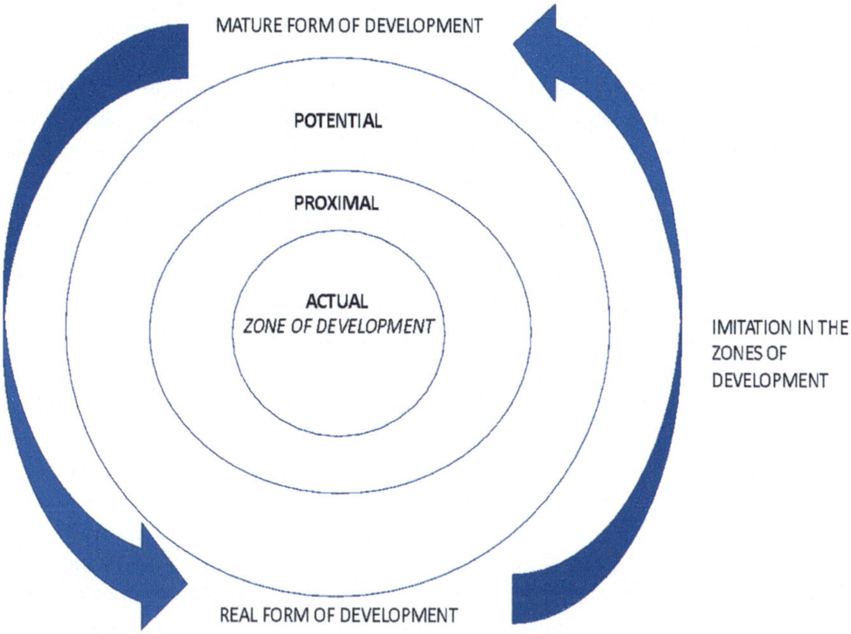

Fig. 19.6 Assessment in a Conceptual PlayWorld

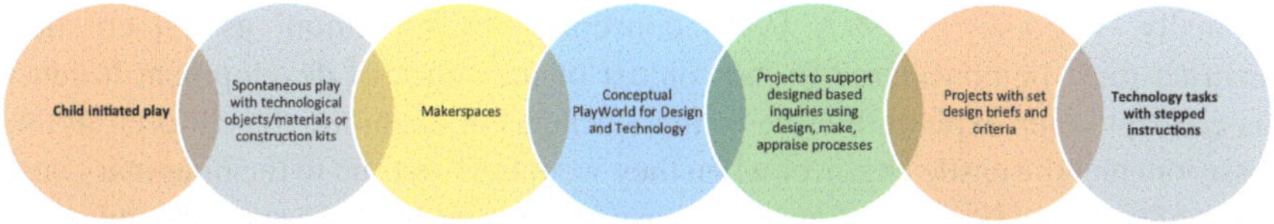

Fig. 19.7 Continuum of the orienting principle guiding design and technology

what a child knows (actual zone of development) and can do when working and designing on their own. However, we can also capture that fuzzy zone between what they can do on their own or what might be in their possible sphere of proximal developmental understanding. That is, when children are playing together with others and are engaged in design and technology with some understanding, but not with proficiency. It is with others acting as play partners, that we can visibly see children being 'stretched' into this proximal zone of development. Play gives this proximal context. By joining in the play and experiencing the drama of the problem to solve, they have a role in the narrative and can make some contribution to the collective design and solutions being proposed and realised through their play. That is, the children are in a dynamic context. Even though they cannot yet achieve the competencies independently, they are nevertheless able to do so with others.

But we can also move beyond formative assessment and into potentive assessment (Fleer, 2015). That is, to assess the context and what is provided for children in the Conceptual PlayWorld environment for technology education. Vygotsky (1994) wrote about the importance of bringing into a child's world examples of the development that a community wishes a child to acquire. In citing the example of language learning of infants, he illustrated how important it was for a child to be in a language-rich environment. Even though they cannot yet understand the words as an infant, they do so over time. Bringing this logic to assessment in technology education, it can be argued that in a Conceptual PlayWorld, the need for design and technology is amplified by the teachers who bring into that Conceptual PlayWorld context the learning they wish for children—such as introducing YouTubes as mature forms of technological principles or using the castle engineer (technical assistant in the role) to share with the children how a drawbridge works. Whilst the children in the first, second or third iteration of being in a Conceptual PlayWorld may not understand the technological principles in the way a technologist might, the children gain the idea of an engineered solution and associated language to bring to future designing of solutions. That is, children in the future can draw on a design, make, and appraise framework when thinking technologically about the problem to hand. This is different to other approaches to assessment because in this theorisation of what the environment affords, the assessment includes a dimension of potential learning. This is theoretically consistent with Vygotsky's (1994) premise of having in the child's environment mature forms of what they want a child to be able to be, learn, or do from the very start of their development.

Finally, Vygotsky's (1997, 1998) conception of 'imitation' as a theoretical construct must be considered in the context of the zones of development because this concept helps explain the role of the teacher. For example, the children in this classroom met the castle engineer when they went back in time to the imaginary situation of Sherwood Forest. The engineer gave insights into engineering principles that they imitated in their own actions—pulley system. These playful experiences were mirrored in the children's designed solutions and eventual collective prototyping. Not as reproductions, but rather as agentic productions through a series of designed solutions to problems that arise in their play. Figure 19.6 shows imitation as a relation between children's present or actual understandings of the iterative process of design, make, and approach in technology education, and the mature form of this process presented by the teachers as part of the collective play and learning in technology education.

19.6 Conclusion

This chapter focused on the problem of the relation between imagination in play and imagination in design in a STEM context for supporting technology education. It was determined that, first there were three periods of pedagogical activity in design that supported the technological solution of a 'grabby' hand machine for retrieving the treasure from the castle. These pedagogical activities were:

1. Collective problem formulation.
2. Researching and developing designs to deepen play.
3. Collective design solutions.

This result is consistent with what is already known about design thinking (Benson & Treleven, 2011; Sung & Kelley, 2019), design-based learning (English et al., 2020), and design in technology (Williams, 2000). There is an obvious relation between collective problem formulation and collective design solutions in a Conceptual PlayWorld, but what is new is how imagination as a psychological function (Vygotsky, 1998) appeared to be resourcing children's designs and deepening their play, which in turn supported their conceptual learning in technology education. This was shown in how a *play problem* could be turned into a *problem to be researched,* which in turn supported a *collective design solution with technological outcomes.* Imagination fuelled this development and transformation. This finding suggests that in a Conceptual PlayWorld, the learning of design and technology becomes personally meaningful to children. The drama of the story, the evolving play narrative, and the designed solutions appeared to enrich and orient children towards technological play in a STEM context of the story of Robin Hood. Moreover, this orientation was shown to evolve over time through dramatic problem situations that needed resolution for the play to continue. Therefore, the play problem in a Conceptual PlayWorld can orient children towards wanting to learn design and technology concepts.

Second, the results of the children's drawings of their designs showed how imagination in play supported imagination in design. Whilst studies into children's play in STEM contexts have revealed the usefulness of the play resources (Parker-Rees, 1997) and a play pedagogy for design (Rogers et al., 2000) to support technological thinking (Sung & Kelley, 2019), imagination in design has not previously been systematically studied as part of children's play narrative. Therefore, the development of children's imagination in design appeared to be a core outcome of the pedagogical activities that the children participated in.

Third, a Conceptual PlayWorld that foregrounds design and technology is different to the pedagogical structures supporting projects for design thinking (Benson & Treleven, 2011; Sung & Kelley, 2019) and design-based learning (English et al., 2020). We found that imaginary play was shown to be the overarching framework of the children's activities and the basis of design and technology planned by the teachers. In Fig. 19.7, different pedagogical frameworks are shown in order to illustrate this point theoretically.

On the left side of the continuum play is the motivating condition for driving the children's learning, whilst as you progress to the right side, it is the concepts in design and technology and the associated project/process/tasks that act as the beginning point for framing the children's thinking about what they are doing.

Whilst this figure captures some of the diversity of models of teaching in design and technology in the early years where play may be emphasised more, it is easy to see on this continuum how in the later years this dimension of play can be lost.

Therefore, the implications for teachers of students in the latter part of primary and perhaps even secondary school is to consider using play as an orienting framework for technology education. It has been noted by Bergen (2009, 425) that 'Successful adults reporting on their extended pretence, construction, and gameplay as children give evidence to the long-lasting, enjoyable, and creative benefits of play. Thus, enhancing playful learning at every level of education, including high school and college programs that prepare professionals for scientific, mathematical, and engineering fields, is warranted'. A Conceptual PlayWorld gives teachers a planning model to create motivating conditions for learning in design and technology where child agency is high and the learning of concepts becomes personally meaningful.

Overall, it can be concluded that under the conditions of a Conceptual PlayWorld of Robin Hood, a *play problem* brings into the teacher's pedagogy and children's actions design thinking and design processes that support technological learning. This approach brings together imaginary play and imagination in design to amplify technological learning. This adds to what is known about effective models of teaching in technology education, where design in imaginary play supports thinking and acting technologically.

Acknowledgements Acknowledgements Research assistance for data collection in this paper was provided by Sue March (field leader), Fatema Taj Johora, and Junqian Ma and data organisation by Freya Fleer-Stout and Ainslie Holland. Special acknowledgement is made of the two teachers who participated in the study, Rebecca and Oriana. Funding from the Australian Research Council Discovery Grant [DP140101131] and from the Australian Research Council Laureate Fellowship Scheme [FL180100161] the research and writing of this chapter.

References

Benson, C., & Treleven, T. (2011). Designerly thinking in the foundation stage. In C. Benson & J. Lunt (Eds.), *International handbook of primary technology education* (pp. 137–150). Sense Publishers.

Bergen, D. (2009). Play as the learning medium for future scientists, mathematicians, and engineers. *American Journal of Play, 1*(4), 413–428.

Crismond, D. P., & Adams, R. S. (2012). The informed design teaching and learning matrix. *Journal of Engineering Education, 101*(4), 738–797. https://doi.org/10.1002/j.2168-9830.2012.tb01127.x

English, L. D., Adams, R., & King, D. T. (2020). Design learning in STEM education. In C. J. Johnson, M. J. Mohr-Schroeder, T. J. Moore, & L. D. English (Eds.), *Handbook of research on STEM education* (pp. 76–86). Routledge.

Fleer, M. (2015). Developing an assessment pedagogy: The tensions and struggles in re- theorising assessment from a cultural-historical perspective. *Assessment in Education: Principles, Policy and Practice, 22*(2), 224–246. https://doi.org/10.1080/0969594X.2015.1015403

Fleer, M. (2017a). Scientific Playworlds: A model of teaching science in play based settings. *Research in Science Education.*https://doi.org/10.1007/s11165-017-9653-z

Fleer, M. (2017b). Digital playworlds in an Australia context. In T. Bruce, M. Bredikyte, & P. Hakkarainen (Eds.), *Routledge handbook of play in early childhood* (pp. 289–304). Routledge Press, Taylor and Francis Group.

Fleer, M. (2018). Conceptual Playworlds: The role of imagination in play and imagination in learning, *Early Years*.https://doi.org/10.1080/09575146.2018.1549024

Fleer, M. (2020). Studying the relations between motives and motivation—How young children develop a motive orientation for collective engineering play. *Learning, Culture and Social Interaction, 24*.https://doi.org/10.1016/j.lcsi.2019.100355

Fleer, M. (2021). The genesis of design: learning about design, learning through design to learning design in play. *International Journal of Technology and Design Education*.https://doi.org/10.1007/s10798-021-09670-w

Kelley, T. R., & Sung, E. (2017). Sketching by design: Teaching sketching to young learners. *International Journal of Technology and Design Education, 27*(3), 363–386. https://doi.org/10.1007/s10798-016-9354-3

Parker-Rees, R. (1997). Making sense and made sense: Design and technology and the playful construction of meaning in the early years. *Early Years, 18*(1), 5–12. https://doi.org/10.1080/0957514970180102

Rogers, G., & Wallace, J. (2000). The wheels of the bus: Children design in in an early years classroom. *Research in Science and Technology Education, 18*(1), 127–136. https://doi.org/10.1080/713694960

Sung, E., & Kelley, T. R. (2019). Identifying design process patterns: a sequential analysis study of design thinking. *International Journal of Technology and Design Education, 29,* 283–302. https://doi.org/10.1007/s10798-018-9448-1.

Vygotsky, L. S. (1966). Play and its role in the mental development of the child. *Voprosy Psikhologii, 12*(6), 62–76.

Vygotsky, L. S. (1994). The problem of the environment. In J. Valsiner & R. van der Veer (Eds.), *The Vygotsky reader* (pp. 347–348). Blackwell.

Vygotsky, L. S. (1997). The history of the development of higher mental functions (M. J. Hall, Trans.). In R. W. Rieber (Ed.), *The collected works of L. S. Vygotsky* (Vol. 4). New York: Plenum Press.

Vygotsky, L. S. (1998). *The collected works of L.S. Vygotsky: Child Psychology* (M. J. Hall, Trans.; R. W. Rieber, Ed. Vol. 5). Plenum Press.

Vygotsky, L. S. (2004). Imagination and creativity in childhood. *Journal of Russian and East European Psychology, 42*(1), 7–97.

Williams, P. J. (2000). Design: The only methodology of technology. *Journal of Technology Education, 11*(2), 48–60. https://files.eric.ed.gov/fulltext/EJ604487.pdf.

Marilyn Fleer holds the Foundation Chair of Early Childhood Education and Development at Monash University, Australia. She was awarded the 2018 Kathleen Fitzpatrick Laureate Fellowship by the Australian Research Council and was a former President of the International Society of Cultural-historical Activity Research (ISCAR). She was presented with the 2019 Ashley Goldsworthy Award for Outstanding leadership in university–business collaboration and was recently elected as a fellow of the Australian Academy of Social Sciences and inducted into the Honour Roll of Women in Victoria as a change agent.

SPRINGER NATURE

GPSR Compliance

The European Union's (EU) General Product Safety Regulation (GPSR) is a set of rules that requires consumer products to be safe and our obligations to ensure this.

If you have any concerns about our products, you can contact us on ProductSafety@springernature.com

In case Publisher is established outside the EU, the EU authorized representative is:

Springer Nature Customer Service Center GmbH
Europaplatz 3
69115 Heidelberg, Germany

The manufacturer's authorised representative in the EU is Springer Nature Customer Service Centre GmbH, Europaplatz 3, 69115 Heidelberg, Germany. If you have any concerns regarding our products, please contact ProductSafety@springernature.com

Printed and bound by CPI Group (UK) Ltd, Croydon, CR0 4YY
01/10/2026
02225850-0013